iPhone in Action

iPhone in Action

INTRODUCTION TO WEB AND SDK DEVELOPMENT

CHRISTOPHER ALLEN
SHANNON APPELCLINE

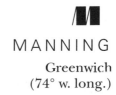

MANNING

Greenwich
(74° w. long.)

For online information and ordering of this and other Manning books, please visit www.manning.com. The publisher offers discounts on this book when ordered in quantity. For more information, please contact

> Special Sales Department
> Manning Publications Co.
> Sound View Court 3B   fax: (609) 877-8256
> Greenwich, CT 06830   email: orders@manning.com

Manning Publications Co.
Sound View Court 3B
Greenwich, CT 06830

Development Editor   Tom Cirtin
Copyeditors: Liz Welch, Andy Carroll
Typesetter: Gordan Salinovic
Cover designer: Leslie Haimes

ISBN 193398886X
Printed in the United States of America
2 3 4 5 6 7 8 9 10 – MAL – 13 12 11 10 09

contents

preface xv
acknowledgments xvii
about this book xix

PART 1 INTRODUCING iPHONE PROGRAMMING 1

1 **Introducing the iPhone 3**

1.1   iPhone core specifications 4

iPhone input and output specifications 5 ▪ *iPhone network
specifications 6* ▪ *iPhone browser specifications 7* ▪ *Other iPhone
hardware features 8*

1.2   How the iPhone compares to the industry 8

The physical comparison 9 ▪ *Competitive internet viewing 9
Mobile web standards 10* ▪ *The rest of the story 10*

1.3   How the iPhone is unique 10

1.4   Understanding iPhone input and output 12

Output and iPhone viewport 12 ▪ *Output and orientations 12
Input and iPhone mousing 13*

1.5   Summary 15

2 Web development or the SDK? 16

2.1 Comparing the two programming styles 16

2.2 A central philosophy: the continuum of programming 18

2.3 Advantages and disadvantages 19

Web development 19 ▪ SDK development 21 ▪ To each program its platform 22

2.4 Stand-alone iPhone development 22

Web development models 23 ▪ SDK development models 24

2.5 Integrated iPhone development 25

Mirrored development 26 ▪ Mixed development 26 ▪ Client-server development 27 ▪ Last thoughts on integration 27

2.6 Summary 27

PART 2 DESIGNING WEB PAGES FOR THE IPHONE 29

3 Redeveloping web pages for the iPhone 31

3.1 The iPhone viewport 32

Making sitewide viewport changes 34 ▪ Making local viewport changes 34 ▪ Viewport properties and constants 35

3.2 Making your web pages iPhone friendly 36

Avoiding missing iPhone functionality 36 ▪ Creating good links 39

Practicing good web work 39 ▪ Fixing common problems 41

3.3 Making your web pages iPhone optimized 43

Detecting the iPhone through USER_AGENT 43 ▪ Detecting the iPhone through CSS 44 ▪ Optimizing with CSS 44

3.4 Manipulating iPhone chrome 45

The three bars 45 ▪ Web clips 46

3.5 Capturing iPhone events 47

One-fingered touches 47 ▪ Two-fingered gestures 48

3.6 Redisplaying web pages 49

The Gmail iPhone pages 50 ▪ The Facebook iPhone pages 51

3.7 Supporting non-iPhone users 52

3.8 Summary 54

4 **Advanced WebKit and textual web apps 55**

    4.1   Introducing the WebKit 56

          New HTML elements 57 ▪ New CSS elements 57

    4.2   CSS transforms, transitions, and animations 59

          The transform function 59 ▪ The transition function 61 ▪ The animation function 64

    4.3   The WebKit database 65

          Loading a database 65 ▪ Running a transaction 65 ▪ A sample database 66

    4.4   Adjusting the chrome 69

    4.5   Recognizing touches and gestures 69

          Accessing events 70 ▪ Converting events 71 ▪ Accessing touches 72 Accessing gestures 74

    4.6   Recognizing orientation 75

    4.7   Upcoming features: CSS gradients and masks 76

          CSS gradients 76 ▪ CSS masks 77 ▪ The Canvas alternative 77

    4.8   Summary 78

5 **Using iUI for web apps 80**

    5.1   Creating your own iPhone UI 81

          The graphical interface 81 ▪ The iPhone data paradigm 83 ▪ Other iPhone design elements 83

    5.2   Getting ready for iUI 84

    5.3   Developing with iUI 85

          The iUI toolbar 86 ▪ iUI lists 87 ▪ iUI dialogs 89 ▪ iUI searches done right with Ajax 90 ▪ iUI panels and rows 91 ▪ iUI buttons 92 iUI attributes 93

    5.4   Creating an iUI back end 94

    5.5   Other iUI tips and tricks 95

          Organizing your code 95 ▪ Improving data listings 97 Compressing iUI 98 ▪ Selecting a different look 98

    5.6   Integrating iUI with other libraries 99

          Using jQuery with iUI 99 ▪ Using iUI with WebKit 100

    5.7   Summary 101

6   *Using Canvas for web apps 102*

6.1   Getting ready for Canvas 103

Enabling Canvas 103 • Ensuring compatibility 103 • Putting it together 104

6.2   Drawing paths 105

Basic path commands 106 • Curve commands 107

6.3   Drawing shapes 110

Drawing rectangles 110 • Writing shape functions 110

6.4   Creating styles: colors, gradients, and lines 112

Color styles 112 • Gradient styles 112 • Line styles 114

6.5   Modifying composition and clipping 114

Global variables 115 • Clipping paths 116

6.6   Transforming and restoring 116

Transformations 116 • State stacking 117

6.7   Incorporating images, patterns, and text 117

Image commands 118 • Pattern commands 119 • Text commands 119

6.8   Putting it together 120

6.9   Applying animation 121

6.10 Summary 123

7   *Building web apps with Dashcode 124*

7.1   An introduction to Dashcode 125

Starting a Dashcode project 126 • The anatomy of Dashcode 126 Running Dashcode projects 130 • Deploying Dashcode projects 130

7.2   Writing Dashcode programs 131

Using library parts 132 • Adding action buttons 134 • Using the list-based Browser template 135 • Working with the stackLayout part 136 • Exploring the rest of Dashcode 139

7.3   Integrating Dashcode with existing libraries 140

Integrating Dashcode with WebKit 140 • Integrating Dashcode with iUI 141 • Integrating Dashcode with Canvas 141 • Deeper integration 142

7.4   Summary 142

8 Debugging iPhone web pages 143

8.1 Using Apache locally 143

8.2 Debugging with your desktop browser 144

*Using Safari 144 ▪ Using Firefox 146 ▪ Using the iPhone
Simulator 148*

8.3 Debugging with your iPhone 149

Using iPhone Debug 150 ▪ Using bookmarklets 150

8.4 Profiling for the iPhone 151

8.5 Summary 152

9 SDK programming for web developers 154

9.1 An introduction to C's concepts 155

*Declarations and typing 156 ▪ Memory management and pointers 157
File structure and directives 158 ▪ Compiling 158 ▪ Other
elements 159*

9.2 An introduction to object-oriented programming 159

Objects and classes 160 ▪ Messaging 160

9.3 The Model-View-Controller (MVC) pattern 161

9.4 Summary 162

PART 3 LEARNING SDK FUNDAMENTALS 165

10 Learning Objective-C and the iPhone OS 167

10.1 Getting ready for the SDK 168

Installing the SDK 168 ▪ The anatomy of the SDK 169

10.2 Introducing Objective-C 171

*The big picture 171 ▪ The message 172 ▪ Class definition 174
Properties 176 ▪ Other compiler directives 178 ▪ Categories and
protocols 179 ▪ Wrapping up Objective-C 179*

10.3 Introducing the iPhone OS 180

The anatomy of the iPhone OS 180 ▪ Windows and views 183

10.4 The iPhone OS's methods 184

*Object creation 184 ▪ Memory management 186 ▪ Event response 187
Life-cycle management 188*

10.5 Summary 189

11 Using Xcode 190

11.1 Introducing Xcode 190

The anatomy of Xcode 191 ▪ Compiling and executing in Xcode 192

11.2 Creating a first project in Xcode: Hello, World! 192

Understanding main.m 193 ▪ Understanding the application delegate 194 ▪ Writing Hello, World! 196

11.3 Creating a new class in Xcode 198

The new class how-to 199 ▪ The header file 199 ▪ The source code file 200 ▪ Linking it in 202

11.4 Other Xcode functionality 202

Adding frameworks with Xcode 202 ▪ Using alternate templates with Xcode 203 ▪ Xcode tips and tricks 204

11.5 Summary 205

12 Using Interface Builder 206

12.1 An introduction to Interface Builder 207

The anatomy of Interface Builder 207 ▪ Simulating in Interface Builder 210

12.2 Creating a first project in Interface Builder: pictures and the web 210

Creating new objects 210 ▪ Manipulating objects graphically 211 Using the inspector window 211 ▪ Working with pictures 213

12.3 Building connections in Interface Builder 214

Declaring an IBOutlet 215 ▪ Connecting an object 215 ▪ Coding with IBOutlets 216

12.4 Other Interface Builder functionality 217

Building other connections 217 ▪ Creating external objects 218 Initializing Interface Builder objects 218 ▪ Accessing .xib files 219 Creating new .xib files 219

12.5 Summary 220

13 Creating basic view controllers 221

13.1 The view controller family 222

13.2 The bare view controller 223

The anatomy of a view controller 223 ▪ Creating a view controller 224 ▪ Building up a view controller interface 225 Using your view controller 226

13.3 The table view controller 231

The anatomy of a table view controller 231 ▪ Creating a table view controller 231 ▪ Building up a table interface 233 ▪ Using your table view controller 238

13.4 Summary 239

14 **Monitoring events and actions 240**

14.1 An introduction to events 240

The responder chain 241 ▪ Touches and events 242

14.2 A touching example: the event reporter 244

Setting things up in Interface Builder 245 ▪ Preparing a view for touches 246 ▪ Controlling your events 248

14.3 Other event functionality 250

Regulating events 250 ▪ Other event methods and properties 251

14.4 An introduction to actions 252

The UIControl object 252 ▪ Control events and actions 252 ▪ The addTarget:action:forControlEvents: method 254

14.5 Adding a button to an application 255

Using addTarget:action:forControlEvents: 255 ▪ Using an IBAction 256

14.6 Other action functionality 257

The UITextField 257 ▪ The UISlider 259 ▪ A TextField/Slider mashup 260 ▪ Actions made easy 261 ▪ Actions in use 261

14.7 Introducing notifications 262

14.8 Summary 262

15 **Creating advanced view controllers 264**

15.1 The tab bar view controller 265

The anatomy of a tab bar controller 265 ▪ Creating a tab bar controller 266 ▪ Building a tab bar interface 267 ▪ Using your tab bar controller 271

15.2 The navigation controller 271

The anatomy of a navigation controller 272 ▪ Creating a navigation controller 273 ▪ Building a navigation controller 274 Using your navigation controller 276

15.3 Using the flipside controller 279

15.4 Modal view controllers 281

15.5 Summary 281

PART 4 PROGRAMMING WITH THE SDK TOOLKIT............ 283

16 **Data: actions, preferences, files, SQLite, and addresses 285**

16.1 Accepting user actions 286

16.2 Maintaining user preferences 288

Creating your own preferences 288 ▪ *Using the system settings 293*

16.3 Opening files 297

Accessing your bundle 298 ▪ *Accessing other directories 299*
Manipulating files 300 ▪ *Filesaver: a UITextView example 301*

16.4 Using SQLite 303

Setting up an SQLite database 304 ▪ *Accessing SQLite 305*
Accessing your SQLite database 305 ▪ *Building a navigation menu*
from a database 306 ▪ *Expanding this example 313*

16.5 Accessing the Address Book 313

An overview of the frameworks 313 ▪ *Accessing Address Book*
properties 314 ▪ *Querying the Address Book 316* ▪ *Using the*
Address Book UI 318

16.6 Summary 322

17 **Positioning: accelerometers and location 324**

17.1 The accelerometer and orientation 325

The orientation property 325 ▪ *The orientation notification 325*

17.2 The accelerometer and movement 326

Accessing the UIAccelerometer 326 ▪ *Parsing the UIAcceleration 327*
Checking for gravity 328 ▪ *Checking for movement 330*
Recognizing simple accelerometer movement 330

17.3 The accelerometer and gestures 333

17.4 All about Core Location 335

The location classes 336 ▪ *An example using location and distance 337*
An example using altitude 340 ▪ *Core Location and the Internet 342*

17.5 Summary 343

18 **Media: images and sounds 344**

18.1 An introduction to images 345

Loading a UIImage 345 ▪ *Drawing a UIImageView 346*
Modifying an image in the UIKit 347

18.2 Drawing simple images with Core Graphics 347

18.3 Accessing photos 349

Using the image picker 349 ▪ Taking photos 349 ▪ Saving to the photo album 350

18.4 Collage: an image example 351

The collage view controller 351 ▪ The collage temporary image view 354 ▪ The collage view 355 ▪ Expanding on this example 356

18.5 Using the Media Player framework 357

The media player class 357 ▪ The volume view 359 ▪ Better integrating the media player 359

18.6 Playing sounds manually 360

Playing simple sounds 361 ▪ Vibrating the iPhone 362 ▪ Playing complex sounds 362 ▪ Other audio frameworks 364

18.7 Summary 365

19 **Graphics: Quartz, Core Animation, and OpenGL 366**

19.1 An introduction to Quartz 2D 367

19.2 The Quartz context 367

Drawing to a UIView 369 ▪ Drawing to a bitmap 370

19.3 Drawing paths 371

Finishing a path 372 ▪ Creating reusable paths 373 ▪ Drawing rectangles 374

19.4 Setting the graphic state 375

Setting colors 375 ▪ Making transformations 376 ▪ Setting clipping paths 379 ▪ Other settings 379 ▪ Managing the state 380

19.5 Advanced drawing in Quartz 381

Drawing gradients 381 ▪ Drawing images 383 ▪ Drawing words 384 ▪ What we didn't cover 385

19.6 Drawing on a picture: an example 386

The photodraw view controller 386 ▪ The photodraw view 388 Expanding on the example 390

19.7 An introduction to Core Animation 391

The fundamentals of Core Animation 391 ▪ Getting started with Core Animation 392 ▪ Drawing a simple implicit animation 392 Drawing a simple explicit animation 393

19.8 An introduction to OpenGL 394

19.9 Summary 394

20 *The web: web views and internet protocols 396*

20.1 The hierarchy of the internet 397

20.2 Low-level networking 397

The CFHost class 398

20.3 Working with URLs 399

Creating an NSURL 399 ▪ Building an NSURLRequest 400
Manipulating HTML data by hand 400

20.4 Using UIWebView 401

Calling up the web view 402 ▪ Managing the web view delegate 403
Thumbnails: a web view example 403 ▪ Google Maps: a Core
Location example 405

20.5 Parsing XML 407

Starting up NSXMLParser 408 ▪ Acting as a delegate 408
Building a sample RSS reader: an XML example 409 ▪ Altitude
redux: a Core Location example 414

20.6 POSTing to the web 416

POSTing by hand 416 ▪ Submitting forms 417

20.7 Accessing the social web 418

Using web protocols 419 ▪ Using TouchJSON 420

20.8 Summary 421

appendix A *iPhone OS class reference 423*
appendix B *External sources and references 427*
appendix C *Publishing your SDK program 429*
index 433

We've both been Apple fans for most of our lives. Shannon fondly recalls early games played on the Apple II and that first Macintosh, which really turned his school news-paper around. Christopher turned his own fandom into a real business with his first entrepreneurial venture, Dreams of the Phoenix, a Mac software company.

Thus, we both had high hopes when we heard rumors of an upcoming iPhone. After the doldrums of the 1990s, Apple was on an upswing, and we were thrilled. We'd already seen how the Apple Airport had revolutionized local area network access and how the Apple iPod had removed the once ubiquitous Sony Walkman from the audio landscape with one fell swoop. We hoped that the iPhone would do the same for the cell phone industry.

It was Christopher who stood in line on that late June afternoon in 2007, purchas-ing one of the 270,000 iPhones sold in those first 30 hours. Despite being a lifelong techie, Shannon had never owned a cell phone before the iPhone, but he got his on June 29 too, and since then it's been his constant companion. (He's still more likely to look up a map while on a long bike ride than he is to make a call, but that's part of the beauty of the device—it's many things for many different people.)

We're programmers, so after we had our iPhones in hand, the next step was to program for them. We both came into this world of iPhone programming through the web.

For Christopher, that's because on June 30, 2007, there was no other way to develop for the iPhone. During that initial nine-month gestation period, any iPhone application *had* to go through the web. Christopher was on the cutting edge. He set up the iPhoneWebDev mailing list, where people puzzled out viewports and other special iPhone features, and he quickly learned how to create great-looking apps for this new platform. He was a participant in the first two iPhoneDevCamps and a judge in the Hackathon.

For Shannon, web programming was an obvious first step because he was already programming popular web sites at www.rpg.net and www.xenagia.net, and he wanted to see what it would take to optimize them for his new iPhone. With the results in hand, he did his first writing about the iPhone—a pair of articles on designing web pages for the iPhone, which can still be found at www.iphonewebdev.com/blog.

When we first pitched this book, it was all about iPhone web development, our initial expertise. But in the middle of our pitch, in March of 2008, Apple announced a whole new method of programming for the iPhone: the SDK. We scrambled to revise our outline at once.

We could have thrown out web development entirely and just moved on to the SDK. Many programmers and authors seem to be doing exactly that. And we think that's a big mistake. Web development and the SDK each offer distinct ways of programming for the iPhone. They also each offer distinct advantages.

Want to simply monetize your program without worrying about anything else? In that case, the SDK *is* probably for you. But, if you want to rapidly deploy your program, frequently update it, interact with other users on the Internet, or take advantage of an existing web infrastructure, you might find that web development is the way to go.

Then there's the possibility of hybridization. If nothing else, you probably want to make great iPhone-optimized web pages to talk about your SDK programs—but we also think there are much deeper options for hybridization.

This concept of there being two paths of iPhone development eventually became a cornerstone of our book. Although we have more material on the SDK, we've written a complete introduction to both topics, offering everything that you need to get started in either type of programming. This comprehensive look at iPhone programming methods and our introductory style are what we think makes this book unique. We now welcome you to join us, as we share with you the many lessons we've learned since the iPhone's release, a year and a half ago.

acknowledgments

Any technical book is a massive undertaking, due to the number of people required to make sure that it reads well, looks good, and is technically correct. Thus, we have to thank the entire Manning staff, without whom this book literally would not exist. They did more than just correct our errors and polish our words; they also helped us make integral decisions about the organization and the contents of the book—decisions that we believe improved it dramatically.

In particular, we'd like to thank the three people at Manning who helped us at the most pivotal times: Troy Mott, our acquisitions editor, who initially agreed to take on the book and who stayed with us every step of the way; Tom Cirtin, our book editor, who guided our style and content on a regular basis; and Marjan Bace, our publisher, who offered some of the biggest challenges regarding our content and organization and initiated some of the best improvements.

We'd also like to thank Liz Welch, our original copyeditor. We've worked with good and bad copy editors over the years, and Liz is definitely a good one. Our text sounded so much better after she was done with it that we don't even want to look at our early drafts any more. Beyond that, tech editors are crucial to the success of a book like this, so we want to thank our original tech editor, Robert McGovern, who caught minutia that we weren't even aware of. His encyclopedic knowledge of the newest iPhone releases was awe-inspiring. Further thanks go to copyeditor Andy Carroll and tech editor Larry Whipple, who came in for the latter half of the book, and to Maureen Spencer, who provided a final proofread. Though it's clichéd to say, it's true: the errors that snuck in are ours, but many others were corrected by all of these people. There were many, many more people behind the scenes at Manning who were crucial to the book, and we'd like to thank them all.

Finally, we'd like to thank the reviewers who generously agreed to read our book as we worked on it; they improved the book immensely: Jonathon Hohle, Adilson Oliveira Cruz, Edmon Begolli, Frank Abelson, Gershon Kagan, Jean Tantra, Matt Smith, Nhoel Sangalang, Noel Rappin, Satnam Alag, Scott Shaw, Amos Bannister, Andrea Gazzinaga, Robert Dempsey, Jeremy Anderson, Patrick Peak, Nathan Levesque, Aiden Montgomery, Martijn Dashorst, Premkumar Rajendran, Rama Krishna Vavilala, and Tristan O'Tierney.

Christopher would like to additionally thank Chris Messina for inviting him to be a founder of iPhoneDevCamp and to thank his long-time MacHack and SmartFriends colleagues for their support and assistance.

Shannon would also like to thank Christopher, who got this book started in the first place.

about this book

iPhone in Action is an introductory book, intended to teach the basics of iPhone programming in a tutorial form. It covers the fundamentals of both major styles of programming for the platform: web development and SDK programming.

You can read this book in one of three ways.

We encourage you to read it straight through, from chapter 1 to 20. This will introduce the platform, introduce both major ways to program for the iPhone, offer advice on when one style of programming is more appropriate than the other, and step you through both styles in turn.

If you prefer to read only about web development, you can just read parts 1 and 2. Note that we've included an introduction to Objective-C at the end of part 2 that will get you started on iPhone SDK programming, even if you've never used a compiled programming language before, so we encourage you to keep on going, and see what else you can learn.

If you prefer to read only about SDK programming, you can just read parts 1, 3, and 4. We still encourage you to at least skim the chapters in part 2, because they include advice especially for SDK programmers, showing how lessons learned from the web can apply to the SDK as well.

The audience

This is an introductory book. We've done our best to make it accessible to everyone who might be interested in developing web pages or writing native programs for the iPhone. We think that it'll be especially useful to people who are looking to wholeheartedly dive into the iPhone arena, because it will allow you to choose whether to write web apps or create native applications for each and every project that you work on.

If you want to learn about iPhone web development, you should have a good understanding of web design in general, including HTML, CSS, and JavaScript. You don't need to know any more dynamic languages to create great iPhone web apps. In fact, Apple's powerful Dashcode development platform is built entirely using these three languages.

If you want to learn about iPhone SDK programming, you should have some experience with programming in general. It'd be best if you've worked with C before, but that's not a necessity; if you haven't, you can read our introduction to C in chapter 9, and you should probably expect to do some research on your own to clarify things. There's definitely no need to be familiar with Objective-C, Cocoa, or Apple programming in general. We'll give you everything you need to become familiar with Apple's unique programming style. You'll probably have a leg up if you understand object-oriented concepts, but again it's not necessary (and again, you'll find an introduction in chapter 9).

Roadmap

We've divided this book into four parts, with one covering introductory iPhone concepts, one covering web development, and two covering SDK programming.

Part 1 introduces the iPhone and the styles of programming that are possible for this device.

Chapter 1 explains the details of the iPhone and how it differs from the mobile phones that predated it. It also contains one of our most important concepts: the six unique features that truly make the iPhone stand out from the pack, and which are also of importance to programmers.

Chapter 2 describes the two styles of iPhone programming—web development and SDK programming—and discusses the strengths of each, so that you can make an informed decision about how to program any individual application. It also briefly touches upon the idea of hybridizing the two styles of iPhone programming.

Part 2 includes all of our information about writing and rewriting web pages for use on the iPhone.

Chapter 3 presents the basics of what you can do to redevelop an existing web page for viewing on the iPhone. In the process, it touches upon many of the most important factors concerning iPhone web pages, such as the viewport, the technology limitations, and the event changes.

Chapter 4 covers the first of three web libraries for use on the iPhone. The WebKit is an extension to HTML being worked on by Apple that gives you access to a variety of great features, from implicit animations to a built-in database.

Chapter 5 discusses the issue of how to make web pages that match the look and feel of iPhone native applications. In the process, it introduces a second notable web library, iUI, which creates iPhone-like animations and tables from simple HTML.

Chapter 6 focuses on the third of our iPhone web libraries, Canvas, another HTML extension championed by Apple. This graphical library gives you the ability to draw complex vector graphics and display them on the iPhone.

Chapter 7 turns to Dashcode, an Apple development environment that you can use to create iPhone web apps. With Dashcode's integrated links to the WebKit and Canvas, you can use the lessons of the previous chapters while working in a graphical layout program.

Chapter 8 finishes off our look at web development by touching upon how you can test and debug your iPhone web apps, using a variety of third-party utilities.

Chapter 9 offers a bridge, providing an introductory look at the tools that will be useful for programming in Objective-C, including the C language, the object-oriented paradigm, and the MVC architectural model.

Part 3 introduces the SDK by covering all of the fundamental topics that you'll need in order to program native apps for the iPhone.

Chapter 10 kicks things off by highlighting Objective-C, which is the programming language used on the iPhone, and the iPhone OS, an immense collection of frameworks that make many complex tasks very easy.

Chapter 11 looks at Xcode, the first major tool in the SDK. This integrated development environment does more than just compile your code. It also helps you correct simple errors as you type and provides quick, integrated access to all the iPhone programming documents.

Chapter 12 shifts the focus to Interface Builder, a graphical design environment that allows you to create and place interface objects without writing a single line of code. Interface Builder is a powerful time-saver for programmers and is used throughout the rest of the text as a result.

Chapter 13 covers simple view controllers. The basic view controller is an important building block of the MVC paradigm, dividing control from view, while the table view controller provides an easy way to organize information while matching the standard iPhone look and feel.

Chapter 14 steps back to talk about user interaction. It covers events, which users generate by touching the screen with one or more fingers, and actions, which happen when users interact with a control object like a button or a slider.

Chapter 15 finishes our look at view controllers by examining two more advanced possibilities. The tab bar view controllers allows for modal selection between multiple pages of content, and the navigation view controller adds hierarchy to tables.

Part 4 completes our look at the iPhone by opening up the SDK's toolkit and examining many different features that may be of interest to programmers. At the same time, it also provides more complex programming examples, which should help programmers to develop full-length iPhone projects.

Chapter 16 opens up the SDK toolkit by talking about data. This includes user input, such as actions and preferences; data storage, such as files and databases; and tools that combine input and storage, such as the iPhone's address book.

Chapter 17 highlights two of the most unique features on the iPhone, the accelerometer and the GPS, showing how the iPhone can track movement through space.

Chapter 18 covers another of the iPhone's strengths—media—by showing how to do basic work with pictures, movies, and sounds

Chapter 19 provides an extensive look at graphics, centering on the iPhone's vector graphic language, Quartz 2D. It also offers a brief overview of Core Animation and touches upon OpenGL for the iPhone.

Chapter 20 concludes our tour through the iPhone's toolkit by examining how it can be used to interact with the Internet. This chapter moves through the entire hierarchy of Internet communication, from low-level host connections to URLs, from web views to modern social languages like XML and JSON.

The appendixes contain some additional information that didn't fit with the flow of the main text. Appendix A contains a list of SDK objects and what they do. Appendix B features links for many web sites of note for iPhone programming. Appendix C includes the current information on how to deploy your SDK programs to actual iPhones.

Code conventions and downloads

Code examples appear throughout this book. Longer listings will appear under clear listing headings, and shorter listings will appear between lines of text. All code is set in a `special font` to clearly differentiate it. Class names have similarly been set in our code font; if you might type it into your computer, you'll be able to clearly make it out.

With the exception of a few cases of abstract code examples, all code snippets began life as working programs. Our complete set of programs can be found at http://www.manning.com/iPhoneinAction. There should be two ZIP files there, one each for the web and the SDK programs. We encourage you to try out the programs as you read; they'll often include additional code that doesn't appear in the book and will provide more context. In addition, we feel that actually seeing a program working can greatly elucidate the code required to create it.

Our code snippets in this book all include extensive explanations. We have often included short annotations beside the code, and sometimes we have numbered cueballs beside lines of code, linking the subsequent discussion to the code lines.

In part 2 of the book, all code snippets are basic HTML (with CSS or JavaScript, as appropriate). In the few places where we used PHP as an example of a more dynamic language, we clearly noted it with `<? ?>` brackets. In parts 3 and 4 of the book, all code snippets are Objective-C. We've usually left header files out of parts 3 and 4, as they tend to be quite basic.

In a few cases in this book, we've included content from multiple files in a single listing in order to provide a bigger picture of the program. In these situations, we've divided the contents of the different files within a single listing like this: `::: file #1 :::`

Software requirements

There are no particular requirements for most of the web portion of this book. You just need to be able to design and deploy web pages. However chapter 7 (on Dashcode) and portions of chapter 8 (particularly the discussion of the iPhone Simulator) refer to software that is available only on a Macintosh.

A Macintosh is absolutely required to do SDK development. You'll also need to have the iPhone SDK, but this is freely downloadable as soon as you sign up with Apple, as is described in chapter 10.

Authors online

This book is intended to be an introduction to iPhone programming. Though it covers an extensive amount of information on the iPhone, there's a lot that we couldn't talk about in a single book. Feel free to come chat with the authors online about additional iPhone topics.

Our main hangout is http://iphoneinaction.manning.com. This blog contains the newest noteworthy links we have found, discussions on "missing classes" that we didn't cover in this book, and occasional articles of more weight.

There is also an Author Online forum where you can post comments and ask questions of other readers as well as of the authors at http://www.manning.com/iPhoneinAction.

And we continue to host Christopher's original iPhone forum on web development, which can be found at http://www.iphonewebdev.com.

About the title

By combining introductions, overviews, and how-to examples, the *In Action* books are designed to help learning and remembering. According to research in cognitive science, the things people remember are things they discover during self-motivated exploration.

Although no one at Manning is a cognitive scientist, we are convinced that for learning to become permanent it must pass through stages of exploration, play, and, interestingly, retelling of what is being learned. People understand and remember new things, which is to say they master them, only after actively exploring them. Humans learn in action. An essential part of an *In Action* guide is that it is example-driven. It encourages the reader to try things out, to play with new code, and explore new ideas.

There is another, more mundane, reason for the title of this book: our readers are busy. They use books to do a job or to solve a problem. They need books that allow them to jump in and jump out easily and learn just what they want just when they want it. They need books that aid them *in action*. The books in this series are designed for such readers.

About the cover illustration

The illustration on the cover of *iPhone in Action* is captioned "Russian, Prince of the Cherkeeses." The Cherkess people are an ethnic group living in the Caucus region of Russia. The illustration is taken from the 1805 edition of Sylvain Maréchal's four-volume compendium of regional and national dress customs and uniforms. The colorful variety of Maréchal's collection reminds us vividly of how culturally apart the world's towns and regions were just 200 years ago. Isolated from each other, people spoke different dialects and languages and their place of origin and their station in life were easy to identify—by their language and by their dress.

Dress codes have changed since then and the diversity by region, so rich at the time, has faded away. It is now hard to tell apart the inhabitants of different continents, let alone different regions or countries. Perhaps we have traded cultural diversity for a more varied personal life—certainly a more varied and faster-paced technological life.

At a time when it is hard to tell one computer book from another, Manning celebrates the inventiveness and initiative of the computer business with book covers based on the rich diversity of regional life of two centuries ago, brought back to life by Maréchal's pictures.

Part 1

Introducing
iPhone programming

Apple's iPhone is more than just a new programming platform; it's an entirely new way to think about mobile technologies. Part 1 of this book gets your feet wet by explaining how the iPhone differs from its predecessors.

We'll start off in chapter 1 with a look at the iPhone itself, then follow up in chapter 2 by describing the two ways you can program for the iPhone: by creating web apps and by using the iPhone SDK.

Introducing the iPhone 1

This chapter covers

- Understanding Apple's iPhone technology
- Examining the iPhone's specifications
- Highlighting what makes the iPhone unique

In the 1980s Apple Computer was the leading innovator in the computer business. Their 1984 Macintosh computer revolutionized personal computing and desktop publishing alike. But by the 1990s the company had begun to fade; it was depending on its loyal user base and its successes in the past rather than creating the newest cutting-edge technology.

That changed again in 1996 when founder Steve Jobs returned to the fold. Two years later he produced the first candy-colored iMac, a computer that walked the line between computing device, pop culture, and fashion statement. It was just the first of several innovations under Jobs' watch, the most notable of which was probably 2001's iPod. The iPod was a masterpiece of portable design. It highlighted a simple and beautiful interface, giving users access to thousands of songs that they could carry with them all the time. But the iPod just whetted the public's appetite for more.

By 2006 rumors and speculation were rumbling across the internet concerning Apple's next major innovation: an iPod-like mobile phone that would eventually be called the iPhone. Given Apple's twenty-first century record of technological innovation and superb user design, the iPhone offered a new hope. It promised a new look at the entire cellular phone industry and the possibility of improved technology that wouldn't be afraid to strike out in bold new directions.

Apple acknowledged that they were working on an iPhone in early 2007. When they previewed their technology, it became increasingly obvious that the iPhone would be something new and different. Excitement grew at a fever pitch. On the release date—June 29, 2007—people camped outside Apple stores. Huge lines stretched throughout the day as people competed to be among the first to own what can only be called a *smarterphone*, the first of a new generation of user-friendly mobile technology.

When users began to try out their new iPhones, the excitement only mounted. The iPhone was easy to use and it provided numerous bells and whistles, from stock and weather reports to always-on internet access. Sales reflected the frenzied interest. Apple sold 270,000 iPhones in two days and topped a million units in just a month and a half. Now, a year and a half after the initial release, interest in the iPhone continues to grow. Apple's July 11, 2008, release of the new 3G iPhone and its public deployment of the iPhone software development kit (SDK) promise to multiply the iPhone's success in the future, with even higher numbers of iPhone sales predicted for 2009 and beyond. The 3G managed to hit a million units sold in just three days. We're atop a new technological wave, and it has yet to crest.

But what are the technologies that made the iPhone a hit, and how can you take advantage of them as an iPhone programmer? That will be the topic of this first chapter, where we'll not only look at the core specifications of the iPhone but also discuss the six unique innovations that will make developing for the iPhone a totally new experience.

1.1 *iPhone core specifications*

The iPhone is more than a simple cell phone and more than a smartphone like the ones that have allowed limited internet access and other functionality over the last several years. As we've already said, it's a *smarterphone*. If the iPod is any indication of market trends, the iPhone will be the first of a whole new generation of devices but will simultaneously stay the preeminent leader in the field because of Apple's powerful brand recognition and its consistent record of innovation.

Technically, the iPhone exists in two largely similar versions: the 2007 original release and the 2008 3G release. Each is a 4.7- or 4.8-ounce computing device. Each contains a 620 MHz ARM CPU that has been underclocked to improve battery performance and reduce heat. Each includes 128 MB of dynamic RAM (DRAM), and from 4 to 16 GB of Flash memory. The primary differences between the two devices center on the global positioning system (GPS) and networking, topics we'll return to shortly.

Programmatically, the iPhone is built on Apple's OS X, which is itself built on top of Unix. Xcode, the same development environment that's used to write code for the Macintosh, is the core of native programming for the device. Putting these two elements together reveals a mature development and runtime environment of the sort that hasn't been seen on most other cell phones (with the possible exception of Windows Mobile) and that upcoming smarterphone technologies won't be able to rival for years.

The iPod Touch

A few months after the release of the original iPhone, Apple updated their iPod line with the iPod Touch. This was a new iPod version built on iPhone technology. Like the iPhone, it uses a 480x320 multi-touch screen and supports a mobile variant of Safari.

However, the iPod Touch is *not* a phone. The original version didn't have any other telephonic apparatus, nor did it include other iPhone features such as a camera. The year 2008 saw the release of a new version of the iPod Touch that included an external speaker and volume controls missing from the original 2007 model. Because of its lack of cellular connectivity, the iPod Touch can only access the internet through local-area wireless connections.

The developer advice in this book will largely apply to the iPod Touch as well, though we won't specifically refer to that device.

However, these general specs tell only part of the story. By looking deeper into the iPhone's input and output, its network, and its other capabilities, you'll discover what makes the iPhone a truly innovative computing platform.

1.1.1 *iPhone input and output specifications*

Both the input and the output capabilities of the iPhone feature cutting-edge functionality that will determine how developers program for the platform. We're going to provide an overview of the technical specifications here; later in this chapter we'll start looking at the iPhone's most unique innovations, and then return for a more in-depth look at the input and output.

The iPhone's input is conducted through a multi-touch-capable capacitive touchscreen. There is no need for a stylus or other tool. Instead, a user literally taps on the screen with one or more fingers.

The iPhone's visual output is centered on a 3.5" 480x320-pixel screen. That's a larger screen than has been seen on most cell phones to date, a fact that makes the iPhone's small overall size that much more surprising. The device is literally almost all screen. The iPhone can be flipped to display either in portrait or landscape mode, meaning that it can offer either a 480-pixel-wide or a 480-pixel-tall screen.

The iPhone's output also supports a variety of media, all at the high level that you'd expect from the designers of the iPod. Music in a number of formats—including AAC (Advanced Audio Coding), AIFF (Audio Interchange File Format), Apple Lossless, Audible, MP3, and WAV—is supported, as well as MPEG4 videos. Generally, an iPhone delivers CD-quality audio and high frame rate video.

Although users will load most of their audio and video straight from their computer, the iPhone can play streams at a recommended 900 kbps over wi-fi, but that can be pushed much higher on a good network. Multiple streaming rates always choose the optimal method for the current network interface—which brings us to the question of the iPhone's networking capabilities.

1.1.2   *iPhone network specifications*

The iPhone offers two methods of wireless network connectivity: local area and wide area.

The iPhone's preferred method of connectivity is through a local-area wireless network. It can use the 802.11g protocol to link up to any nearby wi-fi network (provided you have the permissions to do so). This can provide local connections at high speeds of up to 54 megabits per second (Mbit/s), thus making a network's link to the internet the most likely source of speed limits, not the iPhone itself. Everything has been done to make local-area connectivity as simple to use as possible. Passwords and other connection details are saved on the iPhone, which automatically reconnects to a known network whenever it can. Switches to and from local wi-fi networks are largely transparent and can happen in the middle of internet sessions.

The original iPhone uses the EDGE network for wide-area wireless connectivity, falling back on this network whenever local-area wireless access isn't available. The EDGE network supports speeds up to 220 kilobits per second (kbit/s). Compared to old-style modems, which were accessing the early internet just 15 years ago, this is quite fast, but compared to broadband connectivity it's not that good. Although the original iPhones have *already* been phased out, millions of users are still using them, and thus EDGE network speed remains relevant.

The 3G iPhone supports the third-generation of mobile phone standards, which are well developed in Europe but just emerging in the United States. Network speed standards for 3G are loose, with stationary transfer speeds estimated as low as 384 kbit/s or as high as several Mbit/s. A 3G connection should generally be noticeably quicker than EDGE but still not as fast as a local-area network. In addition, 3G iPhones may drop back to EDGE connectivity if there's insufficient 3G coverage in an area.

These network specifications will place the first constraints on your iPhone web development (and will be of relevance to SDK programs that access the internet as well). If you're working in a corporate environment where everyone will be accessing your apps through a companywide wi-fi, you probably don't need to worry that much about how latency could affect your application. If you're creating iPhone web pages for wider use, however, you have to presume that some percentage of your iPhone users will be accessing them via a wide-area wireless network. This should encourage developers to fall back on lessons learned in the 1990s. Web pages should be smaller and use clean style sheets and carefully created images; data should be downloaded to the iPhone using Ajax or other technologies that allow for sporadic access to small bits of data.

> ## The web vs. the SDK
>
> Throughout the book we're going to talk about two major categories of programming for the iPhone: web development and SDK programming.
>
> Web development involves the creation of web pages that work well on the iPhone. These pages use standard web technologies such as HTML, Cascading Style Sheets (CSS), JavaScript, PHP, Ruby on Rails, and Python; iPhone-specific technologies such as the WebKit, iUI, and Canvas; and iPhone-specific tools like Dashcode.
>
> SDK programming involves the design of programs that run natively on the iPhone. These programs are written in Objective-C, compiled in Xcode, and then deployed through the iPhone App Store.
>
> We'll compare the two programming methods in the next chapter; then we'll dedicate part 2 of this book to learning all about iPhone web development and parts 3 and 4 to digging into Apple's iPhone SDK.

Thus far, we've discussed iPhone specifications that are relevant to both web and SDK development. However, there's one additional element that's clearly web only: the browser.

1.1.3 *iPhone browser specifications*

The iPhone's browser is a mobile version of Apple's Safari. It's a full-fledged desktop-grade browser with access to DOM, CSS, and JavaScript. However, it doesn't have access to some client-side third-party software that you might consider vital to your web page's display.

The two most-used third-party software packages that aren't available natively to the iPhone are Flash and Java. There was some discussion of a Java release in 2008, but the SDK's restriction against downloads seems to have put that effort on hold. We'll talk about these and other "missing technologies" more in chapter 3.

Beyond listing what's available for the iPhone Safari browser (and what's not), we'll also note that it works in some unique ways. There are numerous small changes that optimize Safari for the iPhone. For example, rather than Safari's standard tabbed browsing, individual "tabs" appear as separate windows that a user can move between as if they were individual pages.

iPhone's Safari also features unique "chrome," which is its rendition of toolbars. These gray bars appear at the top and bottom of every iPhone web screen. The chrome at the top of each page shows the current URL and features icons for bookmarks and reloading; we'll investigate how to hide this chrome when we look at iPhone optimized web development in chapter 3. The chrome at the bottom contains additional icons for moving around web pages and tabs. It's a permanent fixture on iPhone web pages. This iPhone chrome is more noticeable than similar bars and buttons on a desktop browser because of the iPhone's small screen size.

Having discussed the general capabilities of the iPhone—its input, its output, its network, and its browser—we've hit all of the major elements. But the iPhone also has some additional hardware features that are worthy of specific note.

1.1.4   Other iPhone hardware features

Cell phones feature numerous hardware gadgets—of which a camera is the most ubiquitous. The iPhone includes all of the cell phone standards, but also some neat new elements, as outlined in table 1.1.

Table 1.1   The iPhone is full of gadgets, some of them pretty standard for a modern cell phone, but some more unique.

Gadget	Notes
Accelerometers	The iPhone contains three accelerometers. Their prime use is to detect an orientation change with relation to gravity—which is to say they sense when the iPhone is rotated from portrait to landscape mode or back. However, they can also be used to approximately map an iPhone's movement through three-dimensional space. Could this make the iPhone the next Wii?
Bluetooth	This standard protocol for modern cell phones allows access to wireless devices. The iPhone uses the Bluetooth 2.0+EDR protocol. Enhanced Data Rate (EDR) allows for a transmission rate about triple that of older versions of Bluetooth (allowing for a 3.0 Mbit/s signaling rate and a practical data transfer rate of 2.1 Mbit/s).
Camera	Another de facto requirement for a modern cell phone. The iPhone's camera is 2.0 megapixel.
GPS	The original iPhone doesn't support real GPS, but instead offers the next best thing, "peer-to-peer location detection." This faux GPS triangulates based on the relative locations of cell phone towers and wi-fi networks for which real GPS data exists, and then extrapolates the user's location based on that. Unfortunately the accuracy can vary dramatically from a potential radius of several miles to several blocks; still, it's better than no GPS at all. The 3G iPhone includes a true Assisted GPS (A-GPS), which supplements normal GPS service with cell network information. Although there is a difference in accuracy between the two types of GPSs, they can both be accessed through the iPhone SDK using the same interface.

Of these hardware features, the ones that really stand out are the accelerometers and the GPS, which are not the sort of things commonly available to cell phone programmers. As you'll see, they spotlight two of the elements that make the iPhone unique: orientation awareness and location awareness. However, before we fully explore the iPhone's unique features, it's useful to put the device in perspective by comparing the iPhone to the mobile state of the art.

1.2   How the iPhone compares to the industry

Although the iPhone is an innovative new technology, it also serves as a part of a stream of mobile development that's been ongoing for decades. Understanding the iPhone's place within the industry can help us to better understand how it's differentiated from the rest of the pack.

1.2.1   The physical comparison

Physically, the iPhone is the sort of stunningly beautiful device that you'd expect from Apple. As we already said, it's almost all screen, highlighting Apple's ability to transform expectations for their electronic devices.

More specifically, the iPhone has a much larger screen than most of the last-generation cell phones, which tended to run from 320x240 pixels to 320x320 pixels and thus had as few as half as many pixels to play with as the iPhone. Although they had keyboards that were comparable with the iPhone's on-screen keyboard, their mousing methods were often based around styluses or tiny trackballs or, worse, scrolling wheels.

We expect other cell phones to start catching up with the iPhone's physical specs pretty quickly, but in the meantime Apple has used those specs to create a totally new cell phone experience—starting with its improved internet experience.

1.2.2   Competitive internet viewing

When compared to its last-generation competitors, the iPhone produces an internet experience that is more usable, better integrated, and more constant than the standard mobile experience.

The improvements in usability stem from the innovative specifications that we've already seen for input, output, and networking. On the input side, you no longer have to use a last-generation scrolling wheel to painfully pick your way through links up and down a page. On the output side, pages are displayed cleanly and crisply without being broken into segments, thus allowing for a faster, more pleasant web experience. Finally, for networking, you have the relatively good speed of the EDGE or 3G network combined with the ability to use lightning-fast local-area networks whenever possible. When compared to last-generation phones plagued by molasses-like internet connections, the change is striking.

With such a strong foundation, Apple took the next step and integrated the internet into the whole iPhone experience in a way that last-generation cell phones failed to do. The iPhone includes a variety of standard programs such as a YouTube interface, a stock program, a maps program, and a weather program that all provide seamless, automatic access to the internet. In addition, the SDK provides simple access to the internet for original applications.

All this functionality is supported by a constancy of internet access that is unlike anything the smartphone industry has ever seen. Supplementing its wi-fi access, an iPhone can access the internet through cheap add-on data plans. These plans allow for unlimited data transfer via the web and email. Thus, users never have to think about the cost of browsing the web. The end result is an always-on internet that, as we'll see, is another of the elements that makes the iPhone truly unique.

The Apple iPhone has brought mobile internet browsing out of the closet, a fact that is going to result in notable changes to current mobile web standards.

1.2.3   *Mobile web standards*

Prior to the release of the iPhone, a number of web standards were being developed for smartphones. The .mobi top-level domain was launched in 2006, built on the Wireless Application Protocol (WAP) and the Wireless Markup Language (WML) standard for cut-down, mobile HTML. In addition, the W3C Mobile Web Initiative has begun work on standards such as mobileOK (which is meant to highlight mobile best practices).

It is our belief that the mobile standards—and even the .mobi domain—are for the most part irrelevant when applied to the iPhone. We believe so because the iPhone provides a fully featured desktop-class browser and has vastly improved input, output, and networking capabilities. There *are* best practices for developing on the iPhone, and we'll talk about some of them in upcoming chapters, but they're not the same best practices required for leading-edge designs prior to 2007. As more smarter-phones appear, we believe that the mobile standards being worked on now will quickly become entirely obsolete.

This is not to say, however, that the iPhone is without limitations. It does not and cannot provide the same experience as a desktop display, a keyboard, and a mouse. New mobile standards for smarterphones will exist; they'll simply be different from those developed today.

Before completing our comparison of the iPhone to the rest of the industry, it's important to note that the vastly improved and integrated internet access of the iPhone is only part of the story.

1.2.4   *The rest of the story*

In 2008 Apple released the next major element in the iPhone revolution, the SDK, a developer's toolkit that allows programmers to create their own iPhone applications. Prior to the release of the SDK, most cell phone development kits were proprietary and highly specialized. The open release of the SDK could revolutionize the cell phone industry as much as the iPhone's web browsing experience already has.

Even that's not the whole story. The iPhone is an innovative product, top to bottom. To further highlight how it's grown beyond the bounds of the last-generation cell phone industry, we've identified six elements that make the iPhone truly unique.

1.3   *How the iPhone is unique*

The iPhone's core uniqueness goes far beyond its powerful web browser and its tightly integrated web functionality. Its unique physical form and the decisions embedded in its software also make the device a breakthrough in cell phone technology. Six core ideas—most of which we've already hinted at—speak to the iPhone's innovation. Understanding these elements (summarized in table 1.2) will help you in whatever type of development you're planning.

The idea of an *always-on internet* is something we already touched on earlier. However what's notable is how successful Apple has been in pushing this idea. Huge data transfer rates show that iPhone users are indeed *always-on*. In Europe, T-Mobile reported that their iPhone users transferred 30 times as much data as their regular

Table 1.2   The iPhone has a number of unique physical and programmatic elements that should affect any development on the platform.

Unique Element	Summary
Always-on internet	A well-integrated, constant internet experience
Power consciousness	A device that you can use all day
Location-aware	A device that knows where it is
Orientation-aware	A device that detects movements in space
Innovative input	Finger-based mousing
Innovative output	A high-quality scalable screen

users. Google has also shown a notable uptick among iPhone users, who are 50 times more likely to conduct a search than the average internet user. Looking at overall stats, the iPhone's mobile Safari has already become the top mobile browser in the United States and is quickly moving up in the international market as well. Anecdotal evidence is consistent, as friends talk about how an iPhone user is likely at any time to grab his or her iPhone to look up a word in Webster's or a topic in Wikipedia, showing off how the iPhone has become the encyclopedia of the 21st century for its users.

When Apple announced the iPhone, they highlighted its *power consciousness.* Users should be able to use their iPhone all day, whether they're talking, viewing the web, or running native applications. Despite the higher energy costs of the 3G network, the newest iPhone still supports 5 hours of talking or 5–6 hours of web browsing. Power-saving tricks are built deeply into the iPhone. For example, have you noticed that whenever you put your iPhone up to your ear, the screen goes black to conserve power? And that it comes back on as soon as you move the iPhone away from your ear? Power savings have also been built into the SDK, limiting some functionality such as the ability to run multiple programs simultaneously in order to make sure that a user's iPhone remains functional throughout the day.

Thanks to its GPS (true or faux), an iPhone is *location aware.* It can figure out where a user is, and developers can design programs to take advantage of this knowledge. To preserve users' privacy, however, Apple has limited what exactly programs can do with that knowledge.

Just as an iPhone is knowledgeable of large-scale location changes, it also recognizes small-scale movements, making it *orientation aware.* As we've already noted, this is thanks to three accelerometers within the iPhone. They don't *just* detect orientation; they can also be used to measure gravity and movement. Although some of this functionality isn't available to web apps, sophisticated input can be accessed by SDK programs.

Finally we come to the iPhone's *innovative input* and *output.* Thanks to a multi-touch screen and a uniquely scaled screen resolution, the iPhone provides a different interactive experience from last-generation cell phones, so much so that we've reserved an entire section for its discussion.

1.4   Understanding iPhone input and output

Although an iPhone has a native screen resolution of 480x320 pixels, web viewers won't see web pages laid out at that resolution. An iPhone allows a user to touch and tap around pages in a way somewhat similar to mousing, but it provides notable differences from a mouse interface.

These differences highlight the final notable elements in the story of what makes the iPhone unique.

1.4.1   Output and iPhone viewport

When using the iPhone for most purposes, you may note that it has a 480x320 screen that displays very clearly. This is not a far cry from the 640x480 video displays common on desktop computers in the late 1980s, albeit with more colors and crispness than those early EGA and VGA displays. Thus, as we mentioned when discussing the slower speeds of the wide-area network, we can again fall back on the lessons of the past when developing for the iPhone.

The iPhone's display becomes interesting when it's used to view web pages, because the 480x320 display doesn't show web pages at that size. Instead, by default a user looks at a web page that has been rendered at a resolution of 980 pixels (with a few exceptions, as we'll note when talking about web development). In other words, it's as if users pulled a web browser up on their computer screen that was 980 pixels wide, and then scaled it down by a factor of either 2:1 or 3:1—depending on the orientation of the iPhone—to display at either 480 or 320 pixels wide.

This scaled view is what the iPhone calls a "viewport." As you'll see, viewport size can be set by hand as part of a web page design, forcing a page to scale either more or less when it's translated onto the iPhone. However, for any web page without an explicit viewport command, the 980-pixel size is the default.

Realizing that most pages will scale by a factor of at least two is vital to understanding how web pages will look on an iPhone. In short, everything will be really, really small. As a result, good web development for the iPhone depends on ensuring that words and pictures appear at a reasonable size despite the scaling. We'll talk about how to do that using the viewport command, CSS tricks, and other methods in chapter 3.

And for SDK developers: note this is an issue for you as well, since the SDK's `UIWebView` class scales the screen just like mobile Safari does. We'll see the first example of this in chapter 11.

1.4.2   Output and orientations

We need to consider one other important element when thinking about the iPhone output: its ability to display in two different orientations, 480x320 or 320x480. Each orientation has its own advantages. The portrait orientation is great for listings, while the landscape orientation is often easier to read.

Each of these orientations also shows off the iPhone's "chrome" in a different way. This chrome will vary from one SDK program to another, but it's consistent when view-

Figure 1.1 The iPhone supports two dramatically different views, landscape and portrait. Choosing between them is not just a question of which is easier to read, but also requires thinking about how much of each view is taken up by toolbars and other chrome. Mobile Safari is used here as an example of how much room the chrome takes up in each display.

ing web pages in Safari, and thus we can use the latter as an example of orientation's impact on chrome, as shown in figure 1.1.

One of the interesting facts shown by this picture is that the web chrome takes up a larger percentage of the iPhone screen in the landscape mode than in the portrait mode. This is summarized in table 1.3.

Mode	Chrome % with URL	Chrome % without URL
Portrait	26%	13%
Landscape	35%	16%

Table 1.3 Depending on an iPhone's orientation, you'll have different amounts of screen real estate available.

The difference between the orientations isn't nearly as bad without the URL bar, which scrolls off the top of the screen as users move downward, but when users first call up a web page on the iPhone in landscape mode, they'll only get to see a small percentage of it. You'll see similar issues in your SDK development too, particularly if you're creating large toolbars for your applications.

Despite this limitation of landscape mode, many of the best applications will likely shine in that layout, as the built-in YouTube application shows.

With discussions of viewports and orientations out of the way, we've highlighted the most important unique elements of the iPhone output, but its input may be even more innovative.

1.4.3 Input and iPhone mousing

As already noted, the iPhone uses a multi-touch-capable capacitive touch screen. Users access the iPhone by tapping around with their finger. This works very differently from a mouse.

It's perhaps most important to say, simply, that *the finger is not a mouse*. Generally a finger is going to be larger and less accurate than a more traditional pointing device. This disallows certain traditional types of UI that depend on very precise selection. For example, there are no scroll bars on the iPhone. Selecting a scroll bar with a "fat finger" would either be an exercise in frustration or would require a huge scroll bar that would take up a lot of the iPhone's precious screen real estate. Apple solved this problem by allowing users to tap anywhere on an iPhone screen, then "flick" in a specific direction to cause scrolling.

Another interesting element of the touchscreen is shown off by the fact that *the finger is not singular.* Recall that the iPhone's touchscreen is *multi*-touch. This allows users to manipulate the iPhone with multi-finger "gestures." The "pinch" zooming of the iPhone is one such example. To zoom into a page, you tap two fingers on a page and then push them apart, while to zoom in you similarly push them together.

Finally, *the finger is not persistent.* A mouse pointer is always on the display, but the same isn't true for a finger, which can tap here and there without going anywhere in between. As you'll see, this causes issues with some traditional web techniques that depend on a mouse pointer moving across the screen. It also provides limitations that might be seen throughout SDK programs. For example, there's no standard for cut and paste, a pretty ubiquitous feature for any computer produced in the last couple of decades.

Besides resulting in some changes to existing interfaces, the iPhone's unique input interface also introduces a number of new touches (one-fingered input) and gestures (two-fingered input), as described in table 1.4.

Table 1.4   iPhone touches and gestures allow you to accept user input in new ways.

Input	Type	Summary
Bubble	Touch	Touch and hold. Pops up an info bubble on clickable elements.
Flick	Touch	Touch and flick. Scrolls page.
Flick, Two-Finger	Gesture	Touch and flick with two fingers. Scrolls scrollable element.
Pinch	Gesture	Move fingers in relation to each other. Zooms in or out.
Tap	Touch	A single tap. Selects.
Tap, Double	Touch	A double tap. Zooms a column.

When you're designing with the SDK, many of the nuances of finger mousing will already be taken care of for you. Standard controls will be optimized for finger use, and you'll have access only to the events that actually work on the iPhone. Chapter 14 explains how to use touches, events, and actions in the SDK. Even though some things will be taken care of for you, as an SDK developer you'll still need to change your way of thinking about input to better support the new device.

When you're developing for the web, you'll have to be even more careful. We'll return to some of the ways that you'll have to change your web designs to account for finger mousing in chapter 3. You'll also have to think about one other factor: internet standards. The web currently doesn't have any paradigms for flicks and other gestures, and thus the events related to them are going to be totally new. In chapter 4 you'll meet some brand-new iPhone-specific events, but they're just the tip of the iceberg and it might be years before internet standards groups properly account for them.

1.5   *Summary*

This concludes our overview of the iPhone. Our main goals in this chapter were to investigate how the iPhone differs from other devices, to discover how it's unique on its own, and to learn how those changes might affect development work.

Based on what we've seen thus far, our biggest constraints in development will be the potentially slow network, the relatively small size of the iPhone screen, and the entirely unique input interface. Although the first two are common issues for other networked cell phones, the third is not.

On the other hand, you've also seen the possibilities of many unique features, such as the iPhone's orientation and location awareness, though you won't get to work with these functions until we discuss the SDK.

In the next chapter we'll look at more of the differences between web development and the SDK so that you can better choose which of them to use for any individual development project.

Web development
or the SDK?

There are two ways you can develop for the iPhone. One approach is to write web pages for mobile Safari, using HTML, CSS, JavaScript, and your favorite dynamic language. The other is to write native applications to run directly on the iPhone, using Objective-C and the iPhone SDK.

We strongly believe that each programming method has its own place, and that you should always ensure you're using the correct one before you get started with any iPhone project.

2.1   Comparing the two programming styles

One of the things that might surprise programmers coming to the iPhone for the first time is the fact that web development and SDK programming can produce similar user-level experiences and support much of the same functionality. We've highlighted this in figure 2.1, which kicks off our discussion of the fact that web development and SDK programming are both reasonable alternatives when creating applications for the iPhone.

Figure 2.1 depicts what iPhone developers call a "utility," a two-page iPhone application that contains the actual application on the front page and setup information on the back page. Within the illustration, we've included a snippet of the code that allows the utility to flip between the pages when the info button is pushed. It's done in

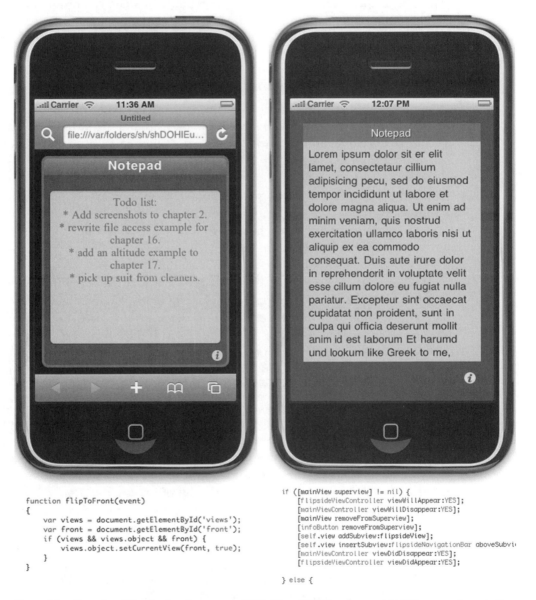

Figure 2.1 Though not identical, web programming (left) and SDK development (right) can produce similar output with similar underlying programming models.

JavaScript in the web example and in Objective-C in the SDK example. Each one produces an attractive rotating animation when the active screen is changed. We'll explain more about how the code for each programming style works in the chapters ahead, but we wanted to give you a preview now.

There's a further similarity between the two programs: each one features on its front page an editable "text view." This view can be used to display text, and can be edited using the iPhone's built-in keyboard.

We'll be talking a lot more about the similarities between the programming styles throughout this book. For now, we mainly want to highlight that neither style is a poor cousin: each has considerable functionality and can be used to create attractive UIs quickly and simply.

2.2   *A central philosophy: the continuum of programming*

Although we think that both styles of iPhone programming are useful, we're well aware that many of you are coming to this book from one of two directions. Either you're a web developer and want to learn how to optimize your web pages for viewing on the iPhone, or you're a programmer and you want to extend your C (or C++ or C# or J2ME) programming experience to the iPhone. We welcome you all, and we're certain that in this book you'll find a great introduction to your style of programming on the iPhone. Even if you've never programmed before but are simply intrigued by this new iDevice that you have, you'll be able to follow this book right through its introductory tutorials. Whichever route you've taken here, we encourage you to read the entire book, because we believe that by understanding—and using—the entire continuum of iPhone programming you'll literally get twice as much out of the experience.

For the *web developer*, we're going to tell you everything you need to know about the specifics of iPhone programming, including digging into some web features that you're probably not familiar with, such as the newest WebKit features, iUI, and Canvas. We hope you'll keep reading from there, as our material on the SDK is all quite introductory, and even if you've never worked with a compiled programming language, you can use this book to move up to SDK programming.

Chapter 9 is the foundation of this transition. It'll provide you with the basis of how a compiled programming language differs from the PHP, Perl, or Ruby on Rails that you might be familiar with. Starting from there, you should be able to learn the same lessons as a more experienced C programmer when we dive into the SDK itself.

For the *SDK programmer*, we're going to provide you with a complete introduction to all of the SDK's coolest features, including all of the unique iPhone features that we've already touched on, such as its GPS, its accelerometers, and its unique input device. However, we hope you won't consider SDK programming the be-all, end-all of iPhone software. We feel there are genuinely places where web development is a better choice.

We'll argue the reasons that you might select web development over SDK right here, in this chapter. Even if you opt not to do any web development, we invite you to at least skim through the web chapters, because we end each with a look at the lessons that web development can teach you about the iPhone as a whole.

Generally, no matter what type of programmer you are, you should think of this book as a toolbox. It's divided into two large compartments, but every tool within has the same goal: creating great iPhone programs. You just need to make sure that you're always using the right tool for the job at hand.

2.3   Advantages and disadvantages

Each of the web and SDK development models has its own advantages and disadvantages. We've summarized the biggest advantages for each model of development in table 2.1.

Table 2.1   Each model of development has its own advantages; for any project, you should use the model that best matches your needs.

Web development advantages	SDK advantages
Ease of development	Sophisticated development environment
Ease of first-time user access	Improved language depth
Rapid deployment	Integration with iPhone libraries
Automated updating	Improved graphics libraries
Access to dynamic data	Ease of continued user access
Access to existing web content	No downloading
Offline server access	Native speed
Integration with external web content	Improved privacy
Access to other users	Built-in economic model

We're going to look at each of these topics in more depth, starting with the web side of things. Afterward we'll offer some more precise suggestions on which programs we think should be developed using each type of programming.

2.3.1   Web development

The bywords of web development are *simplicity, dynamism,* and *globalization.*

Simplicity. Frankly, web development is simpler than using a low-level programming language like C. Although some of the dynamic programming languages are pretty sophisticated, you don't usually have to worry about things like memory management or even (for the most part) object modeling. If you're just outputting plain data, the ease factor in web development goes up by a factor of at least 10 times when compared to doing the same thing using the SDK's tables and other data outputs. Beyond that, when you're done developing a web program, all you need to do is upload your pages to your server. There are no hoops to jump through (other than those your individual company might impose).

It's also a lot simpler for users to begin working with your web program. They're much more likely to randomly view a web page than to pay to purchase your SDK program from the iPhone App Store, and thus it's a great way to attract many more users.

Dynamism. Hand in hand with that is the fact that you can literally update your program at any time. You don't have to worry about when or how your users will get a new program. You just change your server's source code, and the next time your users access the page (or, in the worst case, the next time they update their cache), they'll see all your bug fixes and other changes.

Similarly, you can constantly give users access to your newest data. Whereas data stored within an SDK program is more likely to be static, for a web program a user's view of data changes whenever you update it. This leads us to the next strength...

Globalization. When you create a web-based iPhone program, you become part of a global network that starts at your own server. This has a lot of advantages.

First, it means you can just create one program for use by both iPhone and desktop users (or, at worst, just one back-end, if you decide to program separate front-ends for each class of users). This will improve usability for your users if there's data they want to access (or even update) from both their desktop and their iPhone.

Second, it gives you direct access to your web server. This is particularly important because the iPhone keeps you from running programs in the background due to energy concerns. If you need to keep something running, you can hand it off to your server.

Third, it gives you rapid access to the rest of the web through URLs, Really Simple Syndication (RSS) feeds, and other data links. Granted, these advantages could be mimicked using web access from the SDK. However, if you're depending on the internet anyway, at some point you should just go full out and write your program for the web, allowing you to take advantage of the other strengths listed here.

Fourth, it's also a lot easier for your users to interact with other users, which might be particularly important for chats or multiplayer games.

Looking across the internet, there are numerous examples of superb iPhone web apps programmed by companies who felt that it was the superior medium. Google in particular is developing an entire suite of web apps, mimicking many of their desktop-centric web pages. The Hahlo twitter client is another example of a great iPhone program developed for the web: it makes use of online data but presents it in a tabbed, river format that should look somewhat familiar to iPhone users despite being a web app.

Overall, we feel that web development is the superior programming tool to use when

- You're programming a simple data-driven interface, with little need for the iPhone's bells and whistles
- You're expecting frequent updates to the program's data or to the program itself
- You're depending on the internet for data, users, or other access

2.3.2   *SDK development*

The bywords of SDK development are *sophistication, accessibility,* and *monetization.*

Sophistication. Just as web programs are a lot easier to develop and deploy, the flip side is that SDK programs allow for improved depth. This offers two important advantages.

First, you have greater depth implicit to the SDK itself. The development environment is a wonder of error reporting, complemented by well-integrated documentation and a built-in profiling package. This sophistication is also represented in the programming language, which is a well-considered object-oriented language. Although dynamic web languages are moving in that direction, Objective-C has already been doing OOP for over 20 years. Given that some sophisticated web languages like Java aren't available on the iPhone, the SDK's depth differentiates it that much more from the web.

Second, this depth shows up in the frameworks that you'll have access to when you use the SDK. They'll give you much deeper access to the iPhone's internals than any web page could. Apple has made some unique events, like orientation change, and some multifinger gestures available to web developers through the WebKit, but if you want to use the address book or GPS or want to take a deeper look at the accelerometers, you have to use the SDK. You can also access better graphics when you use the SDK.

Accessibility. Once users buy a program, it's available on their iPhone screen. Although a similar function is available for saving Safari bookmarks, it's likely only a percentage of users will take advantage of it.

That program is also usable wherever a user is, whether that be inside a subway tunnel or in a cell phone dead zone. The iPhone has an always on internet, yet there are inevitably times and places when it's not available—but a native program will be. Even an occasionally connected application might benefit from being native to the iPhone, as they can provide constant access to "old" data.

This goes hand in hand with the fact that the applications will always be running at iPhone speed, not constrained by the internet, a remote server, or some combination of the two.

Finally, because the program is sitting on their iPhone, users might feel more comfortable about using it to save their personal records than they would be if they knew the data was going out to the internet and thus potentially vulnerable (though the iPhone doesn't actually guarantee that its information won't go out onto the net, thus making this a mainly visceral advantage).

Monetization. We don't want to be entirely mercenary, but at the same time we think it's important to note that Apple is making it easy to sell your iPhone SDK programs through their iPhone App Store. Certainly you could depend on advertisements or even subscriptions for a web-developed program, but you don't have to worry about any of that if you write a program using the SDK.

At the time of this writing, the iPhone App Store is just getting started, but it's already shown off some excellent programs that clearly work better using the SDK than the web, primarily due to sophisticated graphics. This mirrors the web programs that we've already seen, like those designed by Google, which are excellent due to their web-based origins.

Overall, we feel that SDK development is the superior programming tool to use when

- You're creating a particularly sophisticated program or suite of programs
- You need to use any function (such as the address book, the accelerometers, the GPS, the camera, or animation) that isn't well supported on the web
- You want to monetize your program but don't have the web infrastructure to do so

2.3.3   *To each program its platform*

At this point you should be able to make your own decisions about which of the two programming platforms will best support the software that you want to write. To supplement your own thinking, we've listed a variety of programs in table 2.2, sorted by the method we'd suggest for programming them.

Table 2.2   **Different programs can each benefit from one of the main developmental models.**

Web programs	SDK programs
Chat programs	Accounting
Data wrappers (general)	Address books and other contacts
Data wrappers (frequently changing data)	Animated graphics
Games (simple multiplayer)	Data wrappers (critical information)
Inventory lists	Games
Schedules (multiperson)	Location-aware programs
Schedules (services)	Photo/graphic programs

The line between web development and SDK programming doesn't have to be as black and white as we make it out here. As you'll see momentarily, models exist for integrating the two types of development. But before we get there, we'd first like to outline our models for programming using just one of the programming packages, as these web development and SDK models will form the basis for most of this book.

2.4   *Stand-alone iPhone development*

The topic of iPhone development isn't just as simple as web versus SDK. We've divided those topics further by highlighting six ways you can develop iPhone pages using the web and two ways you can develop iPhone pages using the SDK. These methods are all summarized in table 2.3, complete with chapter references.

As shown in table 2.3, these models of development ultimately form the skeleton of this book. We'll summarize the methodologies here, and then expand on them in upcoming chapters.

Table 2.3 This book includes details on eight ways that you can program for the iPhone.

Method	Type	References
iPhone incompatible	Web	Brief mentions only
iPhone compatible	Web	Brief mentions only
iPhone friendly	Web	Chapters 3, 8
iPhone optimized	Web	Chapters 3, 8
iPhone web apps	Web	Chapters 4–6, 8
Dashcode	Web	Chapter 7
SDK native apps	SDK	Chapters 10–19
SDK web apps	SDK	Chapter 20

2.4.1 Web development models

We classify web pages into three types:

- Those that haven't received any special development work for the iPhone
- Those that have received normal development work
- Those that have received development work using Apple's Dashcode program

NONDEVELOPED WEB PAGES

The purpose of this book is to talk about how to develop web pages and programs for the iPhone, yet web developers will have a baseline that they're starting from: those web pages that haven't been developed with the iPhone in mind but that are viewed there anyway. We divide these non-developed pages into two general categories.

A web site is *iPhone incompatible* if the web developer has done no work to improve the site for the iPhone and it doesn't work very well. This may be due to a dependence on unsupported plug-ins like Flash and Java. It could also be due to the way in which CSS is used: a web site that uses a microscopically tiny font that's absolutely defined will be unreadable on the iPhone unless a user zooms and scrolls a lot. Very wide columns without viewport tags are another common reason for incompatibility. If you've got an iPhone-incompatible site, that might be why you picked up this book: to figure out how to improve the look and feel of your existing web site for the iPhone.

A web site is *iPhone compatible* if the web developer has done no work to improve the site for the iPhone and it sort of works anyway. It doesn't necessarily look that great, but at least it's usable, and so iPhone users won't actively avoid it. If you've got an iPhone-compatible site, it's probably because you're already making good use of CSS and you understand a core concept of HTML: that it's a markup language in which you can't accurately control how things are viewed.

DEVELOPED WEB PAGES

As an iPhone web developer, however, you want to do better than these undeveloped web pages. That's what the first part of this book is about (though we'll also touch on some SDK tools as we go). We categorize iPhone-specific web development into three types.

A web site is *iPhone friendly* if the web developer has spent a day of time—or maybe less—improving the experience for iPhone users. This involves simple techniques such as using the viewport tag, making good use of columns, using well-designed style sheets, and making use of iPhone-specific links. The basic techniques required to create iPhone-friendly web sites will be covered in depth in chapter 3.

A web site is *iPhone optimized* if the web developers have gone all-out to create pages that look great on the iPhone. They've probably inserted commands that deal with the iPhone chrome and have thought about iPhone gestures. They may link in unique iPhone style sheets when the device is detected. All of this requires a great understanding of how the iPhone works, but also provides a better experience for users. Many times the view that an iPhone user sees on an iPhone-optimized web site may be dramatically different than that experienced by a desktop user; despite the iPhone's claim to be a fully featured browser, there are some things that just don't work as well, and an iPhone-optimized site recognizes that. The slightly more advanced techniques needed to develop iPhone-optimized web sites will also be discussed in chapter 3.

Finally, some web sites may actually be *iPhone web apps*. These are web pages that are intended only to work on the iPhone, and in fact will probably look quite ugly if viewed from a desktop browser. We'll talk about using the functions of Apple's advanced WebKit in chapter 4. Then we'll discuss how to make pages that look like iPhone natives apps in chapter 5, including a look at the iUI library. Finally we'll look at the Canvas graphic library in chapter 6.

DASHCODE PAGES

As part of the SDK, Apple distributes a tool called *Dashcode*. It can be used to package JavaScript and HTML into a format specifically intended for the iPhone. Dashcode can also be used as development platform for many of the web app libraries and presages some of the functionality of the SDK. We'll cover it in chapter 7.

2.4.2 *SDK development models*

iPhone web apps represent a transition for iPhone developers. When you're engaging in simpler types of iPhone development—making existing web sites iPhone friendly or iPhone optimized—you're just considering the iPhone as one of many platforms that you're supporting. However, when creating iPhone web apps, you're instead developing specifically and exclusively for the iPhone.

Some developers will be happy staying with the web development techniques discussed in the first half of this book. Other developers will want to take the next step, to learn the SDK for those programs that could be better developed using that toolkit. Chapter 8 will offer some advice on making the jump from web development to the SDK—even if you've never programmed in a compiled programming language before. Then, the latter half of the book covers the SDK in depth.

THE SDK CONTINUUM

SDK development is even more of a continuum than web programming. As you learn more about Apple's SDK frameworks, you'll gradually add new tools that you can program with. Nonetheless, we've outlined two broadly different sorts of SDK programming.

SDK native apps are those SDK applications that make use only of the iPhone itself, not the internet and the outside world. Even given these restrictions, you can still write numerous complex native programs. We'll start things off with a look at the basic building blocks of the SDK: Objective-C and the iPhone OS (chapter 10), Xcode (chapter 11), Interface Builder (chapter 12), view controllers (chapters 13 and 15), and actions and events (chapter 14). Then we'll delve into the SDK toolkit, talking about data (chapter 16), positioning (chapter 17), media (chapter 18), and graphics (chapter 19).

SDK web apps are those SDK applications that also use the iPhone's always-on internet. In many ways they bring us full circle as we look at the web from a different direction. Although chapter 20 mainly covers how to access the internet using the iPhone, it's also what opens the door to the unique ways in which you can integrate your web and SDK work. That integration can appear in several forms.

2.5 *Integrated iPhone development*

The purpose of this chapter has been to delineate the two sorts of iPhone development—using the web and the SDK. Thus far we've talked quite a bit about what each style of programming does best, and we've even outlined stand-alone development methodologies. We've also touched on the fact that quite often you might want to use the two styles of programming *together.*

This is the only time that we're going to look at these ideas in any depth, but we invite you to think about them as you move through this book, to see how you can use the strengths of both the web *and* the SDK to your advantage. Table 2.4 summarizes our three integrated development methods, highlighting the strengths that each takes advantage of.

We're going to finish this chapter by exploring the three types of integrated development in more depth.

Table 2.4 Writing web programs using both the web and the SDK can let you take advantage of the strengths of both mediums (and all the contents of this book).

Method	Web strengths	SDK strengths
Mirrored development	Ease of first-time user access	Built-in economic model
Mixed development	Any strengths, especially: Rapid deployment Access to dynamic data	Any strengths, especially: Ease of continued access Native speed
Client-server development	Access to dynamic data Offline server access	Improved language depth Integration with libraries Native speed

2.5.1   *Mirrored development*

It's obviously easier to get users to try out a free but limited version of your software than it is to get them to purchase a more complete version. The business model of upgrading users from free to premium versions of software has been used extensively, with "freemium" being the latest buzzword. There are two ways you could create a freemium model for your software.

First, you could do what a lot of developers are already doing and offer a free trial version of your software on the iPhone App Store. This has the advantage of putting the software in the place that people are looking for software, but has the disadvantage that your application could get lost amid the hurly-burly of the store.

Second, you could create a version of your software for the web, using web app technologies. We think this model is particularly useful for those of you who have existing web pages that might already be drawing users to them in more highly targeted ways than the iPhone App Store could. Then, after releasing a limited version of your application over the web using techniques like the WebKit, iUI, and Canvas, you also release a feature-complete version of your application through the App Store using the SDK.

Although we've highlighted the economic reasons for this sort of mirrored development, it's possible that web sites might decide to extend existing web apps to include features not available in their web-based application. If so, then you'll have a clear delineation between what the programs include: the SDK will uniquely include those features that weren't available through the web, like location-aware and orientation-aware data.

2.5.2   *Mixed development*

In a mixed development environment, instead of making the web a subset of your SDK work, you're looking at an overall project and deciding to program some of it using the web and some of it using the SDK. This can be a chaotic methodology if not managed carefully, but it gives you the best opportunity to use the strengths of each sort of development. We find it most likely to work when you have two classes of users or two classes of programmers.

On the user side, a good example might be a situation where you have administrative users and end users. Assume you're managing some data project. The data input methods used by your administrators don't necessarily have to look great. You can develop them quickly using the web and then your administrators can choose whether to input data from their iPhones or from their desktops. Conversely, you want to provide a great user experience for your end users, so you take advantage of the iPhone's native graphical and animation features to output your data in interesting ways.

On the programmer side, you might simply have developers who are more comfortable in either the web or Objective-C arena. A mixed development project allows you to use not only the strengths of the individual programming methods but the strengths of your individual programmers as well.

The exact way in which you do mixed development will depend on the specifics of your project, but there should be real opportunities to take advantage of each programming style's strengths without much redundancy.

2.5.3 *Client-server development*

The final type of integrated iPhone development is the most powerful—and also one that's already in use on your iPhone, whether or not you know it. Client-server development combines back-end web development and front-end SDK development in a fairly typical client-server model. This is effectively what is done by existing iPhone programs such as Maps, Stocks, and YouTube, all of which pull data from the internet and display it in attractive ways while also taking advantage of the iPhone's unique capabilities.

On the one hand, you don't need a lot of the details of web development as presented in this book to create a data back end. Instead you can depend on your existing Perl, PHP, or Ruby on Rails code and use it to kick out RSS or some other data feed that your SDK software can easily pick up. On the other hand, if you're already doing that much development on the web side of things, creating a second web-based interface for iPhone users should be trivial.

Thus, a client-server development environment can give you the excuse to use either of the other integrated development strategies that we suggested.

2.5.4 *Last thoughts on integration*

We know some of you will be gung-ho to create a program that integrates SDK and web work all on your own, but we also recognize that in many larger companies it's likely that different people will undertake different tasks, with some developers doing the web side of things and some instead doing SDK programming.

If this is the case—and particularly if you're creating a suite of programs that integrate SDK and web development alike—don't be afraid to share the knowledge in this book as well as what you learn on your own. The iPhone is a new and innovative development platform. Its unique features create lots of opportunities for interesting work and also some potential gotchas as well. Although we're offering as many lessons as we can, we're sure there's a lot more you'll learn while you're out programming in the real world, and as with all knowledge, the best use is to share it.

2.6 *Summary*

The iPhone encourages two dramatically different methodologies for programming. You can either use traditional web programming techniques—with quite a few special features to account for the iPhone's unique elements—or you can dive into the intricacies of SDK development.

We've divided this book in two: half on web development and half on SDK development. There are two main reasons for this.

First, we wanted it to be an introduction to the entire world of iPhone development. No matter which path you want to take, this is the right book to get you started and to bootstrap you up to the point where you can program on your own. If you're a web developer without much C programming experience, you can even follow the entire path of the book, which will move you straight from the world of the web to the world of the SDK.

Second, we believe that good reasons exist to program in both environments. It's not merely a matter of expertise on the part of the programmer, but instead of capability on the part of the programming languages. There's a lot of simple stuff that's ten times easier to do in HTML, but similarly some complex stuff that's only possible to do using the SDK.

With the understanding of this book's bifurcation clearly in hand, we're now ready to jump into the first compartment of our toolbox; web development, where we'll start to explore what's necessary to make web pages look great on the iPhone device.

Part 2

Designing web pages for the iPhone

Now that you understand the basics of both the iPhone and the two ways to program for it, you're ready to dive into actual development work. Part 2 of this book will cover the first major method of iPhone programming: web development.

You'll start out with the basics in chapter 3, then over the course of the next three chapters you'll learn about three great libraries for building top-quality iPhone web pages: WebKit (chapter 4), iUI (chapter 5), and Canvas (chapter 6). Finally, you'll take a look at Apple's web development environment, Dashcode (chapter 7), and learn how to debug your web pages (chapter 8).

A central concept of this book is the ability to move between web design and SDK programming at will. Chapter 9 is a bridge chapter that highlights SDK concepts for the web developer.

Redeveloping web pages for the iPhone
3

This chapter covers

- Understanding the viewport
- Making pages iPhone friendly
- Making pages iPhone optimized
- Moving toward web apps

As you learned in chapter 2, iPhone-based web apps can give your users great opportunities to leverage the interconnectivity of the internet and to interact with other users. Throughout part 2 we introduce you to lots of tools that you can use to create web pages using web technologies that can be every bit as sophisticated as what you might write using the iPhone SDK.

Before we get there, though, we first want to touch on the fundamentals—those tools that you might use to improve your existing web pages for iPhone users, even before you begin writing totally new iPhone web apps.

The lessons in this chapter should be considered entirely foundational, as they'll be the basis for all the web chapters that follow. They also have wider scope, because you can apply them to web pages that might be viewed on other platforms, not just those for sole use by iPhone users.

Throughout part 2, we'll depend on the three basic tools of web development: HTML, CSS, and JavaScript. Each is crucial to the libraries and tools that we'll be discussing over the next seven chapters, and we presume you already have a solid basis in them. We're only going to touch on deeper programming languages a few times, and in those instances we'll use PHP for our examples, but you should be able to adapt the techniques we describe to whatever language you use for more dynamic web programming.

The Apple docs and web apps

Apple has a lot of comprehensive documentation at its website. These sources are generally more encyclopedic but less tutorial-oriented than what we've written in this book. Because of their more comprehensive nature, we're frequently going to refer you to Apple's docs after we've completed our introduction to a topic.

The Apple docs are generally split into two broad types: web docs and SDK docs. To access either of these, you'll need to sign up for Apple's Apple Developer Connection (ADC) program—which is free (though there are also premium memberships if you're interested).

Web docs. The web docs are located at http://developer.apple.com/webapps/. The most vital document is the "Safari Web Content Guide for iPhone," which contains a lot of general information on developing web pages for the iPhone, much of which is also covered in this book. You'll also find specifics on the WebKit, which is the topic of the next chapter, and Dashcode, the topic of chapter 7.

SDK docs. The SDK docs are available at http://developer.apple.com/iphone/. We'll talk about them in greater depth when we get to the SDK, in chapter 10.

There's also a third category of docs available at Apple—the Mac Dev documentation—but that's a bit beyond the scope of iPhone development.

Before we start redeveloping pages for the iPhone, we'll also point you toward chapter 8. There's info there about setting up a local web server on a Mac and on using a variety of clients for debugging. You might find it useful for following the examples in any of these web development chapters.

3.1   *The iPhone viewport*

The most fundamental concept in any iPhone web work is the viewport. It should be a part of every web page you write for the iPhone, from the simplest web redevelopment to the most complex Canvas-based graphical web app.

We first mentioned the concept of the viewport back in chapter 1. As we explained, though the iPhone has a display resolution of 320x480 (or vice versa, depending on orientation), it maps a much larger "virtual" window to that screen when you run Safari. The default virtual window (or viewport) is 980 pixels wide, which is then scaled down by a factor of approximately 3:1 or 2:1.

Figure 3.1 details what this viewport entails by showing the non-scaled content that can appear in the live area of each of the iPhone's two orientations.

If you choose to stay with the default viewport size, figure 3.1 tells you a lot about how web pages will display to your viewers. In portrait mode, things will appear much as you'd expect, as the viewport will be approximately square; everything above 1090 pixels will appear "above the fold." In landscape mode viewers will see a much abbreviated page, with only the first 425 pixels appearing above the fold.

Fortunately, you're not stuck with the default viewport size. You have two ways to change it. First, any web page served on a .mobi domain and any web page containing mobile XHTML markup automatically uses an alternative default viewport of 320 pixels. Second, you can purposefully change the viewport of any web page by introducing the new viewport metatag. You'll probably do this through a default header that you load across your entire site:

```
<meta name = "viewport" content = "width = 500">
```

Defining a viewport width of 500 would make your web page look as if it appeared in a 500-pixel-wide window before it was scaled onto an iPhone display. It's the simplest sort of viewport command, and probably what you'll do most often.

The remaining question is: why? What's the purpose of using a viewport?

Most of the time, you probably won't have to use the viewport at all. If you call up your web pages on an iPhone, and nothing looks too small, then you're fine. If you instead find out that things *are* small—due either to sitewide decisions or to the content of local pages—that's when you have to add a viewport tag to your web pages. Likewise, if you discover that your page looks really bad in the landscape orientation—due to the small live area—that might be another reason for a new viewport.

Generally, you should look at the viewport as an opportunity. In the world of desktop browsers, you have no idea what size of browser window a user might open, but on the iPhone you can control that exactly.

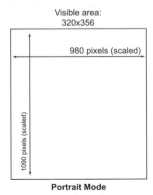

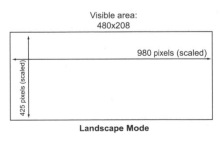

Figure 3.1 The iPhone's viewport allows a much larger web page to be shown, scaled, in the iPhone's display.

About the WebKit

The viewport command is part of the WebKit, an open source application browser engine that offers extensions to the core web standards. WebKit is being used by a number of browser developers, including Apache and Google. More importantly (at least for the iPhone designer) it's the basis of Apple's Safari. This means that a number of WebKit's extensions, not yet broadly available in browsers like Internet Explorer or Firefox, will work on the iPhone. iPhone web developers thus have access to lots of cool gestures and transformations that can give them considerable power using entirely web-based designs.

We're going to cover most of the possibilities of the WebKit in the next chapter. We've opted to cover one WebKit element here—the viewport—because it's crucial to any sort of iPhone-related web design, whether it be a page redevelopment or a full-fledged web app.

Sitewide viewport changes and local viewport changes will each have slightly different causes, and will each require slightly different solutions as a result.

3.1.1 *Making sitewide viewport changes*

Two elements might cause you to make sitewide viewport changes in your global header file: graphics or fonts that are too small.

Graphics are the most common problem. If you use them for navigation or to depict any other critical information, you'll probably have a problem because they're unlikely to be very readable at a 3:1 scale.

Font problems are usually due to absolute values used in the CSS font-size property. Clearly a font set to a small point size is going to be even smaller on the iPhone screen. The better answer is to make changes to your CSS files, which we'll return to shortly. But if you can't for some reason, this may be another reason to change your sitewide headers.

Typically, deciding on a sitewide viewport size will take some fiddling. The exact size of graphics or fonts may force you to select a certain value. If you have a sitewide navigation bar, you'll probably use its width as your viewport size. In the absence of any specific issues, a viewport size of 480 tends to work great. It'll be readable in portrait mode (at a 3:2 scale) and *very* readable in landscape mode (at a 1:1 scale). You won't want to go much lower than that, and you definitely *shouldn't* go all the way down to a 320-wide viewport; a change that extreme will probably make your web page break in other ways, and also ignores the excellent clarity of the iPhone screen.

The goal is to figure that things *will* be smaller on an iPhone than on a desktop browser and find a good compromise within that constraint.

3.1.2 *Making local viewport changes*

Adjusting your global viewport is the first step in making your web pages readable on an iPhone. However, you may also find individual pages that look bad. This situation is most likely to occur on pages that display an individual graphic or applet. The Apple developer pages give an example of a Sudoku applet that appears much too small on an iPhone page because it was only designed to be a few hundred pixels wide. The authors ran into problems with pages that displayed book covers, which tended to max out at 450 pixels wide. In both cases when using a default viewport of 980 pixels, the individual elements appeared up in the top-left of the iPhone screen, much too small and left wasted white space on all sides.

One solution to this problem is to set an individual viewport on each relevant page with a width equal to the known (or calculated) width of the applet or graphic. The other is to use a special `device-width` constant in the metatag, like this:

```
<meta name = "viewport" content = "width = device-width">
```

`device-width` is one of several advanced elements that may be added to the viewport metatag by developers who have more complex sites.

3.1.3 Viewport properties and constants

The iPhone recognizes a total of six viewport properties, as shown in table 3.1. The width is the only viewport property that you will use on most web pages.

Table 3.1 The iPhone recognizes six properties that may be used as part of a viewport metatag to control exactly how an individual web page displays on an iPhone.

Property	Default	Minimum	Maximum	Constants
height	Calculated	223	10,000	device-height, device- width
width	980	200	10,000	device-height, device- width
initial-scale	1.0	minimum-scale	maximum-scale	
minimum-scale	.25	>0	10.0	
maximum-scale	1.6	>0	10.0	
user-scalable	Yes	N/A	N/A	yes, no

We've already discussed how the `height` and `width` properties work: they assume a virtual window of the indicated height or width and then scale appropriately for display on the iPhone. Note that the device-width constant (which we've already met) has a match in `device-height`; you can decide whether you want your web page to fill the width of an iPhone display or its height.

The other four properties all control how zooming works. `initial-scale` determines how much an iPhone zooms when you initially view a page. The default value of 1 fits a web page to the iPhone screen. You might set it to a value smaller than 1 to immediately zoom to a leftmost column to make things more readable to iPhone viewers. But be careful when you use this technique, since it may not be obvious to your users that they're viewing only part of the page.

user-scalable determines whether iPhone viewers are allowed to zoom in and out using pinch-zooming. If it's set to no, then no zooming is allowed. If—as by default—it's set to yes, then users may zoom in down to the minimum-scale value and they may zoom out up to the maximum-scale value. Generally, you shouldn't have to change these last three values for redeveloped web pages, as a viewer choosing how your page looks is what the web is all about. However, if there are good UI reasons for controlling scalability or if you think a page looks particularly bad at certain scales, you may choose to modify them. On the other hand, you probably should turn scaling off for web apps, as those will be programs that you're developing explicitly for viewing on an iPhone.

Note that you can set multiple values in a metatag by separating them either with a comma or a semicolon:

```
<meta name="viewport" content="width=device-height; initial-scale= 0.667">
```

You'll find that we'll keep coming back to viewports in the next several chapters—it's a crucial technique for iPhone web pages.

3.2 *Making your web pages iPhone friendly*

The simplest sort of web page redevelopment involves making your page's "iPhone friendly," which we briefly defined in the previous chapter. In short, this concept involves you taking a day or less of work to do the simple cleanup required to turn existing web pages into web pages that work pretty well on the iPhone.

The basis of an iPhone-friendly page—as with any iPhone-based web page—is a good viewport. Once you've figured that out, you should also look at your pages' technologies and generally consider good web design techniques to make sure your page looks nice. Making a page iPhone friendly is more about fixing problems than about showing off the iPhone's strengths.

3.2.1 *Avoiding missing iPhone functionality*

Although the iPhone is often described as a fully functioning web browser, it's not. In particular, you won't have access to certain third-party plug-ins and you'll discover that many events aren't available on the iPhone.

If you can, avoid using these plug-ins and events. That'll be pretty easy to do when you're creating brand-new web apps, starting in the next chapter. But when you're redeveloping existing web pages, you may find replacing this functionality impossible; nonetheless, it's important to know where your pages will run into problems.

THE MISSING TECHNOLOGIES

In chapter 1 we mentioned the most notable third-party technologies that you won't be able to use on the iPhone: Flash and Java. However, there are several other missing technologies that you might encounter, the most important of which are listed on table 3.2.

The list of unsupported technologies in table 3.2 may well be different by the time this book sees print. The best solution to deal with third-party technologies is *always* to

Table 3.2 Although the iPhone's browser itself is fully functional, some third-party technologies are not yet supported, the most important of which are listed here.

Technology	Comments
Flash	The Flash programming language is widely expected to be supported at some point, though Apple has said nothing official. In the meantime, the WebKit and Canvas offer some weak alternatives for animation, as described in chapters 4 and 6.
Java	Sun announced plans to support Java on the iPhone in 2008, but that's since run afoul of Apple's limitation against downloads in SDK programs. There's currently no word on when and if Java will be supported.
SVG	Scalable vector graphics are not supported. Canvas provides a good alternative, as described in chapter 6.
XSLT	Extensible Stylesheet Language Transformations are not supported.
WML	The iPhone's Safari is not a cut-down, last-generation cell phone browser; thus the Wireless Markup Language is largely irrelevant. However, XHTML mobile profile documents do work at .mobi domains.

check for them. If the technology is not detected, you should ideally deliver the user to an alternative page that displays the same information in a different format. If that's not possible, you should at least deliver users to a page that explains why they can't display the content. Simply displaying a nonworking page is probably the worst alternative of all.

Once you get past those third-party software packages, most things will work correctly in your browser. As we've already discussed, DOM, CSS, and JavaScript are among the advanced web techniques that will work as expected when viewed on an iPhone. However, there's a big asterisk on the statement that JavaScript works, and that has to do with events.

THE MISSING EVENTS

Unfortunately, events won't work quite as you'd expect on the iPhone. Much of this goes back to one of the unique iPhone features that we discussed in chapter 1: its input device. Most specifically, it's important to remember that the finger is not a mouse. Because a finger may or may not be on the screen at any time, your web page suddenly becomes stateless; you can no longer depend on events that presume that the mouse always moves from point A to point B through all the space in-between.

The statelessness of iPhone events causes two traditional categories of events to fail on the iPhone: drags and hovers. Thus, you can no longer allow click and drag (because the iPhone instead uses that gesture for its scrolling) and you can no longer test when the mouse moves over an area (because it doesn't).

The loss of these events is going to affect the way you program using both CSS and JavaScript. In CSS your biggest problem will be with hover styles, which will of course no longer appear, but in all likelihood that won't be a major issue for your web pages. In JavaScript these differences in input cause several specific events to work either differently or not at all, as detailed in table 3.3 (but we also suggest looking at

Table 3.3 Not all JavaScript events work on the iPhone, leaving you with a more restricted palette of options than in a traditional browser.

Functional events	Changed events	Nonfunctional events
form.onreset	formfield.onmousedown	document.onkeydown
formfield.onblur	formfield.onmousemove	document.onkeypress
formfield.onchange	formfield.onmouseout	document.onkeyup
formfield.onclick	formfield.onmouseover	form.onsubmit
formfield.onfocus	window.onscroll	formfield.ondblclick
formfield.onmouseup		formfield.onmouseenter
textarea.onkeydown		formfield.onmouseleave
textarea.onkeypress		formfield.onselect
textarea.onkeyup		window.oncontextmenu
window.onload		window.onerror
		window.onresize

http://www.quirksmode.org/dom/events/ to see if anything has changed by the time you read this book).

It's the changed JavaScript events that bear the most discussion, because they're the most likely to cause you headaches because they *seem* to work.

formfield.onmousedown occurs at an unusual time. Unlike on a desktop browser, the onmousedown event isn't reported until the onmouseup event occurs, making its usage relatively meaningless. This is what breaks the click-and-drag event types that we've already discussed.

formfield.onmousemove, formfield.onmouseout, and formfield.onmouseover are all similarly intertwined. All three always occur in that order when a user clicks on the screen. In addition, if the user clicked on a changeable element, formfield. onmousedown, formfield.onmouseup, and formfield.onclick are also reported immediately afterward. That's a ton of events that actually don't provide much information since they always occur together.

Finally, the window.onscroll event works kind of like formfield.onmousedown, in this case not appearing until after the scroll completes. This is less likely to be a UI issue for the average programmer, but it means that you no longer intercept a scroll before it occurs.

Of the events that just don't work on the iPhone, the formfield.onmouseenter and formfield.onmouseleave events are the most likely to cause problems on your web page. These prevent you from recognizing hover-style events in JavaScript.

Because you can't use these event types, you'll find that many traditional UIs fail. Cut and paste is one of our first losses. Pop-up menus are going to be another casualty since most of them depend on click and drag to work. There *are* workarounds for these: you could develop a new cut-and-paste methodology in which you click on the edges of your text, and you could develop a new pop-up menu methodology in which you click and then release before a menu appears. These are all beyond the scope of the relatively simple web page changes that we're covering here.

We're going to return to the topic of web events on the iPhone twice. Later in this chapter, when we explore iPhone-optimized web pages, we'll highlight the exact events that occur when a user touches or gestures at the iPhone screen. Then in chapter 4 we'll talk about some new iPhone events that are introduced in the WebKit. Your first step in making a web page iPhone friendly will just be working around the event problems that we've highlighted in this section, but if you want to take the next step and rewrite your event model, we'll point you toward those resources.

3.2.2 Creating good links

Events show us once more how finger mousing isn't the same as mechanical mousing. The topic comes up again for basic web designs when you think about how your users select links. The topic is important enough that it's another thing you need to consider when first redeveloping your web pages for the iPhone.

The main problem here is that a mouse pointer typically has a hot spot that's one pixel wide—and a user's finger point is many, many pixels wide. Thus, if you put your links too close together—such as in a column-based navigation bar (navbar) with links one under the other—a user won't be able to select them without zooming in. It's the same story for forms.

Depending on the way you've set up your web page, you may be able to solve this problem instantly. In the case of that columnar navbar, you can just put spaces between your links, and they'll probably look fine on both desktop and iPhone browsers. For more complex setups, including forms, you may need a more wholesale page rewrite—or to create iPhone-specific views, a topic we'll return to when we get to iPhone optimization.

In any case, the point here is to look at your links to see if they're usable on the iPhone, and if not, to fix them with some simple redesigns.

3.2.3 Practicing good web work

If you've defined a viewport, created alternate pages for missing web technologies, and redisplayed any links that were too close together, you've done 90% of the work you need to make your web pages look good on the iPhone. However, before we leave the topic entirely, we'd like to offer our top suggestions for using generally good web practices to make your pages look their best. If you're already an experienced web designer, you've probably got this in hand already, in which case you should skip ahead to iPhone optimization.

GOOD CSS

To make your web pages more accessible on different platforms, we suggest you don't use absolutes in your CSS. Instead, use relative values. For font sizes, use percentages like 80% and 120%, not absolutes like 10pt or 12px.

For font types, allow for a variety of fallbacks, and make sure they include fonts that are available on the iPhone, as listed in table 3.4.

iPhone fonts	Notes
AmericanTypewriter	
Arial	
Arial Rounded MT Bold	
Courier New	Includes Courier
Georgia	
Helvetica	Includes Helvetica Neue
Marker Felt	
Times New Roman	Includes Times
Trebuchet MS	
Verdana	
Zapfino	(Zapfino)

Table 3.4   The iPhone supports a large set of fonts. For an iPhone-friendly page, make sure your CSS files include at least one of these in its standard listing.

Finally, consider carefully how you do any CSS positioning on your web pages. Sure, absolute positioning can make a web page look great, but it'll ensure that your page only works at standard sizes—which means on an iPhone that you'll be forced to use a default viewport size (like 980 pixels) rather than a smaller one that may allow for better scaling of fonts and graphics. Further, there are some quirks with positioning on the iPhone. We've listed our suggestions for using CSS positioning in table 3.5.

Table 3.5   There are four methods that you can use to position elements using CSS—but don't expect them to work quite as you expect on the iPhone.

Type	Definition	Comments
Static	Positioning in normal flow of page	The default behavior
Relative	Positioning relative to the normal flow	Will work on an iPhone, and is the preferred method for more intricate layout in a mixed device environment
Absolute	Positioning relative to the containing block	Will work with an iPhone-specific style sheet, but has more potential to cause problems if used to lay out an entire page due to size differences between different devices
Fixed	Positioning relative to the browser window	Not supported on the iPhone

The biggest surprise here is that fixed positioning is not supported. This is because Apple felt that it did not meld with its new paradigm of zooming web pages. A fixed element just doesn't make sense after you pinch-zoom.

If you're dependent on fixed positioning, though, you *can* use absolute positioning to mimic it. This is a standard web technique that we won't discuss in depth here: you simply create one `<div>` that's the size of your `<body>`, then stick another `<div>` inside that floats to the bottom (or top or whatever) of that top-level `<div>` using absolute positioning.

We'll return to CSS in a bit, when we look at ways you can move your web pages from iPhone friendly to iPhone optimized by creating totally new CSS files intended for use *only* on the iPhone.

GOOD TABLES AND GOOD COLUMNS

Column-based layouts have become a de facto part of web design over the last decade. This sort of design is more important than ever on the iPhone because users can zoom into a specific column using the double-tap feature of the iPhone. While this generally works without additional development work required, a careful developer can make sure that columns are optimized for the iPhone viewer.

First, this means that you should have columns in your web page. You probably already do.

Second, once you have columns, it's important to make sure that your columns logically match the different categories of content that your pages contain. For example, if you have both content and navigation on a web page, you could split those up logically into two different columns. Mixing things together will make your columns less useful for your iPhone readers. This may cause you to rethink using floating tables or embedding tables within tables. On the flipside, you don't want to split up content between multiple columns. For example, having a single story run through several columns on a page probably isn't a good idea as it will force an iPhone user to zoom in, then out, then back in to see everything.

Third, it's important to consider the fact that iPhone viewers may be looking at your pages one column at a time. This means that you need to think even more carefully than usual about what happens when a viewer gets to the end of a column. You thus might want to include some sort of navigation down at the bottom of a column.

3.2.4 *Fixing common problems*

To close up this short section on how to make your web pages more iPhone friendly, we've put together a quick guide for solving common problems, linking together all the suggestions we've offered so far. First, take a look at our iPhone best practices, which we've summarized in table 3.6. Then, if you're having any particular problems, take a look at the individual sections below.

Now let's move on to some of the problems that you may be encountering if you're not following all of these best practices (or maybe even if you are).

GRAPHICS ARE TOO SMALL

This is a classic viewport problem that arises from 980-pixel windows being viewed on a 320-pixel screen. The problem could be solved using a viewport, but we've also listed some other possibilities:

- Decrease the size of the viewport so that it's in the range of 480–640.
- Replace the graphics with corresponding text, especially if you plan to use the graphics for navigation.
- Push the graphic off to a subsidiary page that only shows the graphic and allow the user to click through to that page. Be sure that the subsidiary page shows the graphic at `device-width`.
- Zoom in to the column with the graphics using the `initial-scale` viewport property.

Table 3.6 **Making your web pages iPhone friendly can take just a couple of hours of work, but can result in dramatically improved user experiences for iPhone users. Here are several iPhone best practices that can improve your pages for your users, all summarized from discussions found in chapters 1 and 3.**

Practice	Explanation
Use a viewport	Decide the size of virtual browser that your web pages will support, and lock that in with a viewport command.
Use relative values	Whether you're writing CSS, laying out tables, or doing something else, always use relative values, not absolutes.
Use columns	Lay your pages out in columns whenever you can, and make sure those columns match up to logical data units on your web pages.
Watch your media	Don't use Flash, Java, or other unsupported third-party client software on web pages that you expect iPhone users to access.
Be careful with events	Remember that certain events don't work the same on the iPhone. Don't use click-and-drag events or hover events on iPhone web pages.
Speed up your downloads	Fall back on lessons learned in the 1990s to make faster, leaner web pages that will load quickly over the EDGE and 3G networks.
Separate your links	Put spaces between your links to make it easy for users to click on them with their fingers.
Avoid scrollable framesets	Because of the lack of scroll bars on the iPhone, framesets—which would be individually scrollable on the desktop—provide a subpar experience on the iPhone.
Use words, not pictures	Don't use graphics that just repeat words, as they slow down loading and may be very small on an iPhone screen.
Follow standard best practices	Although we feel that last-generation mobile best practices are already outdated, you should still follow more general web-based best practices. Make sure that your web pages validate and are clean to provide the optimal experience for all of your users.

WORDS ARE TOO SMALL

This also tends to be a viewport-related problem, and is likely to be less common than the graphic issue because a web client is already trying to adjust fonts to make them a reasonable size. Here are some possible solutions:

- Decrease the size of the viewport so that it's in the range of 480–640.
- Adjust your CSS to use relative values instead of absolute values.
- Zoom in to the column with the text using the `initial-scale` viewport property. Note that you probably want to do this if you're highlighting the core content of a page.

COLUMNS ARE TOO BIG

This is more a result of other issues than a problem in and of itself. You'll probably see it when you try to decrease the size of the viewport—to address one of the previous problems—and you discover that a column refuses to shrink. Solutions include the following:

- Use relative numbers (e.g., 50%) to define column widths, not absolute numbers (e.g., 750).
- Reduce the size of any pictures that may be forcing your column to stay wide. For example, if you're running forums and you allow users to post pictures to those forums, you may need to limit the width of pictures that users can post or alternatively to resize them on the fly.
- Allow large pictures to flow into nearby columns. If you're using tables, you might do this using a `colspan` attribute. For example, you might have a page where a 728x90 banner ad sits above the top of your content. For a desktop page, this arrangement probably works fine, but if you're using a smaller viewport for iPhones, you'll need that banner ad to flow into the next column, even if that's off the page on your iPhone display. If you need to flow off the page, in addition to setting the `colspan` attribute for the ad column you'll need to set the viewport's `initial-scale` property to something less than 1.0, which will keep the right-hand side of the ad from showing up on your iPhone screen.

3.3   *Making your web pages iPhone optimized*

In the previous section we explained how to make your web pages look better on the iPhone using a bare minimum of redevelopment. Our goal there was to fix problems and to thus give your iPhone users the same experience as your desktop users.

In the remainder of this chapter, we're going to take the next step. We'll look at more extensively redeveloping pages to work specifically for the iPhone. We won't yet be using any iPhone-specific libraries, but we'll explore some techniques and designs that will only work on the iPhone, and you may end up branching your web pages as part of this optimization process. To kick things off, we need to find out when our users are actually using an iPhone.

3.3.1   *Detecting the iPhone through USER_AGENT*

The idea behind iPhone optimization is to redevelop web pages so that they work great on the iPhone. To begin this process, you must know when a user is browsing from an iPhone. The easiest way to do this is—as is typical in web design—by looking at the user agent. Listing 3.1 shows the best way to do this using a PHP example.

Listing 3.1   Checking the user's agent to see when an iPhone is browsing

```
<?
if (ereg("Mobile.*Safari",$_SERVER['HTTP_USER_AGENT'])) {
    $iphone = 1;
}
?>
```

There are, however, a few caveats in using this method. First, it may not be Apple's preferred method. It was originally undocumented, and though it's documented now, it could easily change. At the least, you should make sure this detection appears in a global header file where you can easily modify it in the future.

Second, it's restrictive. There will soon be other smarterphone devices that have functionality similar to that of the iPhone. You may wish to use a broader net to catch all the smarterphone fish, but we leave that up to the specifics of your own website.

3.3.2   Detecting the iPhone through CSS

Depending on the precise dynamic language you use, you may have other ways that you prefer to use to detect which browser your users are using. Even CSS has its own methods that can be used to detect some browser capabilities.

We note this in particular because this method was originally Apple's only supported way for detecting iPhone usage. Listing 3.2 shows how CSS can recognize an iPhone and thus apply a different style sheet.

Listing 3.2   Applying style sheets using media detection

```
<link media="only screen and (max-device-width: 480px)" href="small.css"
type= "text/css" rel="stylesheet">
<link media="screen and (min-device-width: 481px)" href="large.css"
type="text/css" rel="stylesheet">
```

Besides being supported, Apple's method also has the interesting side effect that it will apply your small-device style sheet to any small screen that views your website, which may give you instant compatibility with future smarterphone devices.

This second method of detection leads right into our first major iPhone optimization topic; if you're going to take the time to improve your website for the iPhone—beyond the day-or-less that we suggested creating a friendly site would take—then your CSS files are the right place to start work.

3.3.3   Optimizing with CSS

The easiest way to improve your web page's readability on the iPhone is to start with your existing big-screen style sheet and then create a new small-screen style sheet that makes everything a little bigger. Table 3.7 offers some suggestions that we have found work well if you haven't made any changes to the native viewport size. Your individual

Element	Changes
Fonts with relative values	Increase 20%–30%
Fonts with absolute values	Increase 2–3 points
Select menus	Increase 20%–30%
Input boxes	Increase 20%–30%

Table 3.7   To make a web page viewable on an iPhone screen, you should create an alternative style sheet and increase the size of all your elements in that style sheet.

website will probably be different; ultimately you'll need to view and review your pages until you find a happy medium.

Of the CSS elements noted, select menus deserve a short, additional note. They're a great input type to use on iPhone-friendly or iPhone-optimized pages because of the built-in support for <select>s on the iPhone, which automatically pop up a large, easy-to-read widget. You should use them whenever you can.

For text-based CSS elements, we'll also note that there's an alternative that will allow you to change *all* of your text-based CSS elements at once. This is done with the -webkit-text-size-adjust property, which is a part of Apple's WebKit. You could easily implement it without doing any more WebKit work, but we've nonetheless left its discussion for the next chapter.

3.4   *Manipulating iPhone chrome*

Thus far we've offered up some standard techniques for differentiating iPhone and desktop viewers. Now that you know when a user is browsing with an iPhone, you can start writing specific code for that situation. Let's begin by looking at a simple optimization you can do by examining the iPhone's chrome and the limited ways in which you can adjust it using standard web techniques.

3.4.1   *The three bars*

The iPhone chrome consists of all those elements that appear at either the top or the bottom of an iPhone page. There are different types of chrome used on the various iPhone programs, but for the mobile Safari web browser, there are just three, as summarized in table 3.8. You may wish to again refer to figure 1.1 for their placement on the screen.

Table 3.8   Three different bars full of buttons and inputs appear on your mobile Safari screen on the iPhone.

Chrome	Functionality	Size
Status bar	Displays overall iPhone status: network connectivity, battery charge, and current time	320x20 or 480x20
URL bar	Displays the web page title and major web page functions: the URL bar, search button, and reload button	320x60 or 480x60
Bottom bar	Displays web page navigation functions: back and forward, bookmark buttons, and tab navigator	320x44 or 480x32

Each of the three bars shown in table 3.8 works slightly differently. The status bar is a permanent fixture of every iPhone page. Users probably don't even notice it because of its omnipresence, but nonetheless it takes up 20 pixels at the top of the screen. The status bar can appear in two different colors, black or gray, but we won't be able to control that until we get to SDK development; for web pages, the status bar is always gray.

The bottom bar is similarly stuck to the bottom of the page. It's a more obvious intrusion because it only appears on web pages. Many web developers have hoped for a way to get rid of the bar or replace it, but so far no methods have been made available. If you want a site-specific bottom bar, you have to use absolute positioning, as we mentioned earlier when discussing CSS.

The URL bar appears at the top of every web page you view, but it scrolls off the top of a page as a user moves downward, recovering those 60 pixels of space. Like the bottom bar, there's an advanced WebKit method that will get rid of this chrome. You can also exert some control over it using normal web methods. You can automatically push the URL bar off the top of the screen with a `window.scrollTo(0, 1)` command in JavaScript. This command must be delayed until the page has loaded sufficiently. Listing 3.3 shows one way to do this.

Listing 3.3   Scrolling the URL bar chrome off the screen with JavaScript

```
<script type="application/x-javascript">
if (navigator.userAgent.indexOf('Mobile Safari') != -1) {      Adds delayed
    addEventListener('load',hideURLBar,false);              ⟵    function
}

function hideURLbar() {            Scrolls
    window.scrollTo(0, 1);         window
}
</script>
```

The code for our URL scroller is very simple. The `hideURLbar` function does the scroll, but it isn't executed until the page is loaded.

Before you use this functionality be warned that, as with your work with the `initial-scale` property of the view screen, the result will not be entirely intuitive for the user. On the one hand, the user might be confused by seeing the page suddenly jump after it loaded. On the other hand, the user might be confused about where the URL bar went to, since it went away without the user doing anything. Some developers like the ability to instantly give their iPhone users another 60 pixels of space "above the fold," but these UI difficulties must be considered.

Of course, the iPhone does *lots* of automated stuff like this. The manner in which it swaps between two orientations as you twist your iPhone around isn't too different from an automatic scroll. So, perhaps users will get used to iPhones moving stuff around for them as they use the device more.

3.4.2   *Web clips*

Although not quite chrome, web clips represent another way in which mobile Safari provides you with unique functionality. Simply, a web clip is an icon that you can use to represent a web page on the iPhone.

As a developer, you just have to create a simple icon to represent your website. The web clip icon should be a 60x60 PNG graphic. Apple also suggests using "bold shapes and pleasing color combinations." You don't have to worry about the rounded corners or gloss that define Apple iPhone icons, because the iPhone will take care of all that for you.

Once you've uploaded your icon to your server, you can specify it with a link of type `apple-touch-icon`, like this:

```
<link rel="apple-touch-icon" href="/apple-touch-icon.png" />
```

Now you can encourage users to add your web page to the home screen of their iPhone, which they do by hitting the plus sign (+) in their bottom bar, then choosing Add to Home Screen. Your web page will appear with a short name and the icon that you designed, laid out in the standard iPhone manner.

Chrome defines the ways in which an iPhone *looks* unique, but as we've already discovered, it also *acts* unique. We've already looked at many events that were different or unavailable on the iPhone. Now that we're working on iPhone-specific code, we can examine the flipside: events that are totally new to the iPhone.

3.5 *Capturing iPhone events*

Earlier in this chapter we showed how some events don't work the same on the iPhone because of its unique input method. At the time, our task was simply to figure out which events to avoid when designing iPhone-friendly pages.

Now we're ready to look at the flipside of user input by examining many of the standard iPhone gestures and seeing how those touches turn into JavaScript events. If you prefer to instead capture touches by hand, we refer you to the next chapter, which discusses some WebKit-specific events that do exactly that.

3.5.1 *One-fingered touches*

Table 3.9 gets us started with iPhone events by looking at the different one-fingered touches (which we previously encountered in chapter 1) and the events that they create.

Table 3.9   **The iPhone recognizes several unique one-fingered touches, some of which correlate to normal web events.**

Touch	Summary	Events
Bubble	User views info bubble with touch-and-hold gesture on clickable elements.	(None)
Flick	User scrolls page with a one-fingered flick.	`onscroll`
Tap / Nothing	User touches to emulate a mouse, but doesn't click on a clickable element.	(None)
Tap / Click	User touches to emulate a mouse and clicks on a clickable element, and it doesn't otherwise change.	`mousemove` `mouseover` `mouseout`

Table 3.9   The iPhone recognizes several unique one-fingered touches, some of which correlate to normal web events. *(continued)*

Touch	Summary	Events
Tap / Change	User touches to emulate a mouse and clicks on a clickable element, and the content changes, as with a select box.	`mousemove` `mouseover` `mouseout` `mousedown` `mouseup` `click`
Tap, Double	User zooms into a column with a rapid double tap.	(None)

The most important thing to note here that the iPhone doesn't give you access to certain events. In particular, you can't see the two pure interface-centric one-fingered touches: zooming and info bubbles.

Of the accessible touches described in table 3.9, the flick is the only one that's somewhat iPhone-specific, but it's just mapped to a normal scroll event. You're not likely to need to know too often when your users are scrolling your page, but if you want to, simple JavaScript lets you do so:

```
<script type=" application/x-javascript">
window.onscroll = function() {
   alert("A SCROLL has occurred");
}
</script>
```

In this example the `window.onscroll` function automatically detects whenever a flick occurs, just as it does a normal scroll on a desktop platform. Some developers have hoped for the ability to determine the length of a flick, but thus far Apple has not made that functionality available.

3.5.2   *Two-fingered gestures*

The iPhone also supports two two-fingered gestures. These are functions that truly distinguish the iPhone, because they would be impossible to emulate easily on a mouse. Table 3.10 summarizes them.

Table 3.10   The iPhone's unique two-fingered gestures generate additional events, some of which can again be seen through normal web-based events.

Gesture	Description	Events
Two-finger flick/scrollable	User pans the screen with two fingers inside a scrollable element	`mousewheel`
Two-finger flick/not scrollable	User pans the screen with two fingers not inside a scrollable element	`onscroll`
Pinch	User zooms in or out with a two-fingered pinch	(None)

We've already met the pinch-zoom functionality, and perhaps it's not too surprising that we can't see its event, given that we haven't been given access to any zoom events so far. The two-fingered flick, on the other hand, is new. It may, in fact, be new to most of your users too, as it's one of the more secret functions of the iPhone. However, if you can trust your users to use it, you can take advantage of this function to accomplish some interesting things.

First, two-fingered flicks give you an option to implement framesets. You might recall that we cautioned against them earlier because one-fingered flick scrolling doesn't work right with them. But if you can train your users to do two-fingered scrolling, they'll be able to easily scroll individual frames.

Second, they allow you to create other types of scrollable elements, solely to detect the `mousewheel` event, and thus introduce new functionality to your web page. The following example shows a page with a `<textarea>` which that is scrollable, along with some JavaScript code that will detect a two-fingered scroll in that area:

```
<script type="text/javascript">
window.onmousewheel = function() {
    alert("A two-fingered SCROLL has occurred");
}
</script>
…
<textarea rows=100 cols=100>
</textarea>
```

This might be a nice feature if you want to lock a header and footer on your page but allow users to scroll the content in the middle.

Third, two-fingered flicks are (obviously) the functionality that you need to train your users to use if you're *already* detecting for `mousewheel` events on your web page.

3.6   *Redisplaying web pages*

To date we've talked about how to take your existing web pages and redevelop them for the iPhone. So far it's been a largely additive process. But what if you have web pages that still don't look good, no matter what you do? In this case, consider a different solution, which probably means totally redisplaying your web pages for the iPhone. Granted, this may not be a possibility for all sites. To use this method, you must have good data abstraction built into your site—usually via individual web pages pulling data from an external source (such as MySQL), and then dumping that content into a template.

The first step in redisplaying your pages is simply creating new templates for iPhone users and then selecting them when you detect an iPhone. For optimal iPhone usage, your new templates should use a "river" format. This means that you redisplay your web pages' data in a single column rather than using a multiple-column approach. You should place your most important navigation elements at the top of the page and your least important navigation elements at the bottom of the page. Your content then flows in between those spaces.

However, you should only decide to redisplay your page if it *really* doesn't look good on iPhones. Because the iPhone is a desktop-class browser, you don't *have* to notably cut down your web pages to make them work. Some professional websites have made the mistake of replacing a great user interface that worked well with the iPhone's columnar zoom with a crippled river formatting; in doing so, they've decreased functionality for iPhone users, not increased it.

Given the limitations of a total redisplay, you may wish to consider a more piece-meal approach. In rewriting web pages for the iPhone, we have been more likely to move a single column here or there than to totally rewrite a page. Such a development process is very personal—much as web design itself is. Often it will be a process of deciding which minor bits of information can be removed from a page to allow everything else to be larger and more readable on a comparatively small screen.

Because the real work of page redisplay is so personal, we can't give precise guidelines for how it will work on your individual pages. But we can look at what's been done already and see what lessons those pages suggest. We'll finish our look at iPhone optimization by exploring a couple of notable websites and what they've done.

There's a fine line between creating an iPhone optimization of an existing web page and redisplaying your web page to the point where you've actually created a new iPhone web app (the topic of the rest of part 2). Our next two examples surely cross that line. Gmail and Facebook both provide insightful looks at how you might redisplay your own pages and also offer a bridge to the topics of the next chapters, where you'll be creating iPhone-specific UIs for pages meant to be viewed exclusively on the iPhone.

3.6.1 *The Gmail iPhone pages*

Google's Gmail was one of the earliest websites to redisplay itself for the iPhone, and perhaps for good reason. The Gmail pages as they existed were a definitive example of how hard it could be to read pages on the iPhone. They contained tons and tons of information in a tiny font, and the majority of it was in one huge column that didn't get much more visible when you double-tapped it.

Today, iPhone Gmail users have a different experience. When they first log in, things seem the same; they're presented with a tiny login screen at the top left of their iPhone screen, which could really use a `device-width` viewport. Yet as soon as they hit the first content page, they encounter an entirely redisplayed page. Figure 3.2 shows the same Gmail page side-by-side in Safari on the Mac and in mobile Safari on the iPhone Simulator.

The Gmail interface offers several ideas that developers redisplaying their pages for the iPhone should take note of. Most importantly, the iPhone page has been redeveloped into a river format. Whereas the desktop web page has a column to the left, everything has been incorporated into a single column on the iPhone page by putting navigation at the top of the page (and, though we can't see it here, also at the bottom).

Beyond that, Google has adopted the same look as the iPhone chrome. We'll talk more about how to make your iPhone web app look like a native app in the future

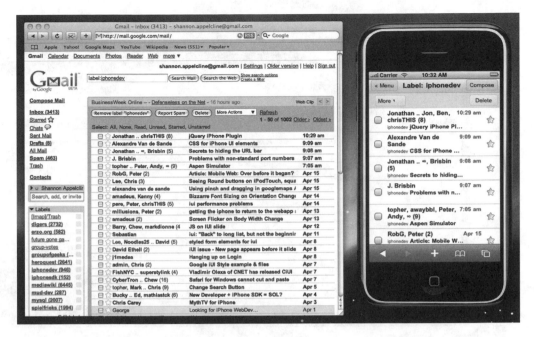

Figure 3.2 Gmail offers a different experience for iPhone users through a page that has been redisplayed to the point where it's become an iPhone web app.

when we cover iUI. Part of this chrome centers on the use of menus that appear only when you click a button, much as is the case with the << Menu button here.

However, the Gmail page also shows some of the limitations of redisplaying your pages on the iPhone. For one, it often hides content at deeper levels; you can no longer look at a list of your Gmail tags without going to a separate web page. In addition, Google has cut back on the amount of information on the screen. This helps you to read things, and also means that you have to download less at a time through a potentially slow EDGE connection, but it also makes it difficult to figure out what's inside a message when you view it from the iPhone. This is because much of the subject gets cut off—a particularly annoying problem for mailing list messages where a mailing list title takes up part of the subject line.

3.6.2   *The Facebook iPhone pages*

Facebook is another website that has been rewritten to redisplay on the iPhone. Their main content pages are much more readable on the iPhone thanks to, once again, a river format. Rather than using the iPhone standard of pushing a menu off to another page, the iPhone Facebook pages create an elegant tabbed interface that works pretty nicely, though it eventually scrolls off the screen, thanks to the lack of absolute positioning on the iPhone.

Rather than repeating our Gmail discussion by looking at those core content pages, we've decided to examine the Facebook login screen, shown in figure 3.3.

Figure 3.3 Facebook's iPhone Optimized login screen makes everything big and easy to use, and gets rid of superfluous content.

If Facebook's normal login page were shown on the iPhone, it'd be impossible to read—*and* impossible to click on the login forms—until you did a columnar zoom on the left-hand column. The iPhone page instead just shows you login widgets that are full sized on your screen. Although it's not shown, they're also perfectly sized to fill an iPhone screen if it's in landscape mode. The downside is that you lose all the extra information from the normal login page. To offset that deficit, the iPhone page gives users the option to hop over to the regular Facebook page, something that more sites should offer.

On the whole, this special Facebook login is much more functional on the iPhone. Further, this type of redevelopment is easy to do, since it just requires recoding a simple form for iPhone usage. If you're clever, you could even do it without recoding the page: a simple viewport command with an `initial-scale` set to zoom in to the left column would have accomplished much the same job as Facebook's wholescale redevelopment.

3.7   *Supporting non-iPhone users*

As we've noted, at some point your iPhone-optimized page will be so iPhone specific that it moves out of the arena of optimization and actually becomes a web app. When this happens, make sure you have a fallback page for non-iPhone users. Just as it's great to have mobile versions of your desktop pages, you should also have desktop versions of your mobile pages.

Lessons for SDK developers

Many of the lessons in this chapter had to do specifically with web design, including good HTML practices and specific attributes and tags that can be used on the Safari web browser. This chapter also gave us our first look at the two biggest innovations we discussed in chapter 1: input and output.

When looking at input, we saw the problems of fat fingers and how events have to be modeled differently because the finger isn't a mouse. These same differences and limitations will appear in your own SDK designs.

When looking at output, we saw the clever way in which the iPhone uses its relatively small screen to model a much larger viewing space. Using a similar model for an SDK program can help you to use the iPhone's architecture to its fullest (and in fact is already used in the SDK's `UIWebView` class).

We also saw our first hints at core iPhone functionality: the chrome and events.

The chrome will be a minor issue when you're doing SDK development. You'll still have a top bar, but you'll have better control over what it looks like. You'll be able to choose whether to have status-like bars or bottom bars depending on the needs of your program. Once you've made a choice, you can just pick the right class of objects from the SDK and drop it into your program.

The events listing in this chapter offer an excellent preview of how iPhone events are different from standard mouse-driven events. iPhone events aren't just about point-and-click; they're about one or more fingers temporarily touching the screen, then moving in different ways. You'll learn about how touches can be accessed—even on the web—starting in the next chapter. In the meantime you should be considering how this new paradigm for input might change how users interact with your programs.

If you have two parallel websites running side by side, you can either automatically forward users from one set of pages to the other depending on their device, or give them the choice to move to the parallel website no matter which device a user is working from. We suggest the latter, as we saw in the Facebook example site.

We recommend this approach because we remain committed to letting users choose their web experience whenever possible, but also because we've already seen iPhone web apps that we felt provided an inferior experience to viewing the desktop web pages. This could well be a personal preference, and that's a great reason to offer a choice. Even if your iPhone pages make the best possible use of an iPhone's unique capabilities, some users may prefer the creakier but probably more fully featured possibilities offered by a true desktop web page.

When you step up to offering a full web app—without any parallel desktop pages—it's polite to at least let desktop users know what's going on. Some websites give a warning as to why an iPhone web app looks so funky on a desktop screen, while others just don't allow users to visit the web app pages from a desktop browser.

Another possibility that we haven't seen yet would be to force a browser resize to a 320x480 screen. Ultimately your choice will depend on what you expect your user base to be—but giving readers more information by providing *some* sort of desktop page is rarely going to be a bad thing.

3.8   *Summary*

Although an iPhone theoretically contains a fully enabled desktop browser, there's no way that a small handheld device could ever provide the complete functionality of a desktop display. Fortunately, there are some easy things you can do to improve the experience for iPhone viewers of your web page.

We've broken our redevelopment suggestions into two parts. First, there are fixes that you need to apply to make your web pages iPhone friendly. Second, there's functionality you can use to directly detect iPhone usage and thus make your pages iPhone optimized. If you follow all the advice in this chapter—adding some simple iPhone variables to your pages, improving your web abstractions, engaging in some iPhone best practices, playing with the chrome, and looking at iPhone gestures—you'll have dramatically improved how your web pages work on an iPhone.

The ultimate in iPhone optimization is creating totally new displays for your web page that format content in a method that's more accessible on the iPhone's smaller screen. If you go too far in this direction you'll end up branching your code base, providing entirely different support for iPhones and desktops. Although that may be further than you want to go if you're just supporting the occasional iPhone user, if you're providing serious iPhone support—and as a reader of this book, you probably are—it may end up being a necessity. This leads nicely into our next topic—iPhone web apps, which are web pages built solely for use on the iPhone.

Advanced WebKit
and textual web apps

In the previous chapter we covered the fundamentals of redeveloping an existing web page for use on iPhones. In the process you learned about important concepts like the viewport, and we discussed a lot of what works—and what doesn't—on Apple's unique mobile platform. We expect, though, that most of you aren't just interested in touching up existing web pages but instead are looking to create totally new programs. Further, after considering the advantages and disadvantages of both web and native development, you've decided that writing a new program using a web language is the best way to go. We're now ready to enter the world of web apps, a topic that will consume the rest of part 2 of this book.

We've identified three ways to create web apps for the iPhone. Each will take advantage of all your existing web knowledge, but each will also connect you with a specific library or Apple browser add-on that will allow you to unlock additional functionality on the iPhone platform.

First, you might choose to build a textual web app, which is an application that is largely built on the fundamentals of HTML itself. You'll be able to supplement it with the advanced features of Apple's WebKit. We'll be discussing this programming method in this chapter.

Second, you might choose to build an iPhone-UI web app, which is an application with a user interface that looks like iPhone's native apps. That will be the topic of chapter 5, where we'll also cover iUI, a programming library meant to make this task easier.

Third, you might choose to build a graphical web app, which is an application that pushes the boundaries of web graphics. That will be the topic of chapter 6, where our discussion will center on Canvas, a graphical add-on introduced by Apple several years ago that is now widely available.

Let's get started with the first of those topics: textual web apps and the advanced functionality of Apple's WebKit. In this chapter we'll explore a wide variety of WebKit options, starting with simple HTML and CSS, and then build on that with advanced functionality such as transformations and databases. We'll conclude with some code that's iPhone only.

4.1   *Introducing the WebKit*

We touched on the WebKit in the previous chapter when we talked about the iPhone's viewport command. As we noted there, the WebKit is a browser engine that underlies several existing browsers. It was originally derived from the Linux browser Konquerer, and is now in wider use on Unix-derived systems.

> **WebKit compatibility note**
>
> Although we're highlighting the WebKit for the advanced functionality it offers to the Apple iPhone, you can make greater use of the features in this chapter that are not iPhone specific. This includes new tags, various sorts of animation, and the client-side database. Most notably, these features should work on Apple's desktop Safari, the brand-new Google Android and Google Chrome platforms, and GNOME's Epiphany browser. Many of these features will become more accessible when they're accepted into new HTML and CSS standards.

It's Apple who has brought the WebKit to the greatest prominence to date, and thus it should be no surprise that it's Apple who is pushing hardest on new features for the WebKit. It's those new features that should excite you as iPhone developers, because they represent totally new functionality that you can use on the iPhone and that isn't yet available on Internet Explorer or Firefox. You'll innately be making your web apps incompatible with those browsers if you start depending heavily on WebKit tricks, but we figure that's the point: using all the tricks you can to produce the best possible web apps for use largely or exclusively by the iPhone.

As you'll recall from chapter 3, there's some information on the WebKit at Apple's Web Apps site. As of this writing, it's pretty rudimentary, but we suspect it'll improve with time. Even now it has somewhat greater breadth than this chapter, covering some special functions that we don't get to here.

We'll offer the warning that all of this functionality is *very* new. Since some of the functions are being offered for future HTML and CSS standards, they could change entirely. Even now they're not entirely consistent. Over the course of writing this chapter we discovered one minor function (a transitioning rotation) that worked on Safari but not mobile Safari and a whole category of functionality (explicit animation) that worked on mobile Safari, but nowhere else. We expect that by the time this book is published, behavior will be more consistent across versions of Safari, but our prime goal has been to document how things will work on the iPhone.

4.1.1   New HTML elements

The WebKit introduces several new HTML elements. We've listed the ones most likely to be useful for iPhone design in table 4.1.

Table 4.1   **WebKit HTML elements give some new basic features for your iPhone web design.**

Tag	Summary
`<canvas>`	WebKit/JavaScript drawing object; discussed in chapter 6
`<marquee>content</marquee>`	Sets content (which could be text or other object) as a horizontally scrolling marquee
`<meta name="viewport">`	Metatag for the iPhone; discussed in chapter 3

There are also some variants of `<pre>` and some alternate ways to do embeds and layers, most of which are being deprecated in HTML 4.01.

4.1.2   New CSS elements

The bulk of the WebKit's new functionality comes from its large set of extensions to the CSS standards. These include the full set of transforms, transitions, and animations that we cover separately later in this chapter, as well as some simpler CSS elements that are summarized in table 4.2.

Table 4.2   **This partial list shows the numerous simple new CSS elements that can be incorporated into your iPhone designs.**

HTML element	CSS properties	Summary
Background	`-webkit-background-size`	Controls the size of the background image.

Box	`-webkit-border-radius`
	`-webkit-border-bottom-left-radius`
	`-webkit-border-bottom-right-radius`
	`-webkit-border-top-left-radius`
	`-webkit-border-top-right-radius`
		Sets the rounded corner radiuses of the box, in length units, either one corner at a time or all using one property.

Box	`-webkit-border-image`	Allows you to set an image as a box border using a somewhat complex syntax, which is explained in Apple's reference pages.

Table 4.2   **This partial list shows the numerous simple new CSS elements that can be incorporated into your iPhone designs.** *(continued)*

HTML element	CSS properties	Summary
Box	`-webkit-box-shadow`	Sets a drop shadow for a box by designating a horizontal offset, a vertical offset, a blur radius, and a color.
Link	`-webkit-tap-highlight-color`	Overrides the standard highlight when a user taps on a link on an iPhone.
Link	`-webkit-touch-callout`	Disables the touch-and-hold info box if set to `none`.
Marquee	`-webkit-marquee-direction`	Controls the direction of the marquee, which can go forward, left, right, up, reverse, or several other directions.
Marquee	`-webkit-marquee-increment`	Controls the distance the marquee moves, in length units.
Marquee	`-webkit-marquee-repetition`	Limits the number of marquee repetitions.
Marquee	`-webkit-marquee-speed`	Sets marquee speed to fast, normal, or slow.

Text	`-webkit-text-fill-color`
	`-webkit-text-stroke-color`
	`-webkit-text-stroke-width`
		Together allows you to differentiate between the interior and exterior of text by setting colors for each and by defining the stroke width using a length unit.

Text	`-webkit-text-size-adjust`	Adds a percentage to increase size of text on the iPhone.

Table 4.2 is not a complete listing, nor does it give you all the details you need to use the properties. There are not only other CSS properties, but also new CSS constants that can be applied to existing properties. Our main purpose is to show you the cooler elements that are available among the WebKit's basic CSS elements, and to encourage you to find out more information at Apple's WebApps reference site.

Figure 4.1 shows how some of the new WebKit CSS properties could be applied to a simple `<div>` to create an attractive box that features rounded corners and a three-dimensional back shadow.

The `<div>` is simply defined with a new class name, as is typical for CSS:

```
<div class="roundedbox">
```

The definition of the roundedbox class then includes several standard CSS properties (to set

Figure 4.1   **New WebKit properties on the iPhone make your pages more attractive.**

the `background-color` and so forth), plus the new `border-radius` and `box-shadow` properties, which appear for the first time in Apple's WebKit.

The code to create the `roundedbox` class is shown in listing 4.1.

Listing 4.1 Using the new CSS web properties to create an attractive box

```
.roundedbox {
    background-color: #bbbbee;
    border: 1px solid #000;
    padding: 10px;
    -webkit-border-radius: 8px;
    -webkit-box-shadow: 6px 6px 5px #333333;
}
```

All of the other new CSS properties could be used in a similar way.

THE IPHONE-SPECIFIC PROPERTIES

Before we finish our discussion of the simpler WebKit CSS properties, we'd like to point out the ones specifically intended for the iPhone. `-webkit-tap-highlight-color` and `-webkit-touch-callout` each give you some control over how links work on the iPhone. It's the last iPhone-specific property, `-webkit-text-size-adjust`, that is of particular note, because it allows you to increase point size by a percentage *only* on the iPhone.

In chapter 3, we talked a bit about adjusting font sizes through multiple CSS files. However, if that's *all* you need to do differently between iPhones and desktop browsers, you can do it with a single line in your CSS file:

```
body {
    -webkit-text-size-adjust: 120%;
}
```

Having now explored the mass of simple additions that the WebKit offers to iPhone developers, we're ready to dive more wholeheartedly into the big stuff, starting with a variety of advanced CSS methods that you can use to manipulate and animate the content of your web page.

4.2 CSS transforms, transitions, and animations

One of the most innovative elements of the WebKit is its ability to manipulate existing HTML objects in various ways, allowing you to accomplish some fancy-looking work entirely in CSS. You have three options: transformation (or static changes), transitions (or implicit animations), and animations (or explicit animations).

4.2.1 The transform function

Transforms allow you to apply various geometric functions to objects in a web page when they're created. There's no animation here (yet), but you have considerable control over exactly what your HTML objects look like.

Each transform is applied using a `-webkit-transform` CSS property:

```
-webkit-transform: rotate(30deg);
```

Several transforms are available, as shown in table 4.3.

Table 4.3   **The WebKit transforms apply to output elements in a variety of ways.**

Function	Argument	Summary
scale scaleX scaleY	Number	Resizes the object
rotate	CSS angle	Rotates the object
translate translateX translateY	Length (or percentage) in X direction, Length (or percentage) in Y direction	Moves the object
skew skewX skewY	CSS angle	Skews the object

The properties in table 4.3 are applied to boxes within CSS. Much like relative positioning, they don't affect layout, so you have to be careful with them.

The following definition could be added to our `roundedbox` class from listing 4.1, turning it into a `wackybox` class:

```
-webkit-transform: rotate(30deg) translate(5%,5%);
```

The result is that your news article appears at an angle, moved somewhat off the screen. Figure 4.2 shows this change, which you can compare to the nontransformed news article that appears a few pages back as figure 4.1. This particular transform isn't that useful if

you want people to read it, but it could be a nice background for a news site or something similar. There are many other things that you can do with transforms, such as setting up banners, printing text at a variety of sizes, and making similar changes on static web pages. Some will be gimmicks, but others can have functional benefits.

Before we leave transforms behind, we'll note that they support one other property, `-webkit-transform-origin`, which can be used to move the origin for scales and skews away from the center of the object.

Although you can do quite a bit with transforms all on their own, their real power appears when you start working with the implicit animation of transitions, which are the next WebKit function that we're going to talk about.

Figure 4.2   Our `roundedbox` **transformed into a** `wackybox`**, which is rotated 30 degrees and translated 5 percent along each of the X and Y axes.**

4.2.2   *The transition function*

A transition is also called an "implicit animation" because it supports animated frames, but you, as a web designer, don't have to worry about how the animation occurs. All you do is define the endpoints.

To define a transition, you place the new –webkit-transition properties in the CSS element that marks the *end* of your animation. You can define what properties to transition (possibly including all of them), how long the transition should last, and how the transition should work. Then, when the block of text changes to the end-point class, a smooth and animated transition will occur. Table 4.4 lists the various transition properties.

Table 4.4   **Transitions let you animate changes of CSS properties.**

Property	Values	Summary
`-webkit-transition-property`	Various properties, including `all`	Defines the property to animate
`-webkit-transition-duration`	Time value, such as `1s`	Specifies how long the animation takes
`-webkit-transition-timing-function`	`ease`, `linear`, `ease-in`, `ease-out`, `ease-in-out`, or `cubic-bezier` (user-defined)	Defines the curve for how the animation occurs; `ease` was `auto` in previous versions of the iPhone OS
`-webkit-transition-delay`	Time value, such as `1s`	Specifies how long to wait to start transition

Unfortunately, not all CSS properties can be transitioned. Apple states that "any CSS property which accepts values that are numbers, lengths, percentages or colors can be animated." This isn't entirely true, because some percentage-based elements such as `font-size` don't yet work. If you want to know whether a CSS property can be transitioned, you can find a list on the Apple website. And there's no harm if you list something that doesn't transition in a property; it just does an abrupt change instead.

You can reduce a transition command to a single line of code using the `-webkit-transition` shorthand property:

```
-webkit-transition: property duration timing-function delay
```

Separate additional transitions after the first with commas.

Traditional websites are already making good use of transitions to highlight page elements when a hover occurs. However, hovers aren't possible on the iPhone, so it's more likely that you'll be making transitions based on a click. Listing 4.2 shows a simple transition of a box to color in the text and background (making it more visible) when you click it.

Listing 4.2   Using transitions to animate changes between styles

```
div {
    -webkit-transition: all 2s;
}
```

```
.clearbox {
   background-color: #bbbbbb;
   opacity: .5;
// Other properties make our box beautiful
}

.visiblebox {
   background-color: #bbbbee;
   opacity: 1;
// Other properties make our box beautiful
}
```

Once you've defined these styles, you just need to add an `onclick` event handler that shifts from one style to the other:

```
<div class="clearbox" onclick="this.className='visiblebox'">
```

This simple transition could easily be built into a more sophisticated interface where a user could make individual elements of your page more or less visible by clicking on those individual elements (or alternatively through some pagewide control panel).

Seeing this simple example highlights why transitions are called *implicit* animation. You don't have to do a thing other than define your endpoint styles and say what you want to animate between them. In this example, your pages enjoy a nice animation from gray to cyan, with increasing opacity, thanks just to your defining two styles (and a transition for all `<div>`s).

However, transitioning between normal styles is just the tip of the iceberg. The coolest thing about transitions is that they can be used with that other new WebKit feature we just discussed: transforms.

TRANSITIONING TRANSFORMS

By putting together transforms and transitions, you can create actual graphical animations of scales, rotates, skews, and translations. In other words, you can make boxes revolve, move, and otherwise change, showing off sophisticated graphical animations, with nothing but CSS definitions (and perhaps a bit of JavaScript to change the styles).

Listing 4.3 shows how to put a transition and a transform together to create an animated thumbnail program.

Listing 4.3   Scaling a picture to create thumbnails

```
::thumbnail.css::
div {          ❶
   -webkit-transition: all 14s;
}

.imagebox {          ❷
  -webkit-transform: scale(.2);
}
::thumbnail.html::
<head>
<title>Thumbnail Viewer in WebKit</title>
<link href="thumbnail.css" type= "text/css" rel="stylesheet">
<meta name="viewport" content = "width = 480">
<script type="application/x-javascript">
```

```
var i = 0;
function animatePic(mystyle) {          ❸
  i = (i + 1) % 2;
  if (i == 1) {
    mystyle.webkitTransform='scale(1) rotate(360deg)';          ❹
  } else {
    mystyle.webkitTransform='scale(.2)';          ❺
  }
}
</script>
</head>
<body>
<div id="mydiv" class="imagebox"
    onclick="animatePic(this.style);">          ❻
<img src="cat.jpg">
</div>
</body>
</head>
```

For the example in listing 4.3, your transforms are enacted in JavaScript. Therefore, your CSS file only needs to do two things. First, you set up all <div>s so that they'll transition ❶, and second, you set the initial scale of your images to be .2 ❷, which means that your images will be read in at full size but then displayed in a much smaller form.

Your <div> is then set up to use the imagebox class and call the animatePic function when clicked ❻. animatePic does all the magic ❸. Every other click, it either scales up and rotates ❹ or scales back down ❺, using the webkitTransform property. The rotate animation is just a graphical flourish to make the animation look more attractive. Figure 4.3 shows this simple transition.

We'll also offer a brief aside on that rotate: for the moment the rotate animation works on Safari, but not on mobile Safari, which is why you don't see the thumbnail twisting in figure 4.3. As we've already explained, we expect the platforms will eventually sync, perhaps by the time this book sees print, but we offer this as a further caveat.

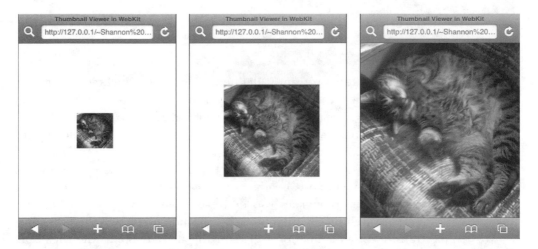

Figure 4.3 A thumbnail (left) is animated by a tap (middle), turning it into a full-page picture (right).

In any case, what's the end result of our work? As long as you're on a setup where you don't mind loading larger pictures, you can thumbnail easily. More importantly, you can see how easy it is to use attractive implicit animations to scale thumbnails on your iPhone by combining the transition and transform properties.

4.2.3   *The animation function*

Transitions support fully featured animations, within the constraints of the CSS properties and the transform abilities. So, how can the WebKit's "explicit" animation improve on that? The answer is by giving you better control over how the animation works.

Listing 4.4 shows an example of animation in use.

Listing 4.4   Using animation keyframes

```
.imagemove {         ❶
  -webkit-transform: scale(.2);

  -webkit-animation-name: 'moveit';        ❷
  -webkit-animation-duration: 5s;        ❸
  -webkit-animation-iteration-count: 1;        ❹
}

@-webkit-keyframes 'moveit' {        ❺

0% {
   left: 0px;
   top: 0px;
  }

 20% {
   left: 0px;
   top: 100px;

 }
 100% {
   left: 0px;
   top: 450px;
   opacity: 0;
  }

}
```

To create an animation, you must define a style ❶ that includes a set of -webkit-animation properties. The three critical ones are -webkit-animation-name ❷, which is the name for the animation; -webkit-animation-duration ❸, which is the length of the animation, and -webkit-animation-iteration-count ❹, which is the number of times to repeat the animation.

The animation itself is defined by a @-webkit-keyframes entry ❺, where the name matches the name you already set. You can define as many frames as you want, setting whatever CSS properties you desire in each frame. Then when the animation occurs, the WebKit will automatically transition among all the individual points. Unlike transitions, an animation will return to its start point when it's done, possibly iterating.

To apply an animation, you just set a box to use the class. If you do this in your regular HTML code, the animation will begin the instant your web page loads. Alternatively, you can change a class name to the animated class when a button click or some other event occurs. For example, if a button in your application requires a double-click, you could start the button shivering when it's clicked the first time to encourage a second click.

Our simple example here just takes a box and shoves it down toward the bottom of the screen, making it disappear as it goes. You could use similar functionality to delete items from a page as a user interacts with it.

The transformations, transitions, and animations are all impressive when put into use. However, there's one more WebKit function that's broadly available on WebKit browsers that we think is even more impressive—and more likely to be useful in your own web apps: the client-side database.

4.3 The WebKit database

WebKit includes a client-side database built on SQLite. This means that you can save much larger amounts on information on your client's machine—which could be an iPhone—than was possible using traditional cookies. This is all done through a set of new JavaScript routines.

4.3.1 Loading a database

A table is loaded into your JavaScript through the openDatabase command:

```
var mydb = openDatabase(databaseName,version,displayName,maxSize);
```

databaseName and displayName are both arbitrary names that you'll select for an individual database. version should currently be set to 1.0, and maxSize is a maximum size for your database in bytes (such as 65536). Here's a typical command for opening a database:

```
var mydb = openDatabase(prefData,'1.0','Preference Data',65536);
```

If a database doesn't exist, the open command will automatically create it. However, this command won't actually create tables in the database, which you must do by hand.

4.3.2 Running a transaction

Activity is run through the database as transactions, which are atomic query units that succeed or fail as a unit. Each transaction may contain one or more queries. A complete transaction looks like this:

```
db.transaction(
   function (transaction) {
      transaction.executeSql(SQL, [array of ?s],
         dataHandler*,errorHandler*);
         // addition transactions
      },transactionError*,transactionSuccess*
);
```

The actual query is done in SQL. You can find good information at www.sqlite.org on the SQL syntax currently supported. Use this site as a reference if you're unfamiliar with the language.

Each of the four handler arguments, marked with an *, is optional, meaning that the simplest call to a transaction just sends the SQL and any arguments. The first two handlers are for individual queries, and the last two handlers are for the transaction as a whole. They each refer to a different function that accepts one or more arguments, as shown in table 4.5.

Table 4.5   **Error and data handlers tell you what SQLite is doing.**

Function	Arguments	Notes
dataHandler	transaction, results	Parses results of successful query
errorHandler	transaction, error	Shows query errors
transactionError	error	Runs if anything in transaction failed
transactionSuccess	N/A	Runs if everything in transaction succeeded

results and error are both fully featured JavaScript objects that give you access to a number of properties and let you know how your SQL query affected your database. The most important ones are listed in table 4.6.

Table 4.6   error **and** results **give you access to SQL responses.**

Property	Summary
error.code	Error code
error.message	Error text
results.rows.length	Number of rows of responses
results.rows.item(i)['name']	Result from column 'name' of row i of your response
results.rowsAffected	Number of rows affected by an UPDATE or DELETE
results.insertId	ID of last INSERT operation

The underlying functionality of the JavaScript database is quite simple—presuming you already know SQL. Thus the question becomes: how can you use it?

4.3.3   *A sample database*

A client-side database will have any number of advantages, all of which you can make use of on the iPhone. However, there's one iPhone-specific trick you should consider: iPhones uniquely might be connected to either a fast network (wireless) or a slow network (EDGE or 3G). So why not give your users the ability to offload their networked data to a local database to improve access time when they're on the road? Listing 4.5 shows a somewhat simplified example that does just that.

Listing 4.5 A database that saves online data to a local resource

```html
<html>
  <head>
<script type="text/javascript" charset="utf-8">

var myDB = openDatabase
    ('bookDB','1.0','Book Listing',65536);        ❶

myDB.transaction(
  function(transaction) {        ❷
    transaction.executeSql('CREATE TABLE IF NOT EXISTS books(id INTEGER NOT
        NULL PRIMARY KEY AUTOINCREMENT,btitle TEXT,bauthor TEXT,
        bpublisher TEXT)',[],nullDataHandler,defErrorHandler);
  }
);

function defErrorHandler(transaction,error) {        ❸

  alert("Error: "+error.message+" (Code: "+error.code+")");
  return true;
}

function nullDataHandler(transaction,results) {        ❹
}

function updateBooks() {        ❺

// Clever code to pull data from your website database goes here

  myDB.transaction(
    function(transaction) {
      transaction.executeSql('DELETE FROM books WHERE 1',
        [],nullDataHandler,defErrorHandler);
      transaction.executeSql('INSERT INTO books
        (btitle,bauthor,bpublisher) VALUES (?,?,?)',
          [btitle,bauthor,bpublisher],nullDataHandler,defErrorHandler);
// other transactions go here, as you read out of your website database

    }
  );

  listBooks();
}

function listBooks() {        ❻
  myDB.transaction(
    function(transaction) {
      transaction.executeSql('SELECT * FROM books WHERE 1 ORDER BY
        bpublisher,btitle',[],bookDataHandler,defErrorHandler);
    }
  );
}

function bookDataHandler(transaction,results) {        ❼
```

```
var dataText = "<table width=280 cellpadding=2>" +
    "<tr bgcolor='" + tablecolor[1] +
"'><td><b>Title</b></td><td><b>Author</b></td><td><b>Pub.</b></td></tr>";

  for (var i=0 ; i < results.rows.length ; i++) {          ❽
    var thisRow = results.rows.item(i);          ❾
    dataText = dataText + "<tr bgcolor='" + tablecolor[i % 2] + "'><td>" +
      thisRow['btitle'] + "</td><td>" + thisRow['bauthor'] + "</td><td>" +
      thisRow['bpublisher'] + "</td></tr>";

  }

  dataText = dataText + "</table>";
  document.getElementById("bookList").innerHTML = dataText;
}

</script>
</head>
</html>
<body>          ❿

<p><input type="submit" value="click" onclick="updateBooks();"> to
update your local list of books.          ⓫

<div id="bookList" class="roundedbox">          ⓬
<p><i>Your book data will be listed here.</i>
</div>
</body>
</html>
```

This process generally follows the examples we've already given. You start off by opening your database ❶ and creating your table if it doesn't already exist ❷. Note that you link to your first two query handlers here. Your default error handler ❸, which you use throughout this program, just reports your error, while your default data handler ❹ doesn't do anything because most of your queries won't return any results.

We opted not to use the bigger picture transaction handlers at all. Your individual project should determine whether or not you need them.

Your updateBooks function ❺ will change the most at your individual site, since this is where you need to read the data from your server-side database and dump it into a client-side database. This function just shows an example of placing one item in the database, using question marks to highlight how the transaction's array argument works. You'll doubtless have some type of for loop in a real program to iteratively map your data into the client-side database.

When you update the books, you also list them ❻, which ultimately updates the data on your web page. This is the only SQL transaction that uses a real data handler, a requirement since it returns results.

Your bookDataHandler ❼ shows how easy this is to code. You just iterate through all the rows you get back ❽, each time creating a variable that makes it easier to access the individual items ❾, then using that variable to write a line of HTML code.

The results show up in the body of your program ❿, which includes both the button that gets things started ⓫ and the <div> where your new data is placed ⓬.

The results are shown in figure 4.4, which as you can see make good use of some of the WebKit CSS elements that we highlighted earlier, showing off the great functionality that the WebKit provides you with.

The JavaScript database is the last WebKit element that you can make use of on the iPhone, but it can also be used more broadly. The last couple of items that we'll discuss are instead iPhone specific.

4.4 Adjusting the chrome

In the previous chapter we showed you some simple methods for dealing with the iPhone chrome. We explained how to scroll the URL bar and noted that the status bar and the bottom bar could not be changed. Using the WebKit, you have slightly more control over things, provided that your user is using iPhone OS 2.1 or higher. All you need to do is enter a new metatag on your web app's home page:

Figure 4.4 Data retrieved from a database can then be displayed.

```
<meta name="apple-mobile-web-app-capable" content="yes" />
```

This code doesn't change the web app when it's run through the browser. It's only when a user chooses to save your app to his or her iPhone home page and then calls it back up that things will act differently. When called back up, your app will appear without the URL bar or the bottom bar: only the status bar continues to eat space in your web app.

Because your user will not be able to navigate using the URL bar, you have to be very careful when using this metatag. You should only do so when navigation is totally self-contained within the program—for example, if you've built it with iUI or Dashcode, both topics that we'll return to in future chapters. This metatag is only appropriate for a true web app.

There's one other metatag of note: `apple-mobile-web-app-status-bar-style` can have its `content` set to `default`, `black`, or `black-translucent` to change the way the status bar looks when a user re-enters your program:

```
<meta name="apple-mobile-web-app-status-bar-style"
content="black-translucent" />
```

That's as much control as you have over the iPhone's chrome; now let's move on to the next iPhone-specific topic: touches and gestures.

4.5 Recognizing touches and gestures

In the previous chapter we introduced some rudimentary ways to access events on the iPhone. We showed you how to correlate iPhone-initiated touches with regular JavaScript events such as `mouseup` and `mousedown`. However, now that you're diving deeper into iPhone web work, you'll be happy to know that there's a whole other way to do things. You can access touches and gestures directly.

The touch support built into the WebKit is similar to the gesture support built into the iPhone's native SDK, revealing the programming power that you have no matter which method you use to write your iPhone programs. In both situations, Apple uses two standard abstractions: the touch and the event. A *touch* occurs when a finger comes down on the screen, when it moves across the screen, or when it's pulled up off the screen. An *event* is a collection of touches. It begins when a finger is first placed on the screen and ends when the last finger comes off. All of the touch actions that occur between the one that began an event and the one that ended an event are stored in the same event record.

To facilitate ease of use, the WebKit also introduces an abstraction that you won't find in the SDK: the gesture. A *gesture* begins when two or more fingers touch the screen and ends when there are one or zero fingers left. Like touches, gestures are collected together into events.

4.5.1   Accessing events

Based on its standard touch and gesture models, the WebKit recognizes seven Document Object Model (DOM) event classes, as shown in table 4.7.

Table 4.7   With touches and gestures, you can recognize iPhone touchscreen events.

Event	Summary
touchstart	A finger touches the iPhone.
touchmove	A finger moves across the iPhone.
touchend	A finger leaves the iPhone.
touchcancel	The system cancels a touch.
gesturestart	Two or more fingers touch the iPhone.
gesturechange	Fingers are moved during a gesture.
gestureend	There are one or less fingers left on the iPhone.

Depending on the complexity of input desired, you may use gestures or touches, or possibly both, in your website. We'll look at both gestures and touches in this section, and you'll see some of the amazing things you can do when you combine them with other WebKit functions.

ACCESSING AN EVENT

You can access any of these new touch events by one of two methods, as you can with other events in HTML. First, you can link an event handler as part of an object's HTML definition:

```
<div ontouchstart="myTouchStart(event);">
```

Second, you can load an event handler through JavaScript:

```
element.addEventListener("touchstart",myTouchStart,false);
```

Each of these methods will pass a standard event object to the function being called (myTouchStart in these examples), generally following all the standard rules for how JavaScript event handling works. We include the specifics here, for completeness, and suggest a JavaScript reference if you need any other information on JavaScript event handling.

TURNING OFF DEFAULT BEHAVIOR

If you are writing your own touch or gesture functionality, you'll probably need to turn off some or all of the default behaviors of Safari's UI. For example, if you're accepting touches, you won't want the iPhone to scroll when a user touches the element in question. Similarly, if you're accepting gestures you won't want the iPhone to pinch-zoom while a user is trying to manipulate the page. You turn off these behaviors by running the preventDefault() method in the appropriate event handlers:

```
function myTouchStart(event) {
    event.preventDefault();
}
```

You'll usually need to run preventDefault() method for touchstart and touchmove events if you're looking at touches. If you're looking at gestures, you'll probably have to run preventDefault() for gesturestart and gesturechange events.

4.5.2 *Converting events*

Whether you're using the touch or gesture events, you're going to need to convert those events into individual touches in order to use them. You accomplish this by accessing a number of properties of the event object, as listed in table 4.8.

Table 4.8 Event properties mainly contain lists of touches.

Property	Summary
target	The target object that generated the touch.
changedTouches	An array of all the most recently changed touches on the page. Usually contains just one touch.
targetTouches	An array of all the current touches for a target element.
touches	An array of all the touches on a page.

Note that changedTouches, targetTouches, and touches each contain subtly different lists of touches. targetTouches and touches each contain a list of fingers that are currently on the screen, but changedTouches lists only the last touch that occurred. This is important if you're working with the touchend or gestureend events. In both cases there will no longer be fingers on the screen, so targetTouches and touches should be empty, but you can still see the last thing that happened by looking at the changedTouches array.

Because the touch properties all produce arrays, you can use JavaScript array functions to access them. That means that `event.touches[0]` will return the first touch and that `event.touches.length` can be used to count the number of touches currently being stored.

We'll see this all in use in the next couple of examples.

4.5.3   Accessing touches

Once you're looking at an individual touch, by using a variable like `event.touches[0]` or `event.targetTouches[0]` you can access additional properties of that touch as shown in table 4.9.

Table 4.9   Touch properties contain specific information about a touch.

Property	Summary
clientX or clientY	X or Y location relative to the current browser screen (absent any scroll offset for the overall web page)
identifier	Unique identifying number for the event
pageX or pageY	X or Y location relative to the overall web page
screenX or screenY	X or Y location relative to the user's overall computer screen (which is of limited use)
target	The target object that generated the touch

The majority of these properties tell you where on a page a touch occurred, using a variety of different X,Y coordinate systems. Putting this together with the basic event information we've already discussed, you can begin to build sophisticated programs that track where a user is touching the screen and take appropriate actions based on that information.

Listing 4.6 shows an example of a simple touch-based program that allows you to drag a color from one box into another.

Listing 4.6   Detecting and measuring touches

```html
<html>
<script type="text/javascript" charset="utf-8">

function colorStart(event) {          ◁─┐

  event.preventDefault();                │  ❶
}                                        │

function colorMove(event) {           ◁─┘

  event.preventDefault();
}

function colorEnd(event) {      ❷

  if (event.changedTouches[0].pageX > 110) {
```

```
      document.getElementById('colorbox').style.backgroundColor
        = event.target.id;
    }
  }
</script>
<style type="text/css" media="screen">        ❸
// Style info goes here
</style>
</head>
<body>
  <div id="red" class="red" ontouchstart="colorStart(event)"        ❹
    ontouchmove="colorMove(event)" ontouchend="colorEnd(event)"></div>
  <div id="green" class="green" ontouchstart="colorStart(event)"
    ontouchmove="colorMove(event)" ontouchend="colorEnd(event)"></div>
  <div id="blue" class="blue" ontouchstart="colorStart(event)"
    ontouchmove="colorMove(event)" ontouchend="colorEnd(event)"></div>
  <div id="colorbox" class="colorbox"></div>        ❺
</body>
</html>
```

What's impressive about listing 4.6 is how little work was required to interpret touch commands. You start off with a set of <div>s—three RGB-colored boxes ❹, which are small boxes filled with the named color, and a special colorbox ❺, which is another small box, this one located at 110x60, which can be filled with the color in question. The layout information for all of these boxes is contained in styles ❸, which we've opted to leave out since they're pretty simple CSS, though we've shown the results in figure 4.5.

Each of the three RGB boxes refers to three touch-related event handlers. When touches start or move ❶, the only thing that happens is that the default behavior is prevented, so that scrolling doesn't occur. All of the code in your program instead occurs in the touchend event handler ❷.

If the touch ends beyond x=110 (which is about where the box to be filled is located), then the background of that box is filled with the color of the box where the touch began. The result is an intuitive interface in which it feels as if you're dragging color from one object to another.

One of the most important things to note in this example is that it's the event handlers from the object where the touch *began* that are used throughout. Even if the touch ended inside the box to be filled, it's the RGB boxes event handler that runs. Similarly, target continues to refer to the object where the event began. A bit non-intuitive, this is nevertheless the most important thing to remember when monitoring touch events through the WebKit.

Figure 4.5 Ready for touchdown! Four boxes set the scene for color dragging.

Unfortunately, unlike with the SDK, there isn't yet a sophisticated manner to measure which object contains a particular touch; we hope to see that in future releases.

4.5.4 Accessing gestures

Having now worked with touches, you'll find the WebKit's gestures are quite easy to use. Essentially, they work identically to touches, but the events trigger only when there are at least two fingers on the screen.

Gestures also have a huge advantage: the WebKit does its best to measure two-fingered pinches and rotations for you. This is done through a pair of two new event properties that only appear for gestures, as described in table 4.10.

Property	Summary
rotation	How much the fingers have rotated
scale	How much a pinch has zoomed in (<1) or out (>1)

Table 4.10 Gestures add two new properties to event objects.

These new properties can allow for simple manipulation of objects using relatively complex gestures. To demonstrate, let's expand our coloring example by allowing the user to scale and rotate the box that's being filled in.

This new example begins by adding a set of four event handlers to the `colorbox`:

```
<div id="colorbox" class="colorbox" ongesturestart="boxStart(event)"
  ongesturechange="boxChange(event)" ontouchstart="colorStart(event)"
  ontouchmove="colorMove(event)"></div>
```

We've reused the touch start and move handlers, because they just prevent the default behavior, and thus ensure that scrolling won't occur after users put their first finger on the screen. The gesture handlers, which are all new, are shown in listing 4.7.

Listing 4.7 Using WebKit gestures to model pinches and rotations

```
function boxStart(event) {          ❶
  event.preventDefault();
}

var origAngle = 0;              ❷
var origScale = 1;

function boxChange(event) {          ❸

  event.preventDefault();

  event.target.style.webkitTransform = 'scale(' + event.scale + origScale
    + ') rotate(' + event.rotation + origAngle + 'deg)';      ❹
}
```

The `boxStart` handler ❶ is more of the same: you again turn off default behavior, here to prevent accidental pinch-zooming of the overall web page. It's the `boxChange` function ❸ that contains your actual code. First, you set some variables for the default angle and scale of the box ❷. Then, you use the `webkitTransform` property (which we met earlier) to scale and rotate the box based on the gesture ❹.

This complex function is easy to write because the WebKit provides you with the movement information of the gesture, rather than you having to figure it out yourself.

4.6 *Recognizing orientation*

The iPhone supports two different types of gestures. Touching the screen is what more immediately comes to mind when you think about user input, but moving the iPhone around—as measured by the accelerometers—is another way in which users can manipulate their iPhone. If you need precise accelerometer data, you'll have to design using the SDK (as discussed in chapter 17), but with the WebKit you can at least recognize orientation changes.

The `orientationchange` event notifies you when a user has rotated the iPhone after your web page has loaded. Besides just notifying you that an orientation change has occurred, the iPhone maintains a special `orientation` property in the `window` object that advises you of the iPhone's current orientation, as described in table 4.11.

Table 4.11 `window.orientation` **always reflects the current orientation of an iPhone device.**

`window.orientation` **Value**	**Description**
0	Portrait view.
90	Landscape view, turned counterclockwise.
–90	Landscape view, turned clockwise.
180	Portrait view, flipped over. Not currently supported.

Listing 4.8 shows how simple it is to detect an orientation change and take an action based on it.

Listing 4.8   An updating web page that always displays `window.orientation`

```
<head>
<meta name="viewport" content="width = 200">
<script type="text/javascript">
window.onorientationchange = function() {        ❶
   document.getElementById("orAnnounce").innerText
      = window.orientation;         ❷
}
</script>
</head>
<body>
<span class="orAnnounce" id="orAnnounce">
<script type="text/javascript">
document.write(window.orientation);        ❸
</script>
</span>
</body>
```

The example in listing 4.8 is simple: whenever the program detects an orientation change ❶, it prints the new value of `window.orientation` on the screen

❷—either 0, 90, or –90. This modifies the orAnnounce <span>, which is set to the starting value at startup ❸. Changing a CSS file or some other minor element would be equally easy. If you'd like to see a more attractive, graphical version of this, we point you toward the first example of chapter 7, where we use Dashcode to create a more attractive orientation reporter.

If you wanted to, you could do even more than some simple outputs or CSS changes. The authors maintain a sample chat program at http://www.iphonewebdev.com/chat/ that totally redesigns itself. In portrait mode it shows only a chat window, but in landscape mode it also shows a list of users online. You can see a bit less of the conversation in landscape mode, but what's there is equally readable thanks to the increased width of the screen.

Alternatively, you could use an orientation change as a standard user-input device. A drum machine, for example, could offer a beat whenever the phone is rotated.

The orientation event is the last major WebKit element at the time of this writing. There's also some neat upcoming stuff that we want to highlight, since it may be available by the time this book sees publication.

4.7 Upcoming features: CSS gradients and masks

The WebKit is constantly growing, and in upcoming years you'll be able to build an increasing number of great features into your iPhone web pages that won't be available to non-WebKit browsers. It can sometimes take a while for new features to make it from the WebKit "nightly builds" into an actual release from Apple, however.

This is the case with two interesting new graphical features that were announced in April 2008: gradients and masks. We have faith that both of these new CSS properties will soon be available in Safari and on the iPhone, but as of this writing they're not yet available. Therefore, we're just going to cover them in passing, with no guarantees that the properties will be quite the same when they actually appear.

If you'd like to use either of these properties, you should check on the internet to see if they've yet made it into an Apple build, and if they've changed any since the early documents that we've referenced (which come from http://webkit.org/blog, the top resource for information on new and upcoming WebKit features).

4.7.1 CSS gradients

CSS gradients will give you the opportunity to embed blended colors on your web page. This feature should be particularly useful on the iPhone, since gradients are already a part of the look and feel of iPhone home page icons, and will make individual programs feel more like native iPhone programs.

The Surfin' Safari WebKit blog states that gradients will use the following syntax:

```
-webkit-gradient(<type>, <point> [, <radius>]?, <point> [, <radius>]? [,
<stop>]*)
```

The type might be linear or radial, while the points define the two edges of the gradient. Color stops then define where colors change, with each stop including a value from 0 to 1 and a color.

These gradients can currently be applied to `background-image`, `border-image`, `list-style-image`, and `content` properties.

The Surfin' Safari blog also offers the following example of how a linear gradient should work:

```
.linear {
   background: -webkit-gradient(linear, left top, left bottom,
      from(#00abeb), to(#fff), color-stop(0.5, #fff),
      color-stop(0.5, #66cc00));
}
```

This particular example uses lots of shorthand to make gradients simpler. For example, the phrases `left top` and `left bottom` are shorthand for the endpoints of the gradient, and `from` and `to` are shorthand for color stops that use those same endpoints.

4.7.2 CSS masks

A mask is a black-and-white shape that you use to show only part of an image. In a mask, alpha values of 0 show nothing; alpha values of 1 display the image content. In other words, it's a way to clip an image.

WebKit masks can be used in a variety of ways. The properties:

- `-webkit-mask`
- `-webkit-mask-origin`
- `-webkit-mask-composite`
- `-webkit-mask-attachment`
- `-webkit-mask-image`
- `-webkit-mask-box-image`
- `-webkit-mask-clip`
- `-webkit-mask-repeat`

can all be used to mask an underlying image in different ways.

`-webkit-mask-image` should provide the simplest masking:

```
<img src="yourimage.png" style="-webkit-mask-image: url(yourmask.png)">
```

`-webkit-mask-box-image` can provide some interesting border masking, if used correctly:

```
<img src="yourpic.png" style="-webkit-mask-box-image: url(anothermask.png)
75 stretch;">
```

Again, since this functionality has not yet made it into Apple builds, check online to find out how and when this functionality can be used.

4.7.3 The Canvas alternative

Although we can't yet fully document these new features, we wanted to point them out as things that you should keep an eye on, because they should soon be available. Because we can't predict when the features will become available in Safari, though, we'll mention one alternative for these functions: Canvas.

We'll be talking about Canvas in chapter 6. It's a vector-based graphic design program that can be used on a variety of browsers, including Safari and mobile Safari. Among Canvas's features are gradients (which work almost identically to the gradients in the WebKit) and masks (which you create by drawing the masking paths by hand). If you must use gradients or masks in your iPhone web app, and mobile Safari doesn't yet include them, consider Canvas as an alternative.

The downside will be that you can't integrate Canvas into a web page in quite the same way you can CSS. It's a separate element, built around the <canvas> tag, rather than a CSS property, which means you'll need to use some layered positioning or something similar to work it into your web page. However, if you need the functionality, then at least you have an alternative that will give you access to it.

Lessons for SDK developers

Engineers at Apple have been leading the charge in the design of the WebKit. It's here that we really start to see commonalities between WebKit and SDK development patterns. Sometimes they're close enough to help you bridge the gap between the two styles of programming, and other times they're far enough apart to cause confusion.

Database programming based on SQLite is the first feature that we'll see repeated in the SDK. Whereas the WebKit's abstraction for the database is already advanced, cleanly breaking out responses from input, the SDK still depends on SQLite's native API; as a result, if anything, database programming is easier in the WebKit than in the SDK at the time of this writing. The SDK's SQLite is covered in chapter 16.

The WebKit's touch event handling is another element that we'll see closely mirrored in the SDK. Both use the same architecture of calling handlers when touches start, change, or end. They group touches into events in slightly different ways: in particular, the fact that a touchend event contains no touches in the WebKit will confuse a SDK programmer, since under the SDK an event will always contain the touches that triggered it. The WebKit's gesture abstraction saves programmers some time; it's not available in the SDK, though SDK programmers will discover that they have access to a lot more information about how and when gestures occur.

Despite these subtle differences, the big picture stays the same and will help programmers model similar user inputs in both web and native apps. The SDK's event handlers are discussed in chapter 14.

We'll see Apple programming patterns and methods two more times in the web part of the book: in chapter 6, when we look at Apple's Canvas library, and in chapter 7, when we investigate their Dashcode developmental platform.

4.8   *Summary*

The WebKit represents some of the quickest changing technology available for web development on any platform. It's the basis for the iPhone's Safari, and that means that you can expect web development on the iPhone to become an increasingly good option as time goes on, further adjusting the balance of development choices that we discussed in chapter 2. The fact that we couldn't fully document some of its features just emphasizes how quickly things are changing.

As an iPhone developer, you'll probably be most excited by the iPhone-specific features that have been implemented. The touch and orientation event handlers provide direct access to some of the unique iPhone features that we've highlighted in this chapter. Some of the "normal" WebKit features are pretty great too. We think you should particularly consider how transitions and the built-in JavaScript database can change the way you program web pages.

When you're building iPhone web apps with the WebKit features, you're still very much building text-based web-centric applications. That's not the only way to build iPhone web apps. As you'll see over the next two chapters, there are two other models that you can use for your web app designs. Next up are iPhone-UI web apps, which look just like iPhone-native apps.

Using iUI for web apps 5

Creating web apps with the WebKit gives you a lot of power and allows for a lot of diversity. But it doesn't address a real concern: what if you want to create web apps that have the same look and feel as native applications on the iPhone?

The reasons for doing so are obvious. You can take advantage of lessons that users have already learned by using Apple's standard iPhone user interfaces. The question of how to do so, however, is slightly more complex.

Certainly you could do so using the WebKit's extensions atop HTML, and we're going to give you the opportunity to do so here by dissecting what makes up the iPhone interface. Unless you have specific needs, however, creating your own interface is probably overkill.

Fortunately, we have another solution to this problem: a third-party library exists that you can use to model the iPhone interface. It's called iUI, and it'll be the focus of much of this chapter. But before we get into iUI, let's see what the iPhone interface looks like.

5.1 Creating your own iPhone UI

The iPhone UI has quite a few unique (and distinctive) features. If you want to model its UI inside a web app, you must consider all of them. We'll provide some guidance by highlighting the most important UI features in this section.

We'll start out by looking at the iPhone's graphic interface, but the UI goes far beyond that. The iPhone's UI also depends on a data-driven content paradigm and the usage of those unique design choices that we highlighted back in chapter 1.

5.1.1 The graphical interface

As shown in figure 5.1, the standard look of an iPhone application is quite distinctive, and thus can be easily modeled in your own web pages.

Figure 5.1   The iPhone has a unique look and feel that should be emulated in iPhone web apps.

What follows are a number of simple guidelines that suggest how to easily model the iPhone's graphical interface.

- *Match the look and feel*—When you're building your own iPhone-UI web apps, the best thing you can do is match the general look and feel of the built-in iPhone programs. This ap-proach has two benefits. First, it gives your users a leg up on using your program, because they're already familiar with the interface. Second, it takes advantage of the work that Apple programmers have already done to figure out what works on the iPhone. The downside is that there isn't a totally consistent interface on all the iPhone applications, a rarity for Apple. The core iPhone programs—the iPod, Mail, and the various phone accessories—do the best job of presenting a consistent UI, but some of the other programs look slightly different. You'll ultimately need to decide how much of the look and feel you want to model, and how much you want to change it to match your own site's branding, a topic we'll return to toward the end of this chapter.

- *Use chrome*—It's most important to match the iPhone's chrome. Put navigation and control bars to the top and the bottom of your App's web pages. You might want to match the iPhone's gray coloring, or you might want to match the colors of the rest of your site.

- *Put action buttons in your chrome*—In the chrome, it's not just the navigation that's important, but the action buttons as well. For example, several applications put a "+" action in their chrome to generate new content. Safari also includes search and bookmark actions.

- *Create sliding menus*—Menus of options should typically be display on subpages, not as a part of your web app's main display. You might click an action button to pop up a menu. Alternatively, a list of options that fills the screen might each have a > symbol that leads to its own subpage, as with the iPod's "more" pages. Though we call these *sliding menus*, for the way they animate on the iPhone, you probably won't mimic that aspect in your web app (unless you use iUI).

- *Include back buttons*—Whenever you slide in a subpage, make sure there's an explicit, labeled button that will bring you back to the page that you left from. Note that the iPhone creates a history stack that could be several levels deep. No matter how deep you go in the structure, you can always return to the top level with back buttons. This is a topic that we'll return to a few times in this chapter.

- *Separate your choices*—Whenever you have a list of options, make sure they're widely separated. The iPhone always includes both spaces and rules between options, giving "fat fingers" plenty of room to fit in.

- *Use the viewport*—If you're creating a web app specifically for the iPhone, you no longer have to worry about creating a full-sized web page that will look good elsewhere. Either set your viewport to 480, or (if you want to be fancy) use either the device-width feature or vary your viewport based on orientation.

- *Don't be afraid to turn off zooming*—Don't be afraid to set `minimum-scale` and `maximum-scale` to 1 in your `viewport` metatag. After all, if you've designed a page specifically for the iPhone, there should be no reason for users to mess with it.

About views

iPhone programs often include multiple pages of information that are all part of the same application. Apple calls these individual pages of content *views* (though the term is also used to refer to individual objects, or subviews, within a page).

Many web apps will be built using the same concept, allowing pages to call up other pages using tab bars, navigational controllers, and other means. As you'll see, the view model is well supported by various tools that you might use when creating iPhone web apps. iUI (which we'll discuss in this chapter) supports it as part of its navigational methods, and Dashcode (which we'll talk about in the next chapter) supports it using a `stackLayout` object.

We'll see even more complex view hierarchies when we begin using the SDK, which is the ultimate model that these other libraries are imitating.

This advice is crucial for putting together an iPhone-UI web app, but it's also primarily visual. When creating iPhone web apps, it's important to think about the base paradigm that underlies your web development.

5.1.2 The iPhone data paradigm

We've already talked quite a bit about how the iPhone is technically different from last-generation mobile phones. We also believe that a core difference exists in the methods that people use to access information on the iPhone.

The desktop internet centers on the paradigm of browsing. Although users will sometimes go to a specific page for certain information, a lot of internet usage instead centers around idle viewing of interesting sites. Because browsing is so important, desktop-centric websites often center on immediately dumping unsorted piles of content on users, hoping to draw them further down into a site.

We believe that these priorities are reversed on the iPhone. You are less likely to attract casual browsers and more likely to attract users who already know what your site is and are visiting it to retrieve specific information. This more data-centric view of web usage has already been verified by the huge jumps that Google saw in iPhone searches. Likewise, some of our sample sites already reflect this paradigm, such as Facebook, which removed most of the information from their iPhone login page.

In other words, it's all about the data.

When you're building your iPhone-UI web apps, think about how the iPhone itself organizes data and try to mimic that. That means using trees of lists. Take a look at the iPod button on your iPhone if this concept isn't immediately clear. When you push it, notice that you get no textual descriptions, just a set of options. When you click Composers, you get another list of options. You can dig down from there to a list of songs, and then you can finally listen to a song.

A similar model for your web app will omit most of your website's text and replace it with the core of your data, easily accessible via sliding menus. It's a pretty big paradigm change, and you may initially feel resistant to it, but we feel that it'll give you best results, because it's what iPhone users are used to and want.

5.1.3 Other iPhone design elements

Beyond the display UI and the data paradigm, you should also think about the iPhone's six unique design elements when creating any web app interface.

- The *always-on internet* is a great benefit because it's easy for you to access online pages and data. But your users may sometimes be accessing your pages via a slower web connection, thanks to EDGE. This all suggests Ajax as a great tool for any web app usage because it requires constant access, but at the same time allows redrawing of pages in bits and pieces. When thinking about an always-on internet, also consider the methodology that Apple uses in native apps like Mail. These programs only load small sets of data—such as 25 messages at a time—waiting to load more until you request it.

- *Power consciousness* won't be a huge issue in your web app work, because you'll be limited to what you can do by Safari. But you shouldn't include JavaScript events that will constantly poll the internet or keep Safari active.
- *Location awareness* can't be accessed from a web app.
- *Orientation awareness* is a topic we covered in chapter 4, when looking at the WebKit.
- *Input* was also covered in chapter 4. If you want, you can continue to use the WebKit's orientation and gesture support as part of your iPhone-UI web apps.
- *Output* is the last of the design elements. One of the cool things about the iPhone's Safari output is that it scales, but as we've already suggested, that's something you might want to omit in an iPhone-UI web app.

By now you might have some great ideas for how to turn your website into an iPhone web app, but the idea of coding everything to look like and work like the iPhone can seem daunting. Unlike our discussions of Canvas and the WebKit, there isn't an Apple library that we can point you to (though one is in process, from what we hear). But if you don't mind using third-party libraries, you can use a free software package called iUI.

5.2   *Getting ready for iUI*

iUI is a library of JavaScript and CSS that is intended to mimic the look and feel of the iPhone in web pages. In other words, it's precisely what you need to make an iPhone-UI web app of the sort that we've been describing. Software engineer Joe Hewitt, the author of iUI, said the following about it in his initial blog posting on the topic:

> *First and foremost, iUI is not meant as a "JavaScript library." Its goal is to turn ordinary standards-based HTML into a polished, usable interface that meets the high standards set by Apple's own native iPhone apps. As much as possible, iUI maps common HTML idioms to iPhone interface conventions. For example, the <ul> and <li> tags are used to create hierarchical side-scrolling navigation. Ordinary <a> links load with a sliding animation while keeping you on the original page instead of loading an entirely new one. A simple set of CSS classes can be used to designate things like modal dialogs, preference panels, and on/off switches.*
>
> *Let me re-emphasize that all of this is done without the need for you to write any JavaScript. It is meant to feel as though HTML was the iPhone's own UI language.*
>
> http://www.joehewitt.com/blog/introducing_iui.php (July 11, 2007)

Thus, not only can you expect iUI to model the iPhone for your web app, but you can also expect it to be easy to use.

iUI is a community project that is offered for your free usage. At the time of this writing, there are two versions of the code. The officially released code is probably what you want to use for any live website, but you can also go to the current SVN repository for the rawest, most up-to-date code. Table 5.1 lists the current places that you can access the code.

Code type	Location
Release Bundle	http://code.google.com/p/iui/
Subversion Repository	http://iui.googlecode.com/svn/trunk/

Table 5.1   iUI can be freely downloaded from the internet, either in its officially released form or via a Subversion repository.

Once you've downloaded iUI, you need to place the iui directory in a standard library location for your website so that you can load JavaScript and CSS files from it. Of course, if you're using iUI in any corporate environment, you'll also want to look at LICENSE.txt to make sure that it meets the requirements of your company. Fortunately, iUI's license is a permissive new BSD license that places few restrictions on what you can do with the software.

Barring problems, you should now be ready to start programming a web app in iUI. But we'll offer one last disclaimer before we talk about that: author Christopher Allen is currently one of the three owners of iUI, along with Joe Hewitt and Sean Gilligan. Naturally, this makes us biased toward iUI. But we wouldn't be supporting it if we didn't think it was a superb product for creating iPhone-specific web pages. We're pretty sure you'll agree too.

5.3   *Developing with iUI*

Once you've got iUI installed, you can include it in your code by referencing iUI's JavaScript and CSS files. Listing 5.1 shows a typical iUI header, placed in the <head> section of your HTML file. This is the beginning of a color-selector application that we're going to develop over the next couple of examples; it will allow you to look up your favorite shades of red (and other colors if you want to extend the example on your own).

Listing 5.1   The headers for an iUI page

```
<meta name="viewport" content="width=device-width; initial-scale=1.0;
maximum-scale=1.0; user-scalable=0;">
<style type="text/css" media="screen">@import "iui/iui.css";</style>
<script type="application/x-javascript" src="iui/iui.js"></script>
```

Note that besides the standard include statements for the style and the script, you'll also set a viewport, following the suggestions we offered earlier for iPhone-UI web apps. This viewport is set to the iPhone screen width and also keeps users from scaling the window.

From here you can start building an application using the iUI classes. Table 5.2 shows all the major classes that you'll be encountering. We'll return to each of them in turn as we build up our color selector.

We'll cover each of these in more detail as we step through our example, which we're now ready to dive into, beginning with the all-important iUI toolbar.

Table 5.2   iUI makes about a dozen classes available for you to use in creating iPhone-like web pages.

iUI class	Element	Summary
button	`<a>`	The standard iPhone-UI toolbar button. Appears at top right normally, or at top left with an arrow pointing backward with `id="backButton"`.
leftButton	`<a>`	Moves a button left.
blueButton	`<a>`	Turns a button blue.
grayButton	`<a>`	A page-width gray button. This is the button to use for important links internal to your page.
whiteButton	`<a>`	A page-width white button. This is the button to use for important links internal to your page.
toolbar	`<div>`	The core class for iUI. Anchors the page.
dialog	`<form>`	Creates a standard iPhone UI for data entry.
group	`<li>`	A nonlinked list item, intended to organize links into groups.
panel	body element	Creates a standard iPhone UI for settings.
row	body element	Creates a left- and right-justified set of data.
toggle	`.row`	Creates toggle buttons in a row.

5.3.1   *The iUI toolbar*

Once you're working on the `<body>` of your iUI web page, you can start using the package's unique styles. This will typically begin with a toolbar, as shown in listing 5.2.

Listing 5.2   A toolbar that anchors an iUI page

```
<div class="toolbar">
    <h1 id="pageTitle"></h1>              ❶
    <a id="backButton" class="button" href="#"></a>        ❷
    <a class="button" href="#searchForm">Search</a>         ❸
</div>
```

This is a standard toolbar that will appear almost unchanged in any iUI program that you create. Every toolbar should have a `pageTitle` line ❶. Most will also have a `backButton` line ❷, unless your app contains no subsidiary pages. The `button` line ❸—which is what we called an action button in section 5.1—is the most optional of the three, and will only appear when you have some action that you want a user to be able to take. A few of these elements bear additional discussion.

First, notice that the `pageTitle` and `backButton` lines do not contain any content. This is because iUI's Javascript automatically fills them. The `pageTitle` will be filled with the title of the element that's currently selected. The `backButton` will be filled by the title of the previous element when you move forward a page (and won't be used until then).

Second, be aware the `backButton` allows access to a whole stack of pages—just like on the iPhone itself. No matter how deep you are in your iUI page stack, you'll always be able to get all the way back to your first page.

Third, note that we said "elements" when referring to the individual parts of an iUI screen. Although an iUI web app will look like several pages on an iPhone, it can all be in one file on your web server (using the standard model of views that we've already highlighted). Individual forms, lists, and <div>s are each named. iUI then redraws the page—using simple animations that model the iPhone—whenever a user moves to a new, named element.

5.3.2   iUI lists

Once you've written your (relatively standard) toolbar, you then need to prepare your default element. It's what will get pulled up on the screen the first time a user visits your iUI page. To create your default element, you must write a <div>, form, list, or other element, then give it the `selected="true"` attribute.

This selected element is the only one that will appear when your page is first drawn. All the other elements on your page should be available via links from your default element; they'll only appear when those links are clicked. Figure 5.2 shows the results that you're aiming for. As you can see, you're creating a simple example of what the SDK will call a table view, one of the most frequently used iPhone user interfaces.

This table view is surprisingly easy to create in iUI, as shown in listing 5.3, which continues our example by showing the default element of our color selector. All that's needed to create the table is a simple unordered list.

Figure 5.2   A list-based paradigm is easy to program in iUI, using just the `<li>` elements.

Listing 5.3   iUI's default element

```
<ul id="home" title="Colors" selected="true">        ❶
   <li><a href="#red">Red</a></li>
   <li><a href="#green">Green</a></li>                ❷
   <li><a href="#blue">Blue</a></li>
   <li><a href="http://en.wikipedia.org/wiki/Color"  ❸
      target="_self">Other Colors</a></li>
   <li><a href="#settings">Settings</a>
</ul>
```

As promised, you start off your default element with the `selected="true"` attribute. Note that you also use an `id="home"` attribute ❶. This isn't a requirement, but helps keep things clear.

The rest of the lines in this element are list items holding links. Although you'll frequently be linking to other elements that are part of the same iUI page, as you do in your first three list items ❷, you're creating a fully functional web page and can thus link to anywhere on the web.

Your penultimate anchor line ❸ shows an offsite link. The `target="_self"` attribute is required when you're linking to a complete (usually off-site) HTML page. This is because iUI assumes that external links usually connect to iUI-style page fragments containing only elements. It therefore tries to load them via Ajax if you don't tell it otherwise. We'll explain more about how you might want to use these page fragments and Ajax when we get to iUI tips and tricks, later in this chapter, but for now just be aware that you need to set the `target` attribute to `_self` for normal links.

About Ajax

Ajax stands for Asynchronous JavaScript and XML. It's a technology that can be integrated into web pages, but it's more a set of techniques than an actual language. It combines JavaScript and XML (as you'd expect from the name) with DOM.

The concept behind Ajax is that the web browser exchanges small amounts of data with the web server without redrawing complete pages. Typically, a form sends some data to the server, the server processes it, and then it sends back a command for the browser to update some small part of a page. Ajax is great for data entry where you want to indicate that the server has accepted the data. As it happens, it's really good for low-bandwidth services too, like EDGE.

Once you've written your default element, you can add other elements in a similar manner. To correctly link these elements, you just have to make sure that their ID matches the originating `href` link. Figure 5.3 shows an example of how a subpage might appear when connected to the `href="#red"` link in the default element. It also highlights another bit of iUI chrome: groups.

The code required to create this new element is shown in listing 5.4. As you can see, it's another unordered list, this time using some new attributes.

Note that you're introducing an additional iUI class in this example. The `group` class can only be applied to `<li>` elements. You apply it to list items that are not links. The result will be a category head that you can use to organize several links below it.

Figure 5.3 Groups help programs to organize lists into logical units.

Listing 5.4 Other elements defining subpages in your iUI web app

```
<ul id="red" title="Red Shades">
  <li class="group">Light Reds
  <li><font color="#ff0000"><b>Computer Red (#ff0000)</b></font>
  <li><font color="#ED1C24"><b>Pigment Red (#ED1C24)</b></font>
  <li><font color="#E0115F"><b>Ruby Red (#E0115F)</b></font>
  <li class="group">Dark Reds
  <li><font color="#DC143C"><b>Crimson (#DC143C)</b></font>
  <li><font color="#800000"><b>Maroon (#800000)</b></font>
</ul>
```

We could easily extend this example to also show pages for green and blue color selections, but doing so wouldn't add anything to what you've already learned. Instead, we're going to jump to the next major element of our color selector example: the searchForm.

5.3.3 *iUI dialogs*

The iPhone already has a standard way to display web searches. You can see it when you click the magnifying glass in mobile Safari. A search dialog pops up toward the top of the screen, complete with a cancel button in the chrome. Meanwhile, the rest of the page is grayed out but can still be faintly seen in the background.

This is the look and feel that is mimicked when you use the dialog class in iUI. It's probably what you want to use whenever you're creating a simple search, and you may find it useful for other types of simple interaction as well. Figure 5.4 shows how it will look when used as a part of your web app.

Listing 5.5 shows how to create this dialog as part of our color selector example.

Figure 5.4 A dialog allows users to conduct searches.

Listing 5.5 Creating a search-like form with the dialog class

```
<form id="searchForm" class="dialog" action="search.php">
  <fieldset>                                                ❶
  <h1>Color Search</h1>                                     ❷
  <a class="button leftButton" type="cancel">Cancel</a>     ❸
  <a class="button blueButton" type="submit">Search</a>
  <label>Color:</label>                                     ❹
  <input id="color" type="text" name="color">               ❺
  </fieldset>
</form>
```

Although we listed dialog as a <form> class in table 5.2, it can technically be applied to anything; but it probably doesn't make much sense for anything but a form.

Inside the `dialog` class, the order of the contents is important. You must have a `<fieldset>` ❶, which is used to logically group form elements together and then draw a box around them. Then you must have a level one header ❷, which will appear amid the chrome at the top. Omitting either of these will cause your dialog to display incorrectly.

The buttons ❸ can be set in several configurations. Here you display both cancel and search buttons. This isn't quite how the iPhone does things by default, but it provides a more complete UI, thus trading off usability for conformity. If you prefer to stick with the iPhone standard, replace the pair of buttons with a single Cancel line:

```
<a class="button" type="cancel">Cancel</a>
```

Note that this example also introduces how to apply the iUI button classes. The button class by itself always displays a right-justified gray button, but by applying multiple classes, you can move that button to the left (with `leftButton`) or turn it blue (with `blueButton`). We'll return to all the ways in which buttons can be used later.

After placing your buttons, you can put together a long series of `<label>` ❹ and `<input>` ❺ lines. In each case, the label is placed into the background of the input, showing what should be entered in that box. This example only includes one such pair, but additional ones can be entered identically.

The one thing that this example doesn't show is the actual back end, search.php. As it turns out, that works in a special manner under iUI.

5.3.4   *iUI searches done right with Ajax*

iUI's search functions use Ajax, bringing us back to the idea of Ajax and page fragments, which we touched on when we looked at external links. iUI uses Ajax in a couple of places to replace individual elements rather than redrawing a whole web page. Not only does this make an application faster, but it also cleverly allows iUI to maintain its overall stack of history pages.

This is all easy to do because iUI takes care of the Ajax details for you in a relatively transparent manner. You don't need to know that you're using Ajax; you just need to know that in some situations you must supply fragmentary pages that contain only iUI elements rather than complete pages.

This is the case with iUI-based searches. Your search result pages should not draw full pages, nor should they forward you to full pages. Instead, they should output an individual element, just like those elements that you've been writing to model other iUI subpages.

Listing 5.6 shows what search.php might output after a search for "Maroon".

Listing 5.6   A search page fragment

```
<ul id="search" title="Maroon">
   <li><font color="#800000"><b>Maroon (#800000)</b></font>
</ul>
```

How the back end of your search works is, of course, entirely up to you. All that matters is that the output supplies a fragment, because that's what iUI's Ajax engine expects to see.

5.3.5 *iUI panels and rows*

That leaves us with just one missing item from our iUI color selector example: the settings page. We'll be laying this out with our remaining iUI classes for this example: panel and row (see listing 5.7). These classes are meant to mimic the Settings page found on your iPhone, with its rows of options that you can change.

Listing 5.7 Panels, rows, and toggles forming another user input model

```
<div id="settings" title="Settings" class="panel">          ❶
    <fieldset>          ❷
        <div class="row">          ❸
            <label>Color in Names</label>          ❹
            <div class="toggle" onclick="return;">
                <span class="thumb"></span>
                <span class="toggleOn">ON</span>          ❺
                <span class="toggleOff">OFF</span>
            </div>
        </div>
    </fieldset>
</div>
```

The panel class ❶ works similarly to the dialog class by building data into a `<fieldset>` ❷. The result is an inset panel. If you'd like, you can put an optional header above the `<fieldset>` but still inside the panel class, which will label your panel for you. We've opted not to do so in this example because you only have one panel on the page, but if you had multiple panels, doing so might help your organization.

THE ROW

Within a panel, information is organized into rows ❸. Again we see similarity to the layout of the dialog, because a `<label>` tag ❹ designates data that is pushed to the left of the element. Here, you also have the option of placing data to the right. In the case of this example, you're placing a special toggle class ❺ in that location. The results are shown in figure 5.5.

You'll most often use the panel and row classes in iUI just as you do here, for toggles. Note that the row class can be used elsewhere if you want to mimic the left/right row-based organization of data in other places. The downside is that rows are set up to contain

Figure 5.5 Toggles support the standard preference UI.

buttons and graphics (like the toggle). If you try to put regular text here, it'll end up top justified, and not looking that great.

THE TOGGLE

toggle is a complex and somewhat special class. It can only be used inside rows, and it must include three additional classes: thumb, toggleOn, and toggleOff. The thumb class is left blank, as has been the case with other iUI classes in the past, and will be replaced with the thumb slider graphic. The contents of toggleOn and toggleOff define what the slider says in each of its on and off positions; this example makes the boring decision of the toggles being labeled ON and OFF, but you can modify that however you want.

The toggle class is managed by an onclick JavaScript event handler. The example includes return;, but you'll want to instead call up something that makes an appropriate change based on the change in settings.

Usually a toggle starts in the off position, but if you instead want to set it on by default, you just need to set the toggled attribute in your toggle <div>:

```
<div class="toggle" onclick="return;" toggled="true">
```

If you've been following along on your home computer, by this point you should have a sparsely functional color selector with examples of all the major iUI classes. Having finished designing an iUI color selector, we've also finished our overview of most of the major classes for iUI. There's still one set of classes that could use a bit of additional clarification: the buttons.

5.3.6   *iUI buttons*

There are a total of five different classes for buttons. We've already seen them in use both in the toolbar and the dialog classes, but we have not yet examined them in a more thorough manner. Generally, there are two types of buttons: the small buttons and the large buttons. The small buttons are slightly trickier to use.

The small buttons include three different classes: button, leftButton, and blue-Button. These three classes can be mixed together, with each of them providing slightly different utility. The button class is always required. It defines the general look and shape of a gray button. It also defaults placement of the button to the right. The leftButton can be stacked with the button to instead push the button to the left. The blueButton can be stacked with either button to make it blue instead of gray. These possibilities are all described in table 5.3.

Table 5.3   The three small button classes can be stacked together to place buttons in different locations and to tint them with different colors.

Classes	Summary
button	Gray button to the right
button leftButton	Gray button to the left
button blueButton	Blue button to the right
button leftButton blueButton	Blue button to the left

You should be careful about moving the colors and locations of buttons too far from the iPhone standards. Given those constraints, though, the small button classes provide considerable ability to easily create attractive iPhone-like graphics.

The large buttons, `grayButton` and `whiteButton`, work differently from the small buttons. They're each used singularly, and each will entirely fill the width of an iPhone screen. You can embed them by adding a class to an `<a>` tag, just as with the other buttons:

```
<a class="grayButton" href="#confirm">I'm Sure!</a>
```

You should use these buttons when you want to make a notable impact on your document somewhere other than your toolbar or dialog box. The large buttons are similar in look and feel to the confirmation buttons that you'll often see pop up on an iPhone, so your users should already be familiar with their usage.

Having completed our look at iUI classes, there's only one other thing we want to look at in our overview of IUI: a few special attributes that are used by the JavaScript libraries.

5.3.7 iUI attributes

iUI recognizes a handful of special attributes that can be applied to various tags. These each tend to be related to one of the classes that we've already met (see table 5.4).

Table 5.4 A small set of attributes can help you to further differentiate your iUI-classed page elements.

Tag	Attribute	Related class
`<a>`	`type=cancel`	`button`, `backButton`, `blueButton`
`<a>`	`type=submit`	`button`, `backButton`, `blueButton`
`<div>`	`toggled="true"`	`toggle`
`element`	`selected="true"`	(none)
`element`	`hideBackButton="true"`	(none)
`<body>`	`orient="portrait"` or `orient="landscape"`	(none)

You've already seen the majority of these attributes. The `cancel` and `submit` anchor types both appeared in our dialog box, though they can appear in any type of form. They give plain links the necessary form-based functionality.

The `toggled` attribute appeared in our example of combining panels, rows, and toggle classes, while the `selected` attribute always defines which page element is currently being viewed. They were each covered thoroughly in the appropriate sections. That leaves just two brand-new attributes.

`hideBackButton` is placed on any element, just as `selected` is. You use it when, for whatever reason, you don't want users to have a back button that would let them return to the previous page. If we didn't want the settings page in our color selector app to have a back button, we'd replace the first line of that element with the following:

```
<div id="settings" title="Settings" class="panel" hideBackButton="true">
```

You probably won't want to do this much (if ever) because it breaks the standard iPhone navigation model.

Finally, orient is a <body> attribute that iUI sets for you. It can be used to tell whether the iPhone is in portrait or landscape mode at any time without having to watch for orientation change events. Of course, it's probably just as easy to use the WebKit's orientation property, making this attribute somewhat redundant.

Having thoroughly described how iUI works, we're now going to briefly revisit how iUI page design is done. What we've discussed thus far assumes a simplistic data model for your website, which probably won't be the case, so let's take a step back and see how iUI might be integrated in a more realistic manner.

5.4   *Creating an iUI back end*

With our color selector (with the exception of that search.php page), we assumed that you were working with static data. If you have unchanging data and you're just looking for a handy way to access it on the iPhone, then our suggestions thus far will work fine. And there are plenty of handy applications that you could write for the iPhone using this methodology. Besides our color selector, you might set up a static contact list, an encyclopedia of municipal laws, or a simple template-based manufacturer's site using the same methods.

But static web display hasn't been common since the earliest days of web design. It's much more likely that you've got a dynamic website that uses PHP, Perl, Ruby on Rails, or some other web development language to render live data from a database or some other dynamic source. As you'd expect, you can use that server-side language to output iUI-classed HTML code. We won't dwell on this too much, as it's a simple application of the iUI classes you've learned to the dynamic programming that you already know.

We've offered a quick example of integrating PHP and iUI in listing 5.8. There is a live program that we currently have running at http://index.xenagia.net/iphone-recent.phtml that shows a listing of some of the most recent fantasy, science fiction, and horror book releases (and then links users through to iPhone optimized web pages outside the iPhone web app).

Listing 5.8   iUI will typically be used as part of a dynamic web page

```
<body>
   <div class="toolbar">          ❶
      <h1 id="pageTitle"></h1>
      <a id="backButton" class="button" href="#"></a>
   </div>
   <ul id="home" title="Recent Books" selected="true">        ❷
<?
   $recent = searchRecentDB(25);          ❸
   for ($i = 0 ; $i < sizeof($recent) ; $i++) {
    if ($recent[$i][entrydate] != $lastdate) {        ❹
```

```
?>
        <li class="group">Entered <? echo $recent[$i][entrydate]; ?>
<?
      }
      $lastdate = $recent[$i][entrydate];
?>
        <li>        ❺
            <a href="display-entry.phtml?mainid=<?
            echo $recent[$i][mainid]; ?>" target="_self"><?
            echo $recent[$i][title]; ?></a>
            <i><? echo $recent[$i][authors]; ?></i>        ❻
<?
      }
    ?>
        </ul>
    </body>
    [XHTML?]
```

You will find nothing here surprising. You've got a standard toolbar ❶, which you follow up with a default element ❷, once again using our old favorite, the unsorted list. The main body of this page is taken up by linked list items ❺. It's interesting to note that iUI will place additional content included after an anchor link on its own output line ❻, as shown here. We've also included some group list items ❹, to help break up the page and give some order to what we're showing.

The data lookup ❸ deserves some additional discussion because it purposefully mirrors the methodology of the iPhone by showing data in bite-sized chunks, here 25 items. We'll look at an even fancier way to do this when we get to iUI tips and tricks in a moment.

If we wanted to spend more time on this example, we'd start off by allowing the user to scroll back through more than just the last 25 books. We'd also try to make our search page look a little more attractive. Finally we'd rewrite our book display pages in iUI too. All of that is ultimately beyond the scope of this example, but we do want to underline one fact: using iUI dynamically is very, very easy. Our working example literally took 10 minutes to put together. If you go view the page in an iPhone browser, you can see that those 10 minutes paid off with a simple, attractive interface.

Now that we've looked at how iUI is likely to be used in a real web environment, we've only got one more iUI-related topic to discuss, one that we've been promising for a while: those tips and tricks that you can use to improve your iUI experience.

5.5 *Other iUI tips and tricks*

We're going to finish up our look at iUI by discussing some tips and tricks that are easy to use but that go beyond the needs of many web app developers. These include a more intensive look at Ajax and how it can help you organize your code and your listings; a simple way to include slightly smaller iUI libraries; and some methods to change the standard look and feel of iUI. We'll kick off with that code organization.

5.5.1 *Organizing your code*

You've already seen that iUI changes the way web pages are written by representing many subpages as elements in a single HTML file. But as files gets larger (and thus more unwieldy over EDGE), you have to decide when you should create a new page in iUI rather than a new element. It turns out that there are three ways organize your iUI pages, as summarized in table 5.5.

Organization	Summary
Single file	For small, singular applications
Multiple fragments	For larger, singular applications
Multiple files	For larger multipurpose applications

Table 5.5   **You can keep elements on the same page, maintain fragmentary subpages, or write totally different pages, depending on the app.**

Each of the three methods shown in table 5.5 should be used in different circumstances.

A single file is the standard process that we've been using so far. It keeps things nicely organized on your side of things. It also preserves iUI's history stack, as we've briefly mentioned before.

Multiple fragments is an alternative that you can use if your file is getting too large and thus causing problems over EDGE. You need to start out with a single main file, as usual, but you can then put individual elements in subpages, and call them up through a normal URL. The trick is that the subpage must be a fragment, just like the fragment that we created for our search results. It can't contain headers or any other information.

For example, in our color selector we might put each of our colors on a subpage, like red.html, green.html, and blue.html. Those subpages would then contain only the `<ul>` element in question. When the link on the main page is clicked, iUI will cleverly load your fragment into your main page, preserving the history stack. This is once more done with Ajax.

Multiple files are the last and least desirable option. Using them will break your history stack, which could be disorienting for your users. But if your file is much too large or too complex, you might want to use them. An even better reason would be if you have several somewhat different applications all glued together.

When you separate iUI pages in this manner, you should be careful to take the navigation between your separate pages outside of the normal iUI interface. That way, your users won't have any expectation of being able to page back to previous screens.

The best method for doing so is a tabbed interface that sits on the first page of each of your iUI web apps, allowing users to move between them. It's important that this only appear on the highest-level pages in your web app, because those will be the places where there isn't any history on the stack.

Remember that whenever you link to an external web page that isn't a fragment, you must include the `target="_self"` attribute. Otherwise, iUI will assume you're loading a fragment, and its JavaScript can get in the way of the new page popping up.

Another method for melding together multiple web apps is to create a home page for your site that has iPhone-like buttons on it, similar to those that appear on the iPhone's home page. This has the advantage of being more attractive, but you'll need to create special navigation to get back to it.

5.5.2 Improving data listings

iUI also uses Ajax in one other place, solving another common iPhone UI problem. As we mentioned a few times, when you're listing data on an iPhone you should cut it down into bite-sized chunks to be easily loaded over the wireless network. We saw this demonstrated in the iPhone's mail program and in our own recent books example.

There's an even fancier way to create these partial listings in iUI. If you use the mail program on the iPhone, you'll note that its Load 25 More Messages button doesn't load a new web page (as would be standard for a desktop web page), but instead shows 25 additional messages at the bottom of the same page. This is the same methodology used for YouTube searches, and can generally be considered yet another bit of iPhone-specific chrome.

You can mimic this easily thanks to iUI's next bit of Ajax magic. You just use `target="_replace"` in your "more" `<a>` link. Returning once more to our color selector example, what if we had too many reds for a single page? All we'd need to do is include the following line at the bottom of our page:

```
<li><a href="darkred.html" target="_replace">See More Reds</a>
```

The darkred.html file would be another page fragment—just like those we've already created—but this one only including list items. When a user clicks on the link, the Ajax code entirely replaces the `<a>` link with the fragmentary page. You could continue this process ad infinitum by including a new replacement link at the bottom of each new fragment loaded.

Table 5.6 reminds us of the various ways in which Ajax can be used with iUI, now that we've met them all.

Table 5.6 iUI uses Ajax to provide speed improvements and also maintain a history stack. You need to remember when to include fragmentary files to make it work.

Tag	Attribute	Content	Replaces
`<a>`	(default)	Fragment	iUI content
`<a>`	`target="_replace"`	Fragment	`<a>` link
`<a>`	`target="_self"`	Web page	iUI page
`<form>`	(default)	Fragment	iUI content

Note that except in the case of `target="_replace"`, your old iUI content page doesn't stick around. Instead you're loading a new view. What remains is the chrome—or if you prefer the overall iUI window—and it's through that chrome that iUI remembers where it is and allows the user to move back through previous pages.

Having finished with Ajax, we can now look at one final feature hidden inside iUI: compression.

5.5.3   Compressing iUI

Do you need your pages to be more efficient because everyone is accessing them via EDGE? If so, you can use specially compressed versions of the iUI CSS and JavaScript files that are packaged with iUI. Just include them under the compressed names:

```
<style type="text/css" media="screen">@import "iui/iuix.css";</style>
<script type="application/x-javascript" src="iui/iuix.js"></script>
```

Don't expect the savings from compression to be huge. As of the current version of iUI at the time of this writing, the CSS file decreases from 8k to 6k and the JavaScript file decreases from 10k to 6k. But every byte can help on an EDGE connection, and the only real cost is readability of the files.

5.5.4   Selecting a different look

In the last few sections we've taken advantage of iUI's ability to look and work exactly like an iPhone. We're now going to contradict ourselves by addressing the following problem: what if you have a standard look and feel for your website that you want to replicate with iUI, rather than sticking with the staid and normal iPhone look?

Because iUI is just a library of CSS and JavaScript, changing its look and feel is not only possible, but it's also easy. At the time of this writing, an iUI web page at http://www.ampersandsoftware.com/NHLapp/ showed how this could be done by introducing some bright red and blue colors into its style. All you need to do is to create additional styles for your pages.

The simplest way to do this is to embed a style statement in the particular tag that you want to change. For example, if you wanted to change the `<fieldset>` background from your Settings example to make the background of the inner panel red, you could code it as follows:

```
<fieldset style="background: #ff0000">
```

This particular method is quick and easy, but it won't work well if you want to make a regular change to lots of elements on your page. CSS doesn't support true class inheritance, which would allow you to create a substyle, but it does allow you to stack multiple classes together. We saw this method used with the iUI buttons. You could similarly create a special style for your red-background `<fieldset>`:

```
.rpanel > fieldset {
   background: #ff0000;
}
```

Then instead of calling just the `panel` class, you call both `panel` and `rpanel` when you create this element:

```
<div id="settings" title="Settings" class="panel rpanel">
```

This method is much more viable for doing an app-wide change to the iUI styles, yet doesn't require you to mess with the original iUI CSS file, thus preserving your ability to upgrade iUI in the future.

As for why in the world you'd want your panel box to be red, we'll leave that as an exercise for the reader.

5.6 *Integrating iUI with other libraries*

Before we finish up with iUI entirely, let's see how it can integrate with other libraries, including both the WebKit that was the subject of the previous chapter and jQuery, another library package that we haven't discussed yet.

5.6.1 *Using jQuery with iUI*

The iPhone is a device that has elicited a lot of excitement in the user community, and iUI isn't the only freely available library that's come about as a result. In early 2008 Jonathan Neal released a jQuery library for the iPhone. You can download it from http://plugins.jquery.com/project/iphone.

jQuery is a lightweight JavaScript library that's intended to make JavaScript and DOM manipulation simpler and more intuitive. You can learn more by checking out *jQuery in Action* by Bear Bibeault and Yehuda Katz (Manning, 2008). The iPhone extension adds a few common iPhone-related manipulations to jQuery. You can access them just by including the appropriate JavaScript scripts (jquery.iphone.js, jquery.js) or if you prefer the appropriate compressed scripts (jquery.iphone.min.js, jquery.min.js).

Once you've done that, you can use a handful of new functions. We're not going to go in-depth into jQuery for the iPhone as we did with iUI. Table 5.7 summarizes the contents of the package, and you can find more complete examples in jQuery's iPhone documentation.

Table 5.7 The jQuery iPhone package provides user-accessible JavaScript functions and variables to make your iPhone web app programming simple.

Function	Type	Summary
`disableTextSizeAdjust`	Function	Stops iPhone from resizing text on page
`enableTextSizeAdjust`	Function	Allows iPhone to again resize text on page
`hideURLbar`	Function	Hides the URL bar chrome Can also move a user back to the top of the page
`orientchange`	Function	Accepts a function that will be executed when the iPhone changes orientation
`version`	Variable	Returns the current version of Safari or false if Safari is not being used

iPhone's jQuery is compatible with iUI; web apps have already been created that use both of them. But there is some overlap. In particular, the `hideURLbar` function is not needed in iUI, which already hides the URL bar whenever it's used.

Lessons for SDK developers

Although iUI is a web-only library, it provides some of the best insights into SDK development, primarily because Joe Hewitt did such a careful job of not only mimicking the look and feel of the iPhone's UI, but also its functionality.

The discussions that lead off the chapter, centering on the iPhone UI and the iPhone's architectural paradigms, provide a great overview for your own SDK programming. A lot of the bells and whistles, such as the chrome and the way data is output, will already be laid out for you with the SDK. Thinking about the data-centric focus of iPhone users and the ways in which the six unique iPhone design elements will influence your program design remain important.

Finally, iUI itself really shows many of the features that SDK programs should have as well. Short data bursts, expanding lists, and special search and setting forms will all be part of a well-designed SDK programming experience. We'll even return to the idea of a windows and views as one of the core concepts of the SDK.

One of the nice features of iUI is that it's really easy to quickly knock out new content. It demonstrates one of the greatest strengths of web development versus SDK development, the quick development time, which you should keep in mind when engaging in any iPhone programming.

The rest of the jQuery functions may be useful even if you're using iUI; the ultimate question is whether they're worth adding more than 50k onto your page's download size. If you're not already using jQuery, they're probably not, but if you are, the iPhone-specific functions should be nice icing on your jQuery cake.

5.6.2   *Using iUI with WebKit*

Finally, let's return to WebKit and the question of how to integrate it with iUI. We think the answer to that is that it's definitely easy but possibly unnecessary.

First, it's easy because iUI just consists of some styles and JavaScript that won't impact the way that most web page elements work. If you want to use a WebKit transition, transform, or animation as part of a web page, iUI won't get in your way. In fact, we think some of elements of the advanced WebKit will work great with iUI. For example, you could use a client-side database to create dynamic iUI-based web pages, using methods similar to the PHP dynamic web page that we described earlier.

Second, it's unnecessary because iUI tends to support a specific type of web design: the list-based data paradigm that we saw earlier was central to much iPhone development. If you're following that paradigm, you probably won't need the bells and whistles of the WebKit; if you're not, then you probably won't need iUI. There are situations when you might combine the two, such as creating a list-based application that had fancy end pages that could make use of WebKit features, but we expect those will be the minority of iUI designs.

Nonetheless, there's little technical reason not to combine the two libraries if the opportunity arises to do something cool.

Alternatives to iUI

At the time of this writing, iUI was the only mature library that gave good access to an iPhone-UI for the web. But we occasionally hear word of alternatives either being freely developed or being developed for sale with fancy graphical interfaces. We're not going to try and cover all of these here, other than to say there will probably be more stuff out there by the time this book sees print, but that we expect iUI to be the leader in the area for at least a while to come.

We will, however, make a special mention of Apple's Dashcode, which is a graphical development platform that can be used to lay out many simple iPhone-UI web apps. If you don't need anything other than JavaScript, CSS, and HTML in your web app, you'll probably want to use Dashcode instead of iUI; it's discussed in chapter 7. But for all those more complex programs that require PHP, Ruby on Rails, or another dynamic language, we think that iUI is a great foundation.

5.7 Summary

The iPhone user interface goes beyond just its chrome. It also includes browser methodologies, data paradigms, and the iPhone's unique features that make iPhone browsing a new and different experience. When you're working on an iPhone-UI web app—intended to model the look and feel of the iPhone and to be used primarily or exclusively by iPhone users—you need to consider all these elements so that you can produce a web page whose look and feel will match the UI that iPhone users are expecting.

You may have to program this yourself to meet the needs or requirements of whomever you're producing web pages for. But if you can, you should use iUI, a handy bundle of CSS and JavaScript code. iUI matches not only the chrome and animations of the iPhone, but also some of its basic ideas for how to transport data in efficient ways.

Having now learned about the WebKit and iUI, you should have some great tools to create web pages that mix text and graphics in attractive ways. But what if you want to create pages with much more extensive graphics (and much less text)? For that purpose, there's one more library which can be of great help, Canvas, which is the topic of our next chapter.

6

Using
Canvas for web apps

This chapter covers:

- Learning about Canvas
- Using Canvas to draw simple shapes
- Using Canvas for animations and other complex graphics

We've already discussed two major ways to lay out high-quality iPhone web apps. As we described in chapter 4, you can create primarily text-based applications that use the new HTML extensions of the WebKit. Alternatively, you can use a third-party library like iUI to create web pages that look a lot like iPhone native apps, as we showed in chapter 5.

But what if you want to create graphical web apps for the iPhone, mirroring items like Clock, Stocks, and Weather? At first this might seem a little daunting, because we've already learned that we can't use either Flash or SVG on the iPhone. Fortunately, Apple has a ready-made answer: Canvas.

Canvas is a scalable vector graphics toolkit implemented as an HTML tag, with limited animation functionality accessible through JavaScript, that was created by

Apple. It was originally created to build the Mac OS X dashboard utilities, but Canvas soon afterward made it into the WebKit (and thus into Safari). Apple has continued to support Canvas not just in mobile Safari, but also generally on the iPhone. The aforementioned Clock, Stocks, and Weather utilities were all built using Canvas native to the iPhone, so it will be very simple for you to mimic the same functionality when you're using Canvas on your web pages—you'll be using the exact same tools as the widget designers for the iPhone.

6.1   *Getting ready for Canvas*

Using Canvas on your iPhone is simplicity itself. There's nothing to download and no libraries to link in; it's already part of the WebKit, as we learned in chapter 4. You just need to use Canvas-related markup and commands, which will then be correctly interpreted by any Canvas-compliant browser. In this section, we're going to look at how to enable Canvas and maintain compatibility with other browsers at the same time, and then we're going put it all together in an example. We'll kick things off with the all-important <canvas> tag.

6.1.1   *Enabling Canvas*

The core of Canvas is the <canvas> tag, which defines a panel on your web page that will display Canvas output:

```
<canvas id="mycanvas" width=320 height=356></canvas>
```

The id isn't required, but it's helpful for referring to the Canvas object. The width and height attributes define the size of the Canvas object, just like similar attributes would for an <img> tag. Note that we've chosen a 320x356 canvas, which happens to be the size of the live area of an iPhone display in portrait mode.

   The graphics within the Canvas object will be entirely controlled by JavaScript. To get access to them, you'll need to use JavaScript to define a context for your Canvas object:

```
var canvas = document.getElementById('mycanvas');
var context = canvas.getContext('2d');
```

Note that though we define our context as being of type 2d, there isn't any 3d context (or any other type for that matter). Canvas is expected to expand in that direction in the future.

   Unfortunately, using Canvas isn't *entirely* that simple, and that's because of the many different browsers that exist on the World Wide Web.

6.1.2   *Ensuring compatibility*

Before we go any further, let's stop a moment and talk about compatibility. If you're working on an iPhone web app, you don't have to worry too much about browsers other than mobile Safari. You've probably already built fallbacks into your iPhone web apps so that users of Internet Explorer and other browsers won't be able to access them, as we discussed in chapter 3.

But Canvas applies to much more than just web apps. You could use Canvas on your iPhone-friendly and iPhone-optimized pages too. If you do use Canvas for more than just web apps, you'll need to consider what other browsers Canvas runs on.

Although Canvas was originally an internal Apple language, it has since gained wider acceptance. It has been incorporated into the WebKit and into the HTML 5 protocol, and it has been implemented as part of the Gecko browser engine. This means that it runs not only on Safari and mobile Safari, but also on Firefox version 1.5 or higher, on Opera version 9 or higher, and on the various WebKit clients that we've already discussed. The holdout, as you've no doubt already sussed out, is Internet Explorer, which many of your users will unfortunately be using.

As a result, if your pages might be viewed by IE (or other, older browsers), you should put some compatibility text on your web page, and you must check for the presence of Canvas in your JavaScript.

The compatibility text on your web page is simple. Just put whatever you want IE viewers to see *inside* the Canvas tag. It'll be invisible to users with Canvas, and it'll automatically display to those without:

```
<canvas id="mycanvas" width=300 height=300>
This page is meant to be displayed on a Canvas-compliant browser. Please
   download <a href=http://www.apple.com/safari/download/>Safari</a> from
   Apple, or use a modern version of Firefox.
</canvas>
```

Within your JavaScript code, you can check for the existence of the `getContext` operation before you start running the rest of your Canvas code:

```
If (canvas.getContext) {
   var context = canvas.getContext('2d');
}
```

This will ensure that your JavaScript runs cleanly and warning-free whether your users have access to Canvas or not.

6.1.3   *Putting it together*

Listing 6.1 puts together the basic Canvas setup and compatibility functionality to show what a web page using Canvas really looks like. This should be used as the basis for any of the advanced Canvas work you'll be doing in this chapter.

Listing 6.1   The parts of a basic Canvas page

```
<html>
<head>
 <title>Canvas Test</title>
   <meta name="viewport" content="width=320; initial-scale=1.0; maximum-
   scale=1.0; user-scalable=0;"/>
   <script type="application/x-javascript">
function drawOnCanvas() {                        ◁──┐ Prepares Canvas
 var canvas = document.getElementById('mycanvas');  │ for input
 if (canvas.getContext) {
```

```
      var context = canvas.getContext('2d');
   }
}
  </script>
</head>
<body onload="drawOnCanvas();" leftmargin=0 topmargin=0>
<canvas id="mycanvas" width=320 height=356>
This page is meant to be displayed on a Canvas-compliant browser. Please
   download <a href=http://www.apple.com/safari/download/>Safari</a> from
   Apple, or use a modern version of Firefox.
</canvas>
</body>
</html>
```

Runs when body is loaded, with margins set

Defines Canvas object with a simple tag

This example just puts together everything you've learned so far. On the one hand, you have your JavaScript, now nicely encapsulated in a function. On the other hand, you have your simple Canvas object.

The only thing that's new is what lies between them, a <body> tag. This does two things: First, it sets an onload attribute, which makes sure that the JavaScript doesn't try to work with your Canvas object until it actually exists. Second, it sets some margins so that your perfectly sized (320x356) Canvas object appears at the top left of your display.

This example also includes a viewport metatag, which should by now be standard for any iPhone work you're doing. Besides setting the viewport to a standard iPhone size for easy reading, this tag also prevents users from resizing the page, which has been pretty standard in our web apps.

Now that you've got your basic coding structure in place, you can use it as the foundation for all the additional Canvas work you're going to do in this chapter.

6.2 *Drawing paths*

Canvas builds its drawings around *paths*, which are collections of lines, arcs, and invisible moves between them. You create a new path by describing any number of these lines, and then you finish up the path by deciding how it's going to look, writing out your whole stack of commands in the process. Nothing gets printed to the screen until you dump out everything with a completion command. This is all done with JavaScript commands that you include as part of a drawOnCanvas-like function, such as the one we included in listing 6.1.

All Canvas drawing is done on a two-dimensional grid with an origin at the top left. This is depicted in figure 6.1.

With these fundamentals of Canvas in hand, you can now begin drawing.

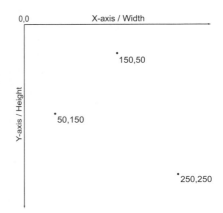

Figure 6.1 Any Canvas object maintains its own two-dimensional grid.

6.2.1   *Basic path commands*

Table 6.1 lists the basic path commands. They're divided into three broad types: creation commands that get you going, draw commands that either draw or move while you're working on a path, and completion commands that are used when you're finishing a path.

Table 6.1   A variety of simple JavaScript commands help you create, draw, and finish basic Canvas paths

Method	Type	Variables	Summary
beginPath	Creation method		Starts a new path
lineTo	Draw method	x,y	Moves the virtual pencil visibly
moveTo	Draw method	x,y	Moves the virtual pencil invisibly
closePath	Draw method		Completes a path by drawing back to the first point
fill	Completion method		Draws a path by filling in the space between visible lines
stroke	Completion method		Draws a path by just drawing the visible lines

Listing 6.2 shows an example of how to use these commands to draw a simple banner. This is just the first step in putting together a Canvas application. Things will get more complex as we learn about additional methods.

Listing 6.2   Simple Canvas commands draw quick two-dimensional shapes

```
var context = canvas.getContext('2d');        ❶

context.beginPath();        ❷
context.moveTo(10,110);        ❸
context.lineTo(10,10);
context.lineTo(40,40);        ❹
context.lineTo(70,10);
context.lineTo(70,110);
context.closePath();        ❺
context.stroke();        ❻
```

You start by repeating the getContext line ❶ from the setup example. The context is important because it's the object that gives you access to all of the drawing methods. For future examples, we'll always assume that we have defined a context by the name of context. After creating the context, you draw a path that defines an image, as shown in figure 6.2.

Any path must start off with a beginPath line ❷. This clears off the drawing stack and resets your virtual pencil to the origin point of 0,0. As a result, most Canvas methods

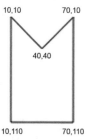

Figure 6.2   This simple banner was drawn with eight path commands.

will follow the `beginPath` with a `moveTo` ❸, to get the virtual pencil to where you want to start drawing without actually drawing anything in between.

For this example, you next use a set of four `lineTo` methods ❹ to draw an M-shape. Because these are lines, they'll display when you complete your path.

The `closePath` ❺ that ends the drawing is entirely optional. It's really just a short-hand way to draw a line between the final point that you explicitly designated and the point that you began drawing at.

But none of this appears on the screen until you use a completion method ❻. You can use `stroke`, as in this example, to just draw the line, or alternatively you can use `fill`, to color everything in. Note that when you use a `fill` command, you don't need a `closePath` command; instead, Canvas will automatically close your shape for you.

6.2.2 Curve commands

Once you've learned about lines, you've just got one other fundamental drawing tool in Canvas: the curve. Three different curve commands are available to you: the arc (which is available through two different methods), the quadratic curve, and the Bezier curve. These are summarized in table 6.2.

Table 6.2   Canvas supplies four methods for drawing curved paths.

Method	Type	Variables	Summary
`arc`	Draw method	x, y, radius, startangle, endangle, anticlockwise	Draws a circle or an arc of a circle
`arcTo`	Draw method	x1,y1,x2,y2,radius	Draws an arc from point to point
`quadraticCurveTo`	Draw method	cpx,cpy,x,y	Draws a quadratic Bezier curve
`bezierCurveTo`	Draw method	cpx1,cpy1,cpx2,cpy2,x,y	Draws a cubic Bezier curve

Each of these curves requires more explanation, because they work slightly differently and the two types of Bezier curves are somewhat complex.

THE ARC

`arc` is a standard circle (or arc) command, and it is the easiest curve method to use. But you need to be slightly careful in its use for two reasons.

First, it steps outside the standard paradigm for the drawing methods. Rather than explicitly defining the endpoints of your arc as data points on your grid, you instead define a center point, a radius, and the endpoints as angles. This makes drawing a circle pretty simple and intuitive, but it can cause problems if you're drawing the arc as part of a stack of paths, in which case you must first move to where your arc will start to avoid leaving behind an unsightly path.

Second, `arc` defines everything in *radians*. If you don't remember your high school geometry, 2π radians is a full circle, the same as 360 degrees. Odds are that you'll be thinking of things in terms of degrees, in which case you'll have to multiply everything by $\pi/180$ in order to convert.

Of the variables only the last, anticlockwise, requires any additional explanation. It's set to either true or false and defines the direction in which the circle is drawn from the start angle to the end angle. Why "anticlockwise" instead of "clockwise," you ask? It's another standard when using radians.

Once you've got these basics, you can draw a circle. The following example draws a 33 radius circle centered at 150,150:

```
context.beginPath();
context.arc(150,150,33,0,360*Math.PI/180,true);
context.fill();
```

You can also use the arc command to draw, well, arcs. The follow example draws a center point and then two arcs around it:

```
context.beginPath();
context.arc(150,150,2,0,360*Math.PI/180,true);
context.fill();

context.beginPath();
context.arc(150,150,20,0,90*Math.PI/180,true);
context.moveTo(185,150);
context.arc(150,150,35,0,90*Math.PI/180,false);
context.stroke();
```

The results of this are shown in figure 6.3, which better shows off some of the functionality we've been talking about.

Both of the arcs in figure 6.3 center around 150,150 with radiuses of 20 and 35 respectively. They both run from 0 degrees to 90 degrees, but the first one goes anti-clockwise, resulting in three-quarters of a circle, while the second goes clockwise, resulting in one-quarter of a circle.

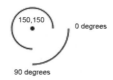

Figure 6.3 Two simple arcs are drawn around a central dot.

Simple calculation tells us that the first arc runs from 170,150 to 150,170 while the second runs from 185,150 to 150,185. If not for the moveTo in between them, a straight line would have been drawn from 150,170 to 185,150 as part of the path that you're drawing. If you'd like to test this out, just input the code, but leave out the moveTo method.

THE ARCTO

Note that there is also a second command, arcTo, which can be used to draw arcs from one point to another. It more closely matches the draw-to paradigm that you've used before, where you draw simple figures connecting one point to the next one.

THE BEZIER CURVES

The two Bezier curves also match this draw-to paradigm: your virtual pencil is on the canvas and you're drawing to another point. But Bezier curves don't necessarily draw very symmetric arcs.

That's pretty much the definition of a Bezier curve. Each one has at least one control point, which defines how the curve changes—whether it's steep or shallow, and over which parts of the curve. The quadratic Bezier curve (quadraticCurveTo) has one control point that connects to both endpoints, and the cubic Bezier curve (bezierCurveTo) has two control points, one per endpoint. If you've ever worked with Adobe Illustrator, those lines that you drag off of the vertices of figures that you've drawn are control points that allow you to make Bezier curves.

Listing 6.3 shows the commands required to draw two Bezier curves.

Listing 6.3   Bezier curves allow for smooth arcs between two points

```
context.beginPath();
context.moveTo(20,200);
context.quadraticCurveTo(20,20,200,20)
context.moveTo(40,300);
context.bezierCurveTo(180,270,150,240,300,40);
context.stroke();
```

Figure 6.4 shows what the output of listing 6.3 looks like. To the left, we have it as it appears on the iPhone screen; to the right, we have a version with the control points and the endpoints drawn in for additional clarity.

We'll offer one final caveat on these Bezier curves: they're tricky to use. The quadratic curve can be used for some nice rounded corners without too much trouble, but figuring out what the cubic curve will look like is entirely trial and error. If you've got a good drawing program that will let you accurately measure the positions of Bezier curves, you might want to use that as your whiteboard; otherwise you'll need to keep inputting control points and seeing how they look on the screen.

Lines and curves may be good, but how can you use them to draw actual stuff? As it happens, Canvas has a very limited selection of more complex shapes that you can draw, forcing you to often fall back on your ingenuity.

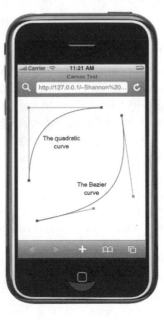

Figure 6.4   The Bezier curves (left) were drawn using the depicted control points (right).

6.3   Drawing shapes

There is only one shape in the standard Canvas library, and that's the rectangle. Beyond that, you can write your own functions to draw other shapes.

6.3.1   Drawing rectangles

You can draw rectangles in three different ways, two of which are closely related to the stroke and fill commands that we've already seen. These possibilities are all described in table 6.3.

Table 6.3   Three rectangle commands allow simpler access to these shapes, without using paths.

Method	Type	Variables	Summary
clearRect	Integrated method	x,y,width,height	Clears the area
fillRect	Integrated method	x,y,width,height	Draws a filled rectangle
strokeRect	Integrated method	x,y,width,height	Draws a rectangle outline

These integrated methods take care of everything for you. There's no need to separately begin a path, then later draw it. Instead, everything is done in one easy method.

The following code would draw one square inside of another:

```
context.fillRect(100,100,150,150);
context.clearRect(125,125,100,100);
context.strokeRect(150,150,50,50);
```

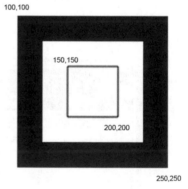

100,100

150,150

200,200

250,250

Figure 6.5   A stack of three rectangles are drawn one atop another.

Note that in each of these method calls, the x,y values define the top left of the rectangle, which is then drawn out from that location. The results are shown in figure 6.5.

We haven't dwelled on it much so far, but shapes in Canvas are drawn one on top of another, in the order of invocation (or at least they are when you use the default composition method, a topic we'll return to). Here, you drew a filled square (using the fillstyle attribute of the context, which we'll also cover in a minute), then cleared the space, and finally drew a stroked square atop it all.

Note that the clearRect command effectively acts as an eraser for a rectangle of space. It'll be useful when you're drawing on top of other drawings, as you did here, or when you're playing with animation down the line.

6.3.2   Writing shape functions

Unfortunately, the rectangle is the only shape that is directly built into Canvas. You can create a circle pretty simply using the arc command, but from there you're entirely on your own. If you're planning to draw a lot of shapes in your Canvas program, you'll probably want to write your own shape functions. Because Canvas does all of its work through JavaScript, this is easy to do, as shown in listing 6.4.

Listing 6.4 An example of a rhombus function

```
function rhombus(context,x,y,length,angle,style) {
    context.beginPath();
    context.moveTo(x,y);

    width = length * Math.sin(angle/2);
    height = length * Math.cos(angle/2);

    context.lineTo(x-width,y-height);
    context.lineTo(x,y-2*height);
    context.lineTo(x+width,y-height);
    context.closePath();

    if (style == 'fill') {
        context.fill();
    } else {
        context.stroke();
    }
}
```

Going back to high school geometry once more (and, to be frank, we had to look it up ourselves), a rhombus is a type of equilateral quadrangle, which is to say a four-sided polygon where all the sides are of equal length.

We've decided to define our rhombuses by the bottom-most point (x,y), the size of the angle just above that point, in radians (`angle`), and the length of one of its sides (`length`). We've also included an option to fill or stroke the rhombus (`style`). Finally, with a bit of trigonometric magic (and, yes, we had to look that up too), we were able to draw a simplistic rhombus (with a very specific orientation) based on those properties.

Here's how our rhombus function could be put to use:

```
rhombus(context,100,100,25,45*Math.PI/180,'fill');
rhombus(context,150,100,25,90*Math.PI/180,'stroke');
```

The results are shown in figure 6.6.

You'll note that the unfilled rhombus is a rotated square, another shape function that you could write for Canvas. The exact shapes you'll want to use in your graphical iPhone web apps will probably vary, but they should be as easy to program as this one.

We've now completed our look at the basic line-drawing functionality in Canvas, so the next question is how to make those lines more attractive.

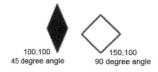

Figure 6.6 Our shape function allows for a variety of rhombuses to be drawn.

6.4   Creating styles: colors, gradients, and lines

Plain black lines aren't going to cut it for an iPhone web app. Fortunately, in Canvas it's easy to modify your simple lines and fills by applying styles and changing other variables.

6.4.1   Color styles

Separate styles can be used to modify the colors of fills and strokes. These properties are summarized in table 6.4.

Table 6.4   **By setting variables, you can choose how your fills and strokes look.**

Property	Type	Value	Summary
fillStyle	Style variable	CSS3 Color	Sets subsequent fills to the color
strokeStyle	Style variable	CSS3 Color	Sets subsequent strokes to the color

Note that both `fillStyle` and `strokeStyle` affect the *following* fill (or stroke) commands. This means that the most recently input style will affect the entire path stack when it's drawn. Earlier ones will be ignored, so if you want to have different shapes with different styles, you'll need to clear the stack after each one with a fill or stroke command.

The actual color definition can be made via most CSS3 definitions. You can use recognized words, #RGB values, rgb values, or even rgba values. Here are four ways to set your fill style to red:

```
context.fillStyle = "#ff0000";
context.fillStyle = "red";
context.fillStyle = "rgb(255,0,0)";
context.fillStyle = "rgba(255,0,0,1)";
```

In the rgba example, the last value is for alpha transparency, set between 0 and 1. If you decrease that value, you'll make your upcoming fills or strokes partially transparent. You'll also meet `globalAlpha`, a global transparency variable, down the line. It'll allow you to change the alpha transparency value of *everything* you're drawing—but the rgba command is more convenient for most usage (at least until you start saving and restoring states).

6.4.2   Gradient styles

Besides colors, you can also produce good-looking gradients in Canvas. These are of particular note because gradients are used throughout the iPhone's user interface. Thus, using gradients will be a notable step toward creating an iPhone-like interface for your graphical web app. Table 6.5 lists the methods required to create gradients; we'll then apply them to fill (and stroke) styles, just like we did with basic colors.

The `createLinearGradient` and `createRadialGradient` methods each define how your gradient will be drawn. With `createLinearGradient` you're defining a

Table 6.5 Fill and stroke styles can include intricate gradients created by a suite of special gradient methods.

Method	Type	Variables	Summary
createLinearGradient	Style creation method	x1,y1,x2,y2	Creates a linear gradient's vector
createRadialGradient	Style creation method	x1,y1,r1,x2,y2,r2	Creates a radial gradient's vectors
addColorStop	Style creation method	position,color	Adds a color to a gradient

simple vector. The x1,y1 point is the start of your gradient (the 0 point for color stops, which we'll get to momentarily) and the x2,y2 point is the end of your gradient (the 1 point). The createRadialGradient method creates a slightly more complex two-dimensional array of vectors. The circle defined by x1,y1,r1 defines the start of your gradient (0), and the x2,y2,r2 points defines its end (1). Because of this complex definition of radial gradients, you can do something that you can't do in most drawing programs: define an asymmetrical gradient.

Once you've got your gradient defined, you can then add color stops to it with the gradient's addColorStop method. Color stops are defined on your gradient between 0 and 1, which are the two endpoints we already met. Like other styles, they use CSS3 colors.

Listing 6.5 shows how a linear gradient is created and then applied to the rhombus function we created earlier; radial gradients are formed similarly.

Listing 6.5 Instead of a color, you can apply a gradient as a style

```
var gradient = context.createLinearGradient(0,100,0,300);    ❶
gradient.addColorStop(0,'rgba(255,0,0,0)');                  ❷
gradient.addColorStop(1,'rgba(255,0,0,1)');
context.fillStyle = gradient;                    ❸

rhombus(context,200,200,50,Math.PI/3,'fill');              ❹
```

When you create a new gradient, you assign its value to a variable ❶. This variable comes with the addColorStop method, the first method we've met that doesn't derive from Canvas's context object. You can use that method to assign as many color steps as you want, though here we've got the minimalist case of just two ❷.

This example uses an rgba method to define its colors, to allow for a nice fadeout, but as we've noted before, you can use any CSS3 color definition. A gradient from one color to a distinct color would work just the same.

You finally apply the gradient to a fill style ❸, and then draw ❹ using that style. All upcoming draw commands will use the same gradient until you change it out for something else. The results are shown in figure 6.7.

Figure 6.7 Gradients allow for attractive coloring.

6.4.3   *Line styles*

Before we finish our look at how to modify lines and fills, we're going to look at one last thing: line styles. These additional variables can be used to really define how your lines look. You probably won't use these nearly as much as colors and gradients, but we nonetheless list them in table 6.6 for the sake of completeness.

Table 6.6   You can get really picky about how your lines look by modifying their styles with a final set of variables.

Property	Type	Value	Summary
lineCap	Style variable	butt,round,square	Defines what the end of a line looks like
lineJoin	Style variable	bevel,round,miter	Defines how two lines come together
miterLimit	Style variable	(number)	Defines the upper limit of when to use a miter join; above that, a bevel join is used instead
lineWidth	Style variable	(number)	Defines the width of a line's stroke

These values are all properties of the Canvas context and can be set accordingly:

```
context.lineCap = 'round';
```

Of the four variables, lineCap and lineJoin will be used pretty infrequently and miterLimit even less so. Most of the time, you'll be happy with the defaults—butt for lineCap, miter for lineJoin, and no miter limit. But lineWidth is of more relevance. This is the value that most drawing programs call the *stroke width*. It's initially set to a unit of 1 pixel, and if you need thicker lines, this is the value to change.

There's still a bit more that you can do to modify your shapes and lines: you can choose *not* to show some of them.

6.5   *Modifying composition and clipping*

You can make your shapes and lines partially invisible three ways: by turning on alpha transparency, by changing the composition method, and by creating clipping paths. These possibilities are summarized in table 6.7.

The global variables are the simplest—but least precise—ways of controlling shapes. Clipping paths use the path functionality that you've already learned to define exactly how your shapes are drawn.

Table 6.7   Two global properties and one method can be used to adjust precisely how your shapes are drawn.

Method or property	Type	Value	Summary
globalAlpha	Global variable	0 to 1	Sets transparency
globalCompositeOperation	Global variable	(numerous)	Sets composite method
clip	Completion method		Creates clipping path

6.5.1 *Global variables*

As the names `globalAlpha` and `globalCompositeOperation` suggest, each of them is a global variable that modifies everything you draw.

    `globalAlpha` sets the transparency level of everything that follows it from fully transparent (0) to fully opaque (1). The default value is 1:

```
context.globalAlpha = .1;
```

`globalCompositeOperation` defines what happens when shapes are drawn one on top of each other. The default value is `source-over`, which means that newer shapes are drawn on top of older shapes, as we saw when we were drawing rectangles (in figure 6.5), and its opposite is `destination-over`. But a variety of more interesting operations may be used to exclude, combine, or otherwise change the way overlapping shapes are composited. Table 6.8 describes these possibilities.

    Just as with `globalAlpha`, all you need to do is set the property to make the change:

```
context.globalCompositeOperation = "darker";
```

Table 6.8 **The `globalCompositeOperation` property changes what happens when you draw shapes on top of each other.**

Value	Type	Summary
source-over	Stacking choice	New shapes are drawn on top of existing content.
destination-over	Stacking choice	New shapes are drawn behind existing content.
copy	New dominant	Only the new shape is drawn.
source-in	New dominant	Only the new shape is drawn, and only where the two overlap.
source-out	New dominant	Only the new shape is drawn, and only where the two don't overlap.
destination-atop	New dominant	The new shape is fully drawn, and the old shape is only drawn where they overlap.
destination-in	Old dominant	Only the existing content is drawn, and only where the two overlap.
destination-out	Old dominant	Only the existing content is drawn, and only where the two don't overlap.
source-atop	Old dominant	The existing content is fully drawn, and the new shape is only drawn where they overlap.
lighter	Blending choice	Where the shapes overlap, the color value is added.
darker	Blending choice	Where the shapes overlap, the color value is subtracted.
xor	Blending choice	Where the shapes overlap, nothing is drawn.

Much as with the composition tools that you find in professional painting programs, if you want to use any of the more complex options like the blending choices, you'll probably need to test things out until you get a result you like.

6.5.2   *Clipping paths*

Clipping paths are much easier to figure out without seeing them first. They work as part of the path-creation tools that we discussed in section 6.2. You start a path with `begin-Path`, and then you use any number of lines, moves, arcs, or Bezier curves to draw. But when you're done, instead of using the stroke or fill methods, you use the `clip` method.

Instead of drawing the path, Canvas uses `clip` to bound what's drawn afterward. The following example shows a circle being used as a clipping path, restricting the square that's drawn under it:

```
context.beginPath();
context.arc(100,250,50,0,360*Math.PI/180,'true');
context.clip();

context.fillStyle = 'gray';
context.fillRect(50,230,100,100);
```

The square is thus drawn only inside the arc. The result is similar to what you can get with the `globalCompositeOperation` values that make one shape dominant over another, but we find clipping to be a more intuitive method. You don't have a bunch of weird names to remember, and it's more obvious that you're creating a specific shape to clip whatever appears after it. The results of this simple clipping are shown in figure 6.8.

Before we finish our look at Canvas basics, we're going to cover two last methods you can use to change the basics of a drawing.

Figure 6.8   A circle clips the filled rectangle drawn after it.

6.6   *Transforming and restoring*

By now you've probably figured out that Canvas is a pretty fully featured scalable vector graphics program. You don't just have basic functionality, but also a lot of subtlety. Here we're going to cover the last two methods that you can use to really change the basics of your drawing: transformations and state stacking.

6.6.1   *Transformations*

Transformations are operations that allow you to change the grid that you're drawing on in various fundamental ways. There are three simple transformation methods, as described in table 6.9.

These methods should all work as advertised. Note that the `rotate` method uses radians, as usual. You can make transformations, change them, and reverse them, as you see fit. The following line, for example, would move our origin to the center of our 320x356 grid:

```
Context.translate(160,178)
```

Table 6.9   Transformations allow you to move your origin or change your grid.

Method	Type	Variable	Summary
translate	Transformation method	x,y	Moves the canvas origin to x,y
rotate	Transformation method	angle	Rotates the canvas by angle
scale	Transformation method	xmult,ymult	Scales the canvas x in width, y in height

This is another area where the next question is probably: *why?* These transformations are solely intended to make your drawing code easier to write. Certain symmetrical shapes might be easier to draw if you center them on the origin, so you might use translate. Similarly, rotate could make it easier to draw a symmetrical array of shapes. Finally, scale could be useful if you want to draw something notably larger or smaller than the rest of your canvas.

6.6.2   State stacking

In these last three sections we've covered quite a few fundamental tools that you can use to create your graphics: fill styles, stroke styles, line styles, global composition variables, global alpha variables, clipping paths, and transformations. Piling all of these Canvas changes together can get confusing; worse, they can really limit the order in which you can do things.

Fortunately, there's a way that you can save the current state of these global variables and later restore them. It uses two methods, as described in table 6.10.

Table 6.10   Save and restore allow you to maintain states for your global variables.

Method	Type	Variable	Summary
save	Style method	(none)	Pushes the current state
restore	Style method	(none)	Pops the last state

All states are maintained in a stack. As noted, save pushes the current state onto a stack, and restore pops the last one off of it. Clearly, you can maintain a long history of states if it's useful. Generally, the best usage of these commands is to save the current state just before you make *any* global change that will be temporary in nature, and then restore when you're done. They're easy to use, and you can see an example of these methods in use shortly, in listing 6.6.

Now that we've finished with all the basics of Canvas, we can move on to a topic that will help our graphical web apps look even more like the native iPhone apps: using images and text.

6.7   Incorporating images, patterns, and text

As you'll see, images are well supported in Canvas. There are a variety of ways to pull in pictures to really rev up your Canvas application. Unfortunately, the same can't be said for text.

6.7.1 *Image commands*

The trickiest part of using an image in your graphical web app is importing the image itself. In order to use an image in Canvas, you'll need to have an Image JavaScript object. The easiest way to get one is to import the image using the `Image` functionality. You'll probably want to use a nice PNG-24 image, which is a great choice to use with Canvas because it supports transparency and an 8-bit alpha channel, meaning that you can put these images straight on top of other objects and still have them look good.

Here's how to import an image from a file:

```
var myImage = new Image();
myImage.src = '/smiley.jpg';
```

You *don't* want these commands to go inside the `onLoad` part of your JavaScript; instead, you want to get that image loading as fast as possible so that it doesn't slow down the drawing of your canvas when you actually use it.

Besides loading an image from a file, you can also use an image that's already loaded on the page. This same functionality can even be used to import a Canvas object! The simplest way to do this is probably with the `document.getElementById` JavaScript method, which can be used in place of an image variable. In the case of importing a Canvas object, make sure that the Canvas object you're importing has been drawn upon first.

Why would you want to duplicate an image already on your page inside your Canvas object? It can help you to control whether all your images are loaded before you get started. Just place the images in `<div>`s that don't display.

Once you've got an image object, you can access it with two methods, as described in table 6.11. The `drawImage` method is listed three times, because it can be called in three different ways.

Table 6.11 **External images can be drawn or used as patterns in Canvas.**

Method	Type	Variables	Summary
drawImage	Draw method	image, x, y	Draws an image at x,y
drawImage	Draw method	image, x, y, width, height	Draws an image at x,y scaled to width,height
drawImage	Draw method	image, slicex, slicey, slicewidth, sliceheight, x, y, width, height	Takes a slice from the original image starting at the image's slicex,slicey, of size slicewidth,sliceheight, and displays it at the canvas's x,y, scaled to width,height
createPattern	Draw method	image, repeat	Creates a pattern variable

The following example shows the three different ways to draw a 150x150 smiley face:

```
context.drawImage(myImage,0,0);
context.drawImage(myImage,150,150,50,50);
context.drawImage(myImage,0,75,150,75,125,125,50,25);
```

The first line draws it at full size. Note that the x,y coordinate is for the top left of the image, so this one is drawn from 0,0 to 150,150.

The second draws it at one-third size from 150,150 to 200,200.

The third `drawImage` method is the most confusing because it rearranges the order of the arguments. Here we're taking a slice of the original image, starting at 0,75 that's 150 wide and 75 tall. In other words, it's the bottom half of our 150x150 image. Then we're drawing it on our Canvas at 125,125, scaled down to a size of 50x25, which is again one-third size because our half of the image is 150x75.

6.7.2 *Pattern commands*

Patterns work slightly differently. They're a new type of fill style (or stroke style) that is made out of an image but that otherwise works the same as colors and gradients. Once again, you must be sure that the image object is fully loaded before calling this method. The following example shows an image being turned into a pattern that's then used when creating our old friend the rhombus:

```
var myPattern = context.createPattern(myImage,'repeat');
context.fillStyle = myPattern;
rhombus(context,100,100,50,Math.PI/3,'fill');
```

The `createPattern` method can take four different repeat values: `repeat`, `repeat-x`, `repeat-y`, or `no-repeat`, each of which repeats the pattern in a different direction (or not at all).

When you're using patterns, you probably want them to be small textures (not like our large smiley face). You may want to use them as a backdrop for your entire canvas, which you can do by patterning a rectangle the size of the canvas. You may even want to use this method to pregenerate some images, rather than drawing them on the fly.

We've got a screenshot of both these sorts of images in the example that finishes off this section, just a page or two hence.

6.7.3 *Text commands*

Now you've learned all the basics of creating stuff in your Canvas app, including shapes, colors, and images. Next you probably want to know how to embed text: unfortunately, you can't, using the Canvas standards. As of this writing, text is not supported, but it looks like it may be soon, because `fillText` and `strokeText` variables were recently added to the Canvas standards. Until those features make it to Safari, there are some alternatives.

First, you can choose to draw your letters by hand, essentially creating new vector fonts. It certainly sounds like a pain in the neck, but it's possible to build up a library of functions for all the letters.

Second, you can play similar tricks by creating images for each letter and then loading those with the `drawImage` method. You can put these all into one image file and then show only the appropriate letter by using the most complex invocation of `drawImage`, which allows you to pull a slice from the original file.

Third, you can use overlaying HTML <div>s—just the sort of absolute positioning that we've argued against elsewhere in this book (but which is perhaps necessary in this situation).

Discussions on this topic are scattered across the internet, alongside individuals' solutions for them. At the time of writing, none of the library solutions were comprehensive enough for us to recommend them here, but take a look around if you need text in your Canvas object and you don't want to write the routines yourself.

With that disappointment behind us, we're going to finish our look at Canvas by seeing how to put it all together.

6.8   *Putting it together*

Having shown lots of stand-alone Canvas methods, we'll now show how a variety of shapes can be painted onto a single Canvas object. Figure 6.9 includes a pair of screenshots that show off most of the Canvas functionality that we've discussed.

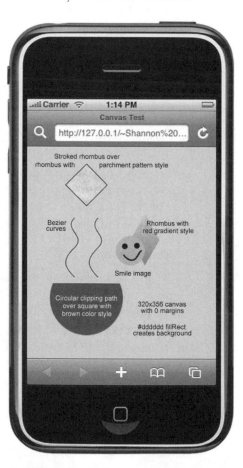

Figure 6.9   Canvas allows for a wide variety of shapes to be created from a small set of primitives, as this combination of images, patterns, curves, and shapes shows.

Here we want to once more note that *order matters*. In this example, the gray background rectangle was drawn first, then everything else on top of it. The smile was drawn after the gradient rhombus that's below it (and here we can see the advantages of using PNG-24, because the blending looks good thanks to that extra alpha channel). The rhombus at the top of the page was similarly drawn as two parts: first a rhombus filled with a pattern, and then a stroked rhombus to give it a clear delineation. Finally, the circular clipping path was the penultimate command, because it clipped everything after it—here, just a rectangle that ends up looking like a bowl thanks to the clipping. fillStyle was also changed a number of times within the example, each time before the next object was drawn.

With that example, we're pretty much done explaining Canvas's basic methods. What we've discussed so far is the core of how Canvas is intended to work, and these are probably the methods you'll use most in making graphical iPhone web apps. But Canvas also does have some limited animation abilities that we'll examine to finish up this chapter.

6.9 *Applying animation*

We're going to start with a caveat: Canvas isn't really intended for animation. You can do it (depending on the good graces of your user's CPU), but unlike with Flash, animation is not a core purpose of the programming language. This means that whenever you do animation with Canvas, it's going to be a lot awkward and a little kludgy—and thus, it's generally something that you should use as spice in your graphical web app, and not as the main course.

You'll have to overcome two main obstacles when animating using Canvas. First, you have to deal with the fact that Canvas doesn't have any animation methods. Worse, once things are drawn on the canvas, they stay on the canvas. That means that whenever you animate, you have to clear your canvas (using clearRect), draw your background, and draw your figure, and then clear your canvas, draw your background, and draw your somewhat changed or moved figure, and so on. You repeat this exercise throughout the animation.

Second, you should note that whenever you're writing to the canvas with a function, *nothing* gets written to the canvas until the function is completed. Thus, you can't use anything obvious like a for loop to do an animation. Instead, you have to draw, end your function, then later call it again with setInterval, setTimeout, or an event handler.

Listing 6.6 shows how to do a simple animation using our old friend the rhombus function.

Listing 6.6   An animation of a changing rhombus

```
function initCanvas() {

    var canvas = document.getElementById('mycanvas');

    if (canvas.getContext) {
        setInterval(animateRhombus,500);        ❶
```

```
    }
}

function animateRhombus() {

    var context = document.getElementById('mycanvas').getContext('2d');

    context.clearRect(0,0,320,356);          ❷

    context.save();          ❸
    context.fillStyle = "#dddddd";           ❹
    context.fillRect(0,0,320,356);
    context.restore();          ❻

    context.save();          ❸
    var gradient = context.createLinearGradient(0,125,0,300);
    gradient.addColorStop(0,'rgba(255,0,0,0)');
    gradient.addColorStop(1,'rgba(255,0,0,1)');
    context.fillStyle = gradient;

    var time = new Date();

    rhombus(context,200,200,50,          ❺
        (time.getSeconds()%10 + 1)/20*Math.PI+.5*Math.PI,
        'fill');
    context.restore();          ❻
}
```

As suggested, this programs kicks off your animation with a setInterval statement ❶. Every 500 milliseconds, you call the animateRhombus function. That routine starts off by clearing your canvas ❷, and then it fills in your background ❹ as required. Note that you make use of the save function ❸, along with restore ❻, so that you don't have to keep track of which styles you've set at which time.

Your actual animation is done with a call to the rhombus routine ❺. You use the system clock to control the animation. This recognizes ten different frames and loads a new one each second. Each frame uses a different value for the rhombus angle, with the result being an animated rhombus that starts off as a square (angle = π radians) and then collapses down toward a line (angle = $.5\pi$ radians).

Clearly, other animations are possible using shapes, lines, or even images. It's easy to imagine an animated version of figure 6.9 where the heat lines above the bowl fluctuate, thanks to changing Bezier curves, or where the smiley face changes by pulling up different image files. Global variables or other means could be used to control the animation if a clock doesn't do the job.

For our purposes, however, the time has come to close the book on Canvas, and to move on to the last aspect of iPhone web development: tools.

Lessons for SDK developers

When you start using the SDK, you're going to have access to a few different drawing libraries. One of them, Quartz, will look a lot like Canvas. Quartz is similarly built around paths, and you'll similarly have to decide whether to stroke or fill paths when you're done with them. You'll also see a lot of other terminology in common, because Apple created both libraries.

There are differences, in part because Quartz is built around Objective-C's unique methods of object-orientation. There are also just enough practical differences to sometimes get you into trouble, such as the fact that the Quartz coordinate system starts at the bottom left, not the top left (though this is partially corrected for on the iPhone).

But there's enough in common that if you learn Canvas, you'll have a big leg-up on Quartz, and vice versa.

6.10 *Summary*

In the previous two chapters, our advice on preparing web apps for the iPhone had a distinctly textual focus. Fortunately, Apple provides a tool that you can use to add great-looking graphics to your iPhone web apps: Canvas. Though its animation functionality is somewhat limited, it does a great job of creating vector-based graphics.

One of the best things about learning Canvas is that its penetration of the browser market is only going to grow. You can already view your iPhone Canvas code on Safari, Opera, and Firefox. As the HTML 5 standard moves toward completion, even Internet Explorer will doubtless catch up with the rest of the field. Something as simple as a `scale` command, Canvas's version of the viewport in a way, could allow you to quickly port your iPhone graphics to those other browsers.

For now, however, Canvas is the last major tool in your iPhone web app toolkit. With graphics, text, and a handful of iPhone UI fundamentals in your back pocket, you should now be able to create whatever great-looking iPhone web apps you imagine.

That doesn't mean that we're entirely done looking at web apps, however. We've finished talking about the great libraries available to you, but there are also some interesting programs available for creating iPhone web pages. The first one we're going to cover is Dashcode, which can offer an interesting (if orthogonal) way to create web apps.

Building web
apps with Dashcode

This chapter covers:

- Introducing the Dashcode development environment
- Programming simple iPhone web apps with Dashcode
- Integrating Dashcode with existing libraries

Thus far, we've talked about a lot of great programming libraries that you can use to create versatile and beautiful web apps for use on the iPhone. The WebKit, iUI, and Canvas each offered different ways to program your web pages.

Although we've looked at several different libraries, we haven't concerned ourselves at all with the tools that you use to construct your web apps. You might be building your pages with Emacs or Adobe GoLive. You might be testing them out with Firefox or Opera. A variety of tools could be used for any of these tasks, but in this and the next chapter we're going to suggest some options that we think are particularly effective. We'll begin in this chapter with a development environment that allows you to create web pages specifically for the iPhone: Dashcode.

We're going to give you plenty of information on Dashcode here. We'll start off with an introductory look at the program and its parts. Then we'll give some specific

Warning: Mac-specific lands ahead

Much of our discussion of developing web pages thus far has not been Macintosh-specific. Clearly, the various libraries will be available no matter what platform you're writing code on. But starting here we're going to have a more Apple-centric focus, because tools are more likely to depend on what computer you're using.

advice on using several of the objects and templates that Dashcode provides. Finally, we'll examine how Dashcode interrelates with the libraries that you've worked with over the course of the last three chapters.

TIP   Some good documentation about Dashcode is available at http://developer.apple.com/webapps/. If you want more information than we've presented in this chapter, read the "Dashcode User Guide."

Let's get started with a look at where Dashcode came from.

7.1   *An introduction to Dashcode*

Dashcode is a development environment that was introduced by Apple in 2006. At that time, it only allowed for the creation of dashboard *widgets*, which are simple web applications built using HTML, CSS, and JavaScript that run under Mac OS X without a browser. Apple's clock, calendar, and calculator are among the applications that have been built using Dashcode.

Apple released Dashcode 2.0 in 2008 as part of the large set of development tools for use with the iPhone. Under this new version of Dashcode, you can create web applications intended to run not on a Macintosh but, instead, on an external website for use with an iPhone.

Dashcode programs are essentially web pages, so all of your experience with HTML, CSS, and JavaScript will continue to be of use. Much of the HTML and CSS will be hidden by Dashcode's graphical user interface, but when you want things to happen, you'll be programming directly in JavaScript.

You could theoretically use Dashcode to program web apps of considerable complexity, but we suggest using it mainly for simple widgets like those found in the native Mac OS X dashboard. For more complex applications, you'll want to have access to a more complex language like PHP or Ruby on Rails, and though you could integrate that functionality with Dashcode work, as we'll explain at the end of this chapter, you'd probably do better to use your standard development environments. Dashcode is really best for small and simple (yet elegant) web apps.

At this time, Dashcode 2.0, which is the version that you'll need to write iPhone web apps, is only available as part of the iPhone SDK. You should thus jump forward to chapter 10 for some information on how to install it. Once you've done so, you can run Dashcode from /Developer/Applications/Dashcode.

7.1.1 Starting a Dashcode project

Once you've started up Dashcode, you'll need to begin a project by selecting a template. Each of these templates comes partially filled in with different starting objects intended to make your development experience quicker and simpler. The various possibilities are summarized in table 7.1.

Table 7.1   Dashcode templates get you started quickly

Template	Summary	SDK equivalent
Browser	A navigation controller that is list-based	Navigation-based application
Custom	A totally blank application	Window-based application
Podcast	An application that displays and plays podcasts	N/A
RSS	An application that reads and displays an RSS feed	N/A
Utility	A flipside controller with two screens	Utility application

Note that for each template we've listed an SDK equivalent template. Apple has used many common techniques for both SDK and Dashcode development, and if you're transitioning from one to another, these equivalents will help guide you; otherwise, you can ignore them.

Generally, you should be able to easily decide which template you want to use based on the needs of your web app. If you're linking to podcasts or RSS feeds, you'll use those specific templates. The Utility template should be used whenever you want a simple one-page application with information or preferences of some sort on a second page, and the Browser template should be used whenever you want to build hierarchies of lists. If none of the templates applies, the Custom template is the right place to start.

7.1.2 The anatomy of Dashcode

When you start a Dashcode project, the main screen will display what your project currently looks like, and it'll also feature a huge variety of buttons and other controls that can be used to build it up. For example, the starting screen for a project based on the RSS template is shown in figure 7.1.

The Dashcode screen is broadly divided into three parts. Above is the *top bar*, which features a few useful buttons. Below and to the right is the *canvas*. This is area where you can see what your web app looks like. To the left is the *navigator*, which gives you access to the entirety of your program (and to some helpful advice). There are three additional screens that aren't initially visible, but which are each quite important: the *source code panel*, the *inspector* and the *Library*.

We'll talk about each of these in turn.

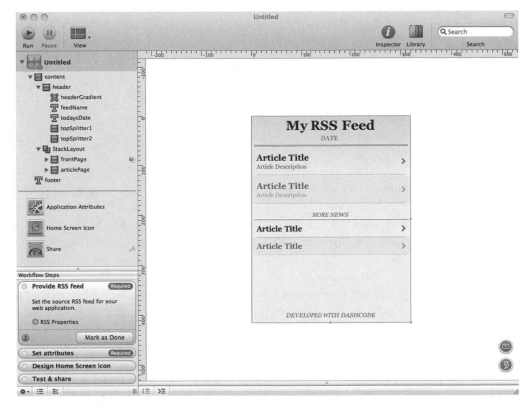

Figure 7.1 Dashcode includes a top bar (top) for important buttons, a navigator (left) for getting around your web page, and a canvas (right) to show off your content.

ABOUT THE TOP BAR

The top bar provides you with buttons to get to a few important pages inside Dashcode. It's how you call up the inspector and the Library. The View menu lets you replace the canvas and source code panels with some of the subsidiary panels that we're not going to dig into in this overview. Finally, the Run button lets you test out web apps as you write them.

The top bar doesn't bear much additional discussion, but it's a great navigational tool inside Dashcode.

ABOUT THE CANVAS

Dashcode's development canvas is the right panel of the main page. It's simple and easy to use.

You can manipulate graphical objects there by dragging them around, and the canvas will try to help you keep things aligned by showing clear blue lines when objects align to the middle or edges of the screen. It'll also sometimes limit where your object can go. For example, if you drag around some of the locations for the article listings in the existing template, you will see limits on positioning based on whether you're using absolute or document-flow positioning for an individual object. This can be adjusted using the inspector for the object, a topic we'll return to shortly. Resizing items is equally simple.

Besides that, you can also change the textual content of most objects within the canvas. For example, double-clicking on My RSS Feed in the RSS example will allow you to change that title on the page.

Just keep in mind that the canvas is a graphical user interface. You can use it to eyeball the placement of objects in your web app and to make quick and easy changes to the content.

ABOUT THE NAVIGATOR

The navigator, at the left of the Dashcode screen, contains links to all of the various objects that exist as part of your app. The large blue button (*Untitled*, in figure 7.1) is where most of those objects are linked in, a topic we'll return to momentarily. Meanwhile, there are a number of additional features in the navigator that we'll cover first.

The *Application Attributes* button gives you access to some of the big-picture stuff, like your web page's name and what your app does when an iPhone is rotated.

The *Home Screen Icon* button lets you create a 60x60 web clip for your application, just like you did for your own web pages in chapter 3. You should make sure to do this for any web app you're writing for the iPhone.

Finally, the *Share* button allows you to deploy your web app. We'll discuss this in section 7.1.4.

Now let's return to the blue button at the top of the navigator. If you look, you'll see that it can be opened to reveal an ever-deepening hierarchy of objects that make up your web page. Three of the first categories that you'll see are the header (which is a `<div>` of text that appears across several pages in your web app), the `StackLayout` (which is an object that contains a listing of all the pages that make up your web app), and the footer (which is a `<div>` of text that appears at the bottom of your web app).

Each of these categories can be opened to reveal additional items. For example, the sample header in the RSS template contains five different objects: a gradient, a title, a date, and two horizontal rules. Clicking on one of these objects will highlight it on the web page and also allow you to easily modify it with the inspector, which we'll return to shortly.

ABOUT THE SOURCE CODE

You can access your web app's source code by calling it up through the View menu in the top bar—this will create a new panel in the bottom right of your main window. Dashcode's programming is all done in JavaScript, so that's what you'll see here. One of the coolest features of Dashcode is that the JavaScript is quite well integrated. As you'll discover when you're hooking up buttons, you can hop straight over to the source code, and Dashcode will even fill in some of the details concerning what code you need to write and how.

ABOUT THE INSPECTOR

You can call up an inspector window by clicking the appropriate button in the top bar. This will open a window that is used to modify specific information for individual objects in Dashcode. For example, the Fill & Stroke inspector is shown in figure 7.2. Here you can manipulate objects more precisely than you can inside the canvas.

The inspector window includes five different tabs that can be used to modify a wide variety of settings. Here they are, from left to right:

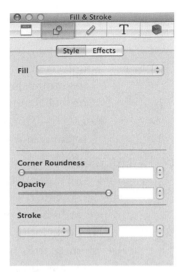

- *Attributes*—Manipulates some of the most basic information about an object, such as the words or images displayed on it.
- *Fill & Stroke*—Changes the background color of an object, and how its corners are rounded; also manipulates simple iPhone effects, such as glass and recess.
- *Metrics*—Modifies CSS positioning information, such as where an object goes, how big it is, and whether it uses absolute or document-flow positioning; also determines how an object resizes, which can be of relevance when the iPhone's orientation changes and the size of the objects need to change.

Figure 7.2 The inspector window allows you to modify individual Dashcode objects.

- *Text*—Changes fonts, colors, spacing, and other text-related settings.
- *Behaviors*—Adds event handlers for simple web events and advanced WebKit events, such as the touch and gesture events discussed in chapter 4.

The changes you make in the inspector window are largely self-explanatory, but we'll look at some of them in-depth—particularly the resizing controls and the event handlers—in our examples of Dashcode programs in the next section.

ABOUT THE LIBRARY

You can call up the Library window by clicking the appropriate button in the top bar. The Dashcode library contains a variety of objects that you can add to your programs, as shown in figure 7.3.

The items that you can add to your Dashcode programs are divided into three broad classes:

- *Parts*—The Parts Library contains all the objects that you might want to add to your program, broadly divided into Buttons, Shapes, Containers, Media, and Text. This includes a variety of attractive graphical objects, some of which are styled to match the look and feel of the iPhone.

Figure 7.3 The Dashcode library gives you access to widgets, code, and your pictures.

- *Code*—The Code Library features rudimentary code snippets, which primarily remind you how to get and set information for various objects. Many of them show you how to extract information from some of the standard Dashcode objects that you'll find in the Parts Library, such as the stack layout. If you're a beginning JavaScript programmer, this will be quite useful; otherwise you will probably only use this tab a few times when you're learning Dashcode.
- *Photos*—This is a built-in interface to iPhoto, giving you quick and simple access to any images in your iPhoto library.

All the library sections are easy to use. To add a new object to your project, drag it onto your canvas. Blue lines will help you center your object, if you so desire. Afterward, you can resize the object or otherwise manipulate it using the canvas or the inspector, as appropriate.

That concludes our look at the parts of Dashcode. We'll be putting this all to actual use momentarily, but first we need to talk about what you can do with a Dashcode project when you're done with it.

7.1.3   *Running Dashcode projects*

At any time, you can test out your current Dashcode project by clicking the green Run button that appears in the top bar. This will run your program inside the iPhone Simulator, a handy iPhone emulator that we'll discuss more completely in the next chapter. Dashcode also gives you access to a sophisticated debugger. If there's a mistake in your JavaScript code, you'll get precise information on what went wrong and how.

7.1.4   *Deploying Dashcode projects*

When you're using Dashcode, you're creating web pages, complete with HTML and CSS files. In order to make them available for use on iPhones, you need to place them on a web server.

This is easy to do. You just click the Share button in the navigator. You'll have the option to deploy your web app to your local file system.

Dashcode will then create a directory containing quite a few files, even for a simple program. The collection of files created for a Dashcode program with just a few buttons is shown in figure 7.4.

We suggest using your local server for testing all programs. We'll talk more about how to do that in the next chapter, which includes guidelines for running an Apache server on your Mac.

Presumably you'll eventually want to move your Dashcode program over to some larger server, but we'll leave the specifics of that final deployment to you.

Now that you've seen the basics of how Dashcode works, you're ready to dive into some actual programming.

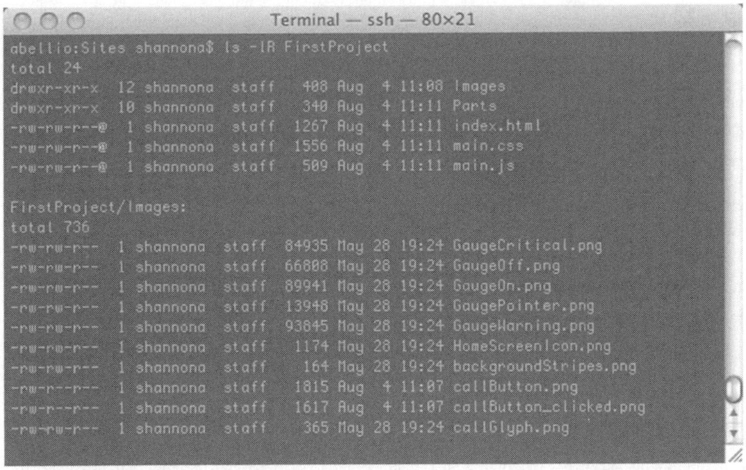

```
abellio:Sites shannona$ ls -lR FirstProject
total 24
drwxr-xr-x  12 shannona  staff   408 Aug  4 11:08 Images
drwxr-xr-x  10 shannona  staff   340 Aug  4 11:11 Parts
-rw-rw-r--@  1 shannona  staff  1267 Aug  4 11:11 index.html
-rw-rw-r--@  1 shannona  staff  1556 Aug  4 11:11 main.css
-rw-rw-r--@  1 shannona  staff   509 Aug  4 11:11 main.js

FirstProject/Images:
total 736
-rw-rw-r--   1 shannona  staff  84935 May 28 19:24 GaugeCritical.png
-rw-rw-r--   1 shannona  staff  66808 May 28 19:24 GaugeOff.png
-rw-rw-r--   1 shannona  staff  89941 May 28 19:24 GaugeOn.png
-rw-rw-r--   1 shannona  staff  13948 May 28 19:24 GaugePointer.png
-rw-rw-r--   1 shannona  staff  93845 May 28 19:24 GaugeWarning.png
-rw-rw-r--   1 shannona  staff   1174 May 28 19:24 HomeScreenIcon.png
-rw-rw-r--   1 shannona  staff    164 May 28 19:24 backgroundStripes.png
-rw-r--r--   1 shannona  staff   1815 Aug  4 11:07 callButton.png
-rw-r--r--   1 shannona  staff   1617 Aug  4 11:07 callButton_clicked.png
-rw-rw-r--   1 shannona  staff    365 May 28 19:24 callGlyph.png
```

Figure 7.4 You'll realize how much work Dashcode does for you when you see all the files it creates, even for a simple program. This terminal window shows just some of the files created.

Saving in Dashcode

It's important to remember that there are two ways to output files from Dashcode.

First, you can (and should) save your Dashcode project. Do this as soon as you get started, using File > Save. Then, in the Share menu, check the box that says "Save project to disk before deploying" to ensure that your saved Dashcode project always matches your current deployment.

Second, you can (and will) deploy your HTML code to a web server when it's done.

It's easy to forget about the Dashcode project when you're outputting HTML code. By checking that box in the Share menu, you'll never have to worry about it, and you'll be sure that your Dashcode project itself is always up to date.

7.2   *Writing Dashcode programs*

Dashcode is ultimately a tool for writing dynamic web pages of light complexity. Not only does it provide you with a great graphical interface, but it also offers you a huge library of complex objects that can each save you hours of programming.

But making use of it is largely up to you. Programming with Dashcode requires knowledge of HTML and JavaScript that go beyond the scope of this book. You may wish to consult a book like *Secrets of the JavaScript Ninja* by John Resig (Manning, 2009) for information on these topics.

But to help give you a leg up on using Dashcode, we've highlighted four of the most important (or complex) topics that you might encounter: using library parts, adding action buttons, using the list-based Browser template, and working with the stack layout.

7.2.1   *Using library parts*

In chapter 4, we used the WebKit to create a simple web program that reported the orientation of a user's iPhone. At the time, we opted to display the information with a textual interface rather than spending the time to put together graphics. Now, with Dashcode at our disposal, we can take advantage of the library parts to display this information graphically with a minimum of work. Table 7.2 shows how to do so, step by step.

Table 7.2   We can create a graphical orientation gauge in just a few minutes in Dashcode.

Step	Description
1. Create a project.	Select File > New Project. Choose a Custom project.
2. Create a gauge.	Drag a gauge from the Parts library to the top center of your Dashcode canvas.
3. Adjust the gauge.	Pop up an inspector window and click on the Attributes tab. In the Values section, change the range to go from 0 to 359, to allow the full range of values. Change the threshold to 0 and warning and critical to 360, to make sure the gauge always remains green. In the Geometry section, change the angles to go from 0 to 359 to match the gauge up with our possible orientation directions. Change the pointer reach to 120% to help it stand out more.
4. Adjust the title and resize.	In the main window, click the Application Attributes button in the Navigator sidebar. Change the Title to Orientation Gauge. Change the viewport to Adjust page width to fit, to keep your gauge from resizing. Open the Metrics inspector for the content object. Change resizing so that the left and right springs are outside of the subwindow, rather than inside.
5. Input code.	Open a source code panel using the View button. Drag Set Gauge Value from the Code library to the source code panel. Adjust it appropriately to set the gauge's value on startup and when the orientation changes.
6. Add a home screen icon.	Design and input a home screen icon using the Home Screen Icon button in the navigator sidebar.
7. Deploy.	Release your new program.

You should be able to run through this complete procedure by following the steps in the table, but the following subsections include some additional information on the more complex steps.

The first three steps—creating the project, creating the gauge, and adjusting the gauge—are all pretty simple. Figure 7.5 shows what your miniature gauge will look like; by the time you're done with this project, the arrow will always point to the top of your iPhone.

Figure 7.5   A simple gauge shows the ease-of-use of the Dashcode parts.

You could also look through the inspector window for the gauge to see if there's anything else that you might want to adjust.

The fourth step, where you adjust the resizing, has a few elements that we haven't previously covered.

RESIZING OBJECTS

Dashcode supports two models of viewports. The default, which it calls "Zoom pages to fit," uses a fixed-size viewport (typically 320 pixels wide) and thus causes an iPhone to zoom in when you move from portrait to landscape mode. The other choice, "Adjust page width to fit," instead sets the viewport to `page-width`, which means that the page content remains at the same size when an orientation change occurs.

The latter results in *resizing*, which we haven't talked about much up to now. That's primarily because it's a pain to deal with unless you have a program helping you out, like Dashcode (or like Interface Builder, later in this book). When an iPhone changes orientation without zooming, the top-level window implicitly changes size, so the program then needs to figure out what to do with its subwindows. Do elements like the gauge maintain their position relative to the center of the page, the left, or the right? The top or bottom? Each of these answers might be correct for a different element on a different page. This is what the Autoresize box of the Metrics inspector is for, as shown in figure 7.6.

In this case, for the content object's Autoresize options, you clicked the horizontal springs inside the box to make them go away, and then you clicked new horizontal springs into existence outside of the box. The result is that when the content object resizes, it keeps the gauge at the middle of the screen. Turning on only one of the right or left springs would have kept it justified in that direction.

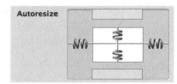

Figure 7.6   The Autoresize box tells a window where its subwindows should go when the window's size changes, usually though an orientation event.

WRITING THE ORIENTATION CODE

As we've previously noted, the Code library returns rudimentary code that will show off the basics of how to access many of the Dashcode objects. Here's what the Set Gauge Value code snippet looks like, with Apple's original comments:

```
// Values you provide
var gaugeToChange = document.getElementById("elementID");
// replace with ID of gauge to change
var newGaugeValue = 100; // new gauge value

// Gauge code
gaugeToChange.object.setValue(newGaugeValue);
```

Once you have that, you just need to place it at an appropriate place in your source code, with appropriate values filled in. Figuring out the ID of the gauge is simple. It's on the Attributes tab of the inspector. After that, you just need to individualize the code sample to reset the gauge value when orientation changes occur. We already saw how to look up the orientation value using the WebKit in chapter 4. Listing 7.1 shows what happens when you put that together with a Dashcode object.

Listing 7.1 Automatically setting a gauge based on orientation

```
function load()        ❶
{
    dashcode.setupParts();
    var gaugeToChange = document.getElementById("gauge");
    var newGaugeValue = (window.orientation * -1 + 180) % 360;      ❷
    gaugeToChange.object.setValue(newGaugeValue);         ❸
}

window.onorientationchange = function() {        ❹

    var gaugeToChange = document.getElementById("gauge");
    var newGaugeValue = (window.orientation * -1 +180) % 360;
    gaugeToChange.object.setValue(newGaugeValue);
}
```

You start by adjusting the existing `load()` function ❶ so that the pointer will show the orientation when the program starts. This is done by massaging the `window.orientation` value ❷ to make the pointer always point toward the top of the iPhone, and then setting the gauge ❸ using the gauge's built-in `setValue` method. Afterward, you do the same thing whenever an orientation change occurs ❹.

As outlined in table 7.2, you then finish your program by creating the home-screen icon and deploying the program, both simple steps. That's all there is to building a graphical orientation detector. It has a few more lines of code than the textual one you wrote in chapter 4, but the improvement in style is stunning, thanks to the built-in functionality of Dashcode.

7.2.2 *Adding action buttons*

We've just seen how easy it is to output to a Dashcode part. It's equally easy to take input from a Dashcode object. As an example, we'll put together a quick application that includes two lozenge buttons and a horizontal indicator between them. This layout is shown in figure 7.7.

Our goal is to make these buttons control the indicator. This would be easy enough to do if you were writing your HTML files by hand—you'd just need to add some `onclick` event handlers to the appropriate objects in your HTML.

It's even easier in Dashcode. All you need to do is open the Behavior tab in the inspector for your button. Here you'll see a list of events

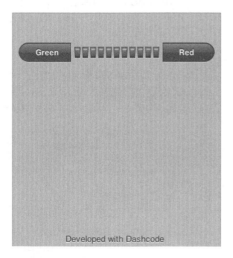

Figure 7.7 A few buttons can be easily added through Dashcode.

(which includes both standard onclick events and the ongesture events that we met in chapter 4) and handlers. To write a new handler, just type the name of a JavaScript function into the Handlers area. The result is shown in figure 7.8.

Typing in a handler does two things. First, it immediately creates the function in your JavaScript file. You can see it by viewing source code. Second, it creates a link from the Behavior inspector, marked by an arrow. Double-clicking

Figure 7.8   The Behavior inspector allows you to quickly assign functionality to buttons and other inputs.

that arrow will take you straight to the appropriate function, making it easy for you (or someone else) to examine your Dashcode project in the future.

At this point, you can write your button code using the JavaScript techniques that you're already familiar with. You'll want to write the decreaseIndicator function that you've already linked in, plus an increaseIndicator function for the other button. The Code library contains two code fragments that can help get you started: Get Indicator and Set Indicator.

Listing 7.2 shows some sample code that could be used to increase the indicator.

Listing 7.2   Modifying an indicator just like other simple objects

```
function increaseIndicator(event)
{
   var indicatorToChange =
      document.getElementById("horizontalLevelIndicator");
   var indicatorValue = indicatorToChange.object.value;

   if (indicatorValue < 11) {
      newIndicatorValue = indicatorValue +1;
      indicatorToChange.object.setValue(newIndicatorValue);
   }
}
```

But the code wasn't really the point of this section, because it's another simple application of JavaScript. Our real goal here was to show you how easy it is to hook up actions to buttons in Dashcode.

7.2.3   *Using the list-based Browser template*

Having now looked at how to use Dashcode parts to output data and accept input, we're ready to dig a bit further into Dashcode, starting with its templates. Each of the templates other than Custom has quite a bit of functionality built into its JavaScript file. Fully investigating all of them is beyond the scope of this book, but we do want to give some particular attention to the Browser template, which allows you to create hierarchical lists of data.

The Browser template closely matches the core data-based paradigm for the iPhone that we highlighted in chapter 5. There we saw it used in iUI's lists; we'll meet it again in the SDK when we work with the navigation controller in chapter 15. Now we're going to look at how Dashcode does lists.

Dashcode manages its lists through a listController object, which contains two methods: numberOfRows returns how many rows a list should contain, and prepareRow sets up an individual row, including its onclick handler. These methods are generally called *data source methods*, which means they define and create the data content of an object. This is a concept that also appears in the SDK.

By default, the list is created from an array named resort, each element of which has a name and a location. The length of this array is used to set numberOfRows, while a row's name is read to create the main output of prepareRow. Additional content from the array is accessed after the onclick handler for a row launches a new page using the detailController.

In order to create your own lists, you'll need to manipulate these functions. Table 7.3 offers suggestions on how to do so.

Table 7.3 You can modify the methods of the listController to create your own list.

Task	Solution
Read from a different array.	Change the array.length call in numberOfRows. Change the array lookups in prepareRow and detailController.
Read from a database.	Make a COUNT(*) call in numberOfRows. Insert SQL lookups into prepareRow and detailController.
Change contents of subpages.	Modify the detailController method.
Go to a different sort of subpage.	Modify the onclick handler in prepareRow.

As of this writing, the listController code included with the Browser template isn't sufficiently generalized to make new code plug-and-play (because the template includes methods specifically needed for its built-in ski resorts example). But the template does do a good job of showing you a functional example of a list, and that should be a good starting point for doing your own coding.

7.2.4 *Working with the stackLayout part*

We're going to finish up our look at Dashcode programs by examining what might be the most complex element of Dashcode: the stackLayout. Though the stackLayout is a pretty major building element that will end up controlling multiple views in many of your programs, it's just another Dashcode part. You can find it in the Container section of the Parts library.

We'll look at the main things you might want to do with a stackLayout.

CREATING A STACKLAYOUT

Many templates will come with their own stackLayouts already in place. But if you want to create a stackLayout, all you need to do is drag it from the Parts library to your canvas. Positioning shouldn't really matter, because it's a virtual object. A stack-Layout will be created with two different views, which is to say, two different pages that can each be filled with different content.

POPULATING A STACKLAYOUT

The two views that are created by default are imaginatively named view1 and view2. Each of these can be used to control a screen full of information.

If you want to add more views to your stack-Layout, go to the Attributes tab of the inspector window for the stackLayout. It includes a Subviews section, as shown in figure 7.9. You can add additional views by clicking the plus (+) button.

Once you've got the right number of views for your web app, you can fill them by clicking on the view to which you want to add content (which will display that view on the canvas), and then dragging new objects to your canvas. The new object will immediately be placed in the appropriate view. With this process you can fill out multiple pages of content.

Figure 7.9 The stackLayout object allows you to add additional views with a GUI.

BUILDING OUTSIDE THE STACKLAYOUT

When you're building multiple pages with your stackLayout, you may want to include information on the top or bottom of every page, such as a header or footer.

To do this, drag a Box container to your canvas, and place it outside and above the stackLayout (for a header) or outside and below the stackLayout (for a footer). You'll be able to add content to these areas of the page, as usual, by dragging and dropping.

MANIPULATING A STACKLAYOUT

When a user first visits a web app that uses a stackLayout, they'll see a page consisting of the first view of your stackLayout plus any header and footer that you created. So, how do you allow users to navigate from one view to another?

You use a few methods that come with the stackLayout part. They are highlighted in table 7.4.

Table 7.4 The stackLayout part contains methods that can be used to manipulate the views.

Method	Arguments	Summary
getAllViews	N/A	Returns an array of the IDs of all views
getCurrentView	N/A	Returns the ID of the current view
setCurrentView	View, Reverse	Changes to the view, with the transition to the view occurring in reverse if the reverse Boolean is set to true
setCurrentViewWithTransition	View, Transition, Reverse	Changes to the view, using the transition variable, possibly reversed

The transitions bear some additional discussion. These are the ways by which one screen changes into another. In the iPhone SDK, these transitions are usually fancy animations involving pages sliding on top of each other, and that's what's being reflected here.

The standard transition is a slide from right to left, which can be reversed using a Boolean argument. But as noted, there's also a `setCurrentViewWithTransition` method that allows you to define a transition object as part of your view change. This transition could be a dissolve, a slide, a flip, a revolve, a swap, or several others, each laid out with specific timing. The Dashcode User Guide (at http://developer.apple.com/webapps/) contains additional information on creating these transitions.

To show how easy it is to manipulate Dashcode's `stackLayout`, we've designed a tab bar that allows a user to move between three views in a `stackLayout`. We would have liked to create a standard iPhone UI tab bar, attached to the bottom of the page, but there weren't any parts that looked quite right. Instead we took advantage of some of the attractive buttons included in the Parts library, as shown in figure 7.10.

In order to mimic this layout, include a left-rounded push button, a push button, and a right-rounded push button. You can choose to put these buttons in one of two places.

First, you could opt to place them in a header. This will allow you to have an identical set of tab bar buttons on every page.

Figure 7.10   A few buttons can make a tab bar.

Second, you could place the buttons at the top of each view. This has the advantage of allowing you to differentiate your buttons, such as by highlighting the button for the current page, but it requires a little extra work. This is the tactic that we took.

In order to lay out your buttons in a row, you'll need to use absolute positioning. You'll also want to change the Fill & Stroke information of each: give the side buttons rounded corners (15px) and the middle button unrounded corners (0px). Finally, you may choose to change the color of the text and the button background for the current page.

Once you've got your graphical elements in place, you just need to link up the buttons to actions using the techniques we've already discussed. Listing 7.3 shows four simple functions that can be hooked up to the buttons to provide navigation among all the pages.

Listing 7.3   Creating a tab bar using the `stackLayout`

```
function gotoPageOneRev(event)
{
    document.getElementById('stackL').object.setCurrentView('view1',true);
}

function gotoPageTwoRev(event)
{
    document.getElementById('stackL').object.setCurrentView('view2',true);
```

```
}
function gotoPageTwoFor(event)
{
    document.getElementById('stackL').object.setCurrentView('view2',false);
}
function gotoPageThreeFor(event)
{
    document.getElementById('stackL').object.setCurrentView('view3',false);
}
```

Each function in listing 7.3 transitions the user to a different page. Good use is also made of the transition functions: when a user increases their page number, the For function is used, which scrolls the page right to left, and when a user decreases their page number, the Rev function is used, which scrolls the page left to right.

USING VARIABLE VIEWS

Before we finish with the stackLayout part, we'd like to highlight one last technique. If you take a look at the Browser template, which we discussed in the section 7.2.3, you'll see that it contains a stackLayout with two views, listLevel and detailLevel. The first view shows the list of items that you see when you load up the program, and the second displays the details of an individual item that you load up when you click on a list entry.

What's interesting about this is that there are a multitude of detail pages, one for each list item. How does the Browser incorporate them all into one view? It does so by rewriting the detail page before it's called up each time. This is all done within the list controller's onclick handler, which is shown in listing 7.4.

Listing 7.4 Updating a single view to look like a multitude of pages

```
var handler = function() {
    var resort = resorts[rowIndex];
    detailController.setResort(resort);          ❶
    var browser = document.getElementById('browser').object;
    browser.goForward(document.getElementById('detailLevel'),resort.name); ❷
};
```

This code fragment doesn't show the details, but it gives enough of the big picture to make our point. First, it calls a special function to rewrite the contents of the secondary view ❶. That's, of course, done using DOM. Second, it calls up the secondary view ❷. The list controller uses the browser.goForward method as an alternative to setCurrentView. We find the stackLayout method more readable, but, as usual, you can find more info on the browser method online.

In any case, the trick is a good one, and it shows how you can use the stackLayout to represent many similar pages with slightly different content.

7.2.5 *Exploring the rest of Dashcode*

We've done our best in this chapter to point out the main features of the Dashcode development environment. We've shown how to output to parts, how to accept input from them, and how to use two of the most complex parts, the listController and the `stackLayout`. We could probably write several more chapters on all of the parts and code available within Dashcode, but Apple's already done the job, so we'll point you one more time toward their "Dashcode User Guide." At the time of this writing, the user guide's appendix B contains an excellent list of parts with special functionality.

To aid your own exploration, table 7.5 lists some of the most interesting parts that you might want to look up.

Table 7.5   Some of the Dashcode parts can provide you with complex functionality.

Part	Summary
Browser	A grouping element that contains the `goForward` navigation method used by the `listController`
Canvas	A `<canvas>` area, usable as discussed in chapter 6
Column Layout	A simple way to lay out side-by-side columns
Edge-to-Edge List	A list like the one used in the Browser template
Quartz Composer	An alternative graphical tool
QuickTime	An area for playing QuickTime media
Rounded-Rectangle List	An alternative form of list, built inside a rounded-rectangle on a page
Stack Layout	A collection of different views

Together, these parts (and many simpler ones) can provide you with considerable power in Dashcode, even when building relatively simple web apps.

That concludes our look at Dashcode's parts and templates. But before we leave this Apple tool behind entirely, we want to address one final question: how does Dashcode relate to what you've already learned about web apps?

7.3   *Integrating Dashcode with existing libraries*

Over the previous three chapters, we've talked about some great libraries that you can use to create iPhone web apps, and in this chapter we introduced the Dashcode development environment. Since these are somewhat orthogonal directions of iPhone development, we want to briefly touch upon how they can be used together. We're going to cover each of the iPhone-related libraries in turn: WebKit, iUI, and Canvas.

7.3.1   *Integrating Dashcode with WebKit*

Apple's advanced WebKit introduces three classes of features: HTML extensions, CSS extensions, and JavaScript extensions.

The new HTML and CSS features will be largely invisible to you inside Dashcode. This does have its downsides. For example, you don't have as granular control over the viewport in Dashcode as when programming by hand; instead, you only have access to a couple of options for how the web page zooms. On the other hand, you

Lessons for SDK developers

As with the WebKit, Dashcode is being managed at Apple and thus shares a lot of its design sense with the SDK. Although Dashcode itself will never be of direct use for an SDK developer (except perhaps for quickly mocking-up an application on the web), the ideas in this chapter are crucial.

First, as noted within the chapter, the graphical code-creation environment of Dashcode has a clear analogue in the SDK: Interface Builder. Both development environments allow programmers to lay out objects using a graphical interface, then to link them with code. Further, both programs include a lot of the same features, including an inspector to look at individual objects and a library containing standard objects. Once you become familiar with either program, the other will be easy to learn as well.

Some of Dashcode's parts, such as the lists, should look quite familiar to SDK developers, as they match up with the same ideas used in the SDK. But it's probably Dashcode's concept of a variety of pages (views) that are managed by a single controller (the `stackLayout`) that is most important for SDK developers. This matches not only the MVC architectural model, which is core to the SDK and which we'll meet in a couple of chapters, but it also reflects how these ideas will be abstracted in the SDK.

can depend on Apple to add new features for you without having to learn the new WebKit code.

Conversely, you'll always have full access to any new JavaScript features inside Dashcode. We already saw how to integrate the WebKit's orientation features. The client-side database is probably another JavaScript extension that you'll want to take advantage of inside Dashcode. Some of the best JavaScript features may even get integrated into Dashcode itself, as is already the case with the `ontouch` and `ongesture` events.

7.3.2 Integrating Dashcode with iUI

The third-person iUI library can't really be integrated with Dashcode. It depends on its own CSS and JavaScript libraries, and these are unlikely to play well with the extensive CSS files generated by Dashcode. Fortunately, iUI and Dashcode can generally be seen as alternatives, as they each provide ways to use web design to create web apps that look like native iPhone apps.

7.3.3 Integrating Dashcode with Canvas

Apple's Canvas graphical extension is the easiest of all the libraries to incorporate into Dashcode. You just place a Canvas part from within Dashcode, and then you can write JavaScript code as usual.

7.3.4   Deeper integration

For any Dashcode project, you could opt for deeper integration by deploying your Dashcode project and then mucking with the source code by hand—adding WebKit HTML, adjusting viewports, linking in iUI libraries, or whatever else you wish.

There is something to be said for this approach. You can use Dashcode to do simple layout for your web pages and then do coding from within your favorite HTML design platform. As it happens, this is the same division of labor that Apple uses in the SDK, dividing the work between two programs, Xcode and Interface Builder.

But unlike Interface Builder, Dashcode isn't really set up for this sort of back and forth work, so we suggest keeping it basic in your Dashcode work. Use Dashcode to create simple web apps that fall within the boundaries of its capabilities. For more complex work, we suggest using your preferred development platform from the start.

7.4   Summary

Dashcode is a new tool for creating iPhone web apps. Instead of creating your applications using a text-dominant developer platform, you can use a graphical user interface that makes the placement of objects within a page simple and intuitive. Dashcode makes life even easier for you by providing access to a variety of "parts," which are preexisting objects with attractive graphical interfaces and predefined behaviors. Though you might not be able to program your most complex pages with Dashcode, there's a lot you can do with the tool, and the results will be quickly produced and attractive.

Besides its front-end development support, Dashcode also includes some sophisticated back-end development tools, such as an iPhone Simulator and a built-in debugger. That's great if you're writing using Dashcode, but what if you're instead writing a larger scale iPhone web app? How do you test and debug your software then? As it happens, there are a variety of good answers, some of which overlap the topics we've already discussed, and all of which are included in the next chapter.

Debugging
iPhone web pages

This chapter covers:

- Installing a local server to aid debugging
- Using a variety of browsers and add-ons
- Profiling iPhone code

Now that you've learned how to code iPhone web pages in a variety of ways, you're probably ready to dive right in. But as we discussed in the last chapter, programming great iPhone web pages isn't just about using the right libraries, it's about using the right tools as well. In chapter 7, we discussed a specific tool, the Dashcode development platform. Now we're going to look at some more general tools that can be used to test and debug a variety of programs. We'll begin with the most fundamental tool of all: the Apache web server.

8.1 Using Apache locally

If you have a Mac OS X computer, you can take advantage of its built-in Apache web server to quickly prototype and test web pages. Setting it up is a simple process, as outlined in table 8.1.

Table 8.1   Setting up your local Macintosh to preview web pages is quick and simple.

Step	Description
1. Start up your web server.	From the Apple menu, choose System Preferences. Click the Sharing icon. Select the Web Sharing checkbox.
2. Share files.	Create files with a plain text editor, such as Emacs, which can be accessed from the Terminal. Move the files to the Sites folder in your directory.
3. Test.	Visit your web pages from your Mac at http://127.0.0.1/~Your Username/Test, test, test.

The benefit of developing and testing your web pages locally is that you can do so without affecting your live pages. Further, since Mac OS X is essentially a Unix system, you can set up your local system to mimic your web server as closely as you'd like. With just a few minutes of work, Shannon was able to set up his own test page, which we've used to double-check a lot of the web code in this book. This was the URL:

```
http://127.0.0.1/~Shannon Appelcline/test.html
```

After you've viewed a web page from your local machine and seen that it generally works, you'll probably want to test things from your iPhone as well (though, as we'll see, the iPhone Simulator is another option). This can be done by replacing the 127.0.0.1 in the previous URL with the actual IP address of your machine. You can find your IP address by clicking Network under System Preferences in the Apple menu. You also might need to make one other change when typing your test URL into your iPhone. If you have a username with a space in it, as Shannon did, you'll need to replace that space with "%20" (the correct symbol for a space in a URL) when you type it into the iPhone. Here is the alternative URL that we used for our test machine:

```
http://192.168.1.100/~Shannon%20Appelcline/test.html
```

Though we haven't talked about any actual debugging techniques yet, using your local Apache web server will make using any debugging techniques quick and easy.

8.2   *Debugging with your desktop browser*

Once you've got a local server in place, you can use your desktop browser to see how your iPhone web pages really work. On the Macintosh you've got three great choices—Safari, Firefox, and the iPhone Simulator. Each has its own debugging advantages.

8.2.1   *Using Safari*

On a Macintosh, your default browser will be Safari (unless you've changed it). Like most modern browsers, Safari comes with built-in development tools. You can find them under the Develop menu. If the Develop menu doesn't appear, you can activate it in the Safari preferences: choose the Advanced tab, and check the "Show Develop menu in menu bar" check box. Once you've done that, you'll see a menu full of cool features, beginning with the Web Inspector.

The Web Inspector is the main element of Safari's debugging system. It'll show you the code underlying your web page, including stylesheets, images, and script files. Most importantly for programmers working with the WebKit, you can also look up the contents of client-side databases, as shown in figure 8.1.

Within the Web Inspector, you'll find two other tools: the Error Console and the Network Timeline. The console shows you errors in JavaScript; it's what you need when you're using Canvas to create graphical web apps. Also note that this console replaces Drosera, the JavaScript debugger that was previously available under Safari. The timeline will list all the individual files used by a page and show you how long each takes to load. It's terrific for when you're trying to decrease load time for mobile devices.

Back on the Develop menu there's one other cool feature, the User Agent utility. It causes the browser to send a different user agent to the server, thus pretending it's a different type of browser. If you want, you can have it pretend to be mobile Safari. It even differentiates between the iPhone and the iPod Touch. We generally suggest using the iPhone Simulator instead (and we'll return to it shortly), but differentiating between those two devices can occasionally be of use.

Though you might be content to stop with Safari, it's by no means the only browser option that you have under Mac OS X.

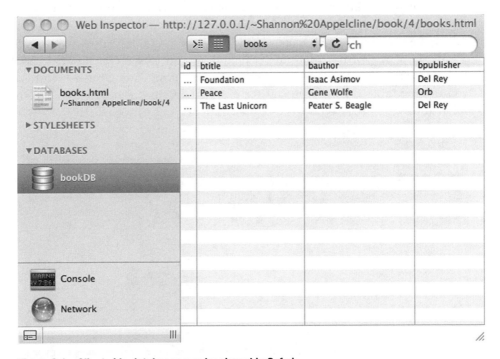

Figure 8.1   Client-side databases can be viewed in Safari.

8.2.2   *Using Firefox*

Though we should perhaps remain loyal to Apple software in this book, we've found Firefox to be an even stronger platform for web page development, mainly due to its robust and well-supported add-on system. Table 8.2 lists the current places to download Firefox and the add-ons that we like best for web development.

Table 8.2   Firefox and the two add-ons can all be downloaded from Mozilla.

Program	Location
Firefox	http://www.mozilla.com/firefox/
DOM Inspector	https://addons.mozilla.org/en-US/firefox/addon/1806
Web Developer add-on	https://addons.mozilla.org/en-US/firefox/addon/60
Firebug add-on	http://www.getfirebug.com/

Firefox ships with a JavaScript Error Console that seems to provide slightly better and more detailed information than Safari, as shown in the comparison in figure 8.2. Prior to version 3 of Firefox, a DOM Inspector shipped with the browser. It provided more architectural info than Safari's similar Web Inspector, but less info on the files that make up a page (and no info on databases!). Now it's available as an add-on. Firefox notably has no built-in tool to let you see how long your page takes to download.

Firefox really shines when you install Chris Pederick's Web Developer add-on. This gives you access to a menu and optional toolbar that provide you with a huge array of information about the web page you're viewing.

We find the forms functions—which allow you to see exactly what variables are holding what values in your forms—and the outline functions—one of which lets you outline table cells, so you can see exactly how they're built—to be the most useful features. There are also CSS-related functions, a variety of validators, and a lot

Figure 8.2   Firefox (left) and Safari (right) both include JavaScript error consoles that you can use to debug Canvas and any other JavaScript work you're doing.

more. And yes, there's a Speed Report too, though it depends upon an external service to work.

USING THE FIREBUG CONSOLE

Firebug is another great add-on for Firefox. It's by iUI developer Joe Hewitt, and it gives you an in-depth look at all of a page's code. Whereas Web Developer is more about how things *look*, Firebug is more about how they *work*. There's some overlap, but the two make a good combination. Besides providing data on things like the DOM, CSS, and what headers look like on the internet, Firebug also provides a great `console` object that gives you a variety of tools for debugging.

After you install Firebug, you can activate it at any time by choosing Open Firebug from among your Firefox tools. When you do that, a panel will appear along the bottom of the screen. Most of the tabs provide you with the information that we've already discussed. It's the Console tab that bears further discussion.

First of all, the command line at the bottom, marked by ">>>", gives you the ability to enter raw JavaScript commands that will be executed immediately. Figure 8.3 shows

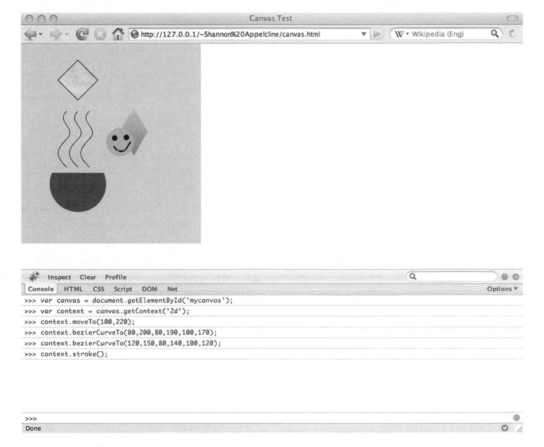

Figure 8.3 The Firebug console allows you to type in JavaScript functions, making Canvas even simpler.

how easy it is to play around with Canvas by adding a new Bezier curve to our final Canvas example from chapter 6.

The Firebug console also has another great feature: it introduces a new `console` JavaScript object that can be used to aid JavaScript debugging in various ways. Some of the most important `console` methods are listed in table 8.3.

Table 8.3   The Firebug `console` methods make it easy to report info and data.

Method	Summary
`console.log`	Writes text to the Firebug console. May include variables for strings (`%s`), numbers (`%i` or `%d`), floating-point numbers (`%f`), and object hyperlinks (`%o`). Variant methods `console.debug`, `console.info`, `console.warn`, and `console.error` similarly write text to the Firebug console, but with different emphasis.
`console.assert`	Tests whether an expression is true, and, if not, writes to the console.
`console.dir`	Creates a listing for an object, identical to the Firebug DOM info.
`console.dirxml`	Creates an XML listing for an object, identical to the Firebug HTML info.
`console.profile`	Encapsulates the JavaScript profiler when used with `console.profileEnd`.
`console.trace`	Creates a JavaScript stack trace.

The `console` methods can be called from within your JavaScript functions. Information will then be reported directly to your Firebug console when you view your page from Firefox:

```
console.log("Starting JavaScript Execution!");
```

Our listing of `console` methods is by no means exhaustive. Firebug also provides over a dozen special functions that are available only from your console command line and that may be used to monitor events, look at objects, profile code segments, and otherwise figure out how your JavaScript is working. An in-depth look at these topics is beyond the scope of this book, but the Firebug website has great documentation on the console and the command-line options alike.

As a result of all these features, Firebug should be your first stop when you're doing Canvas work or any other type of JavaScript coding. But what if you *really* need to see precisely how something looks on an iPhone? If you're working on a Macintosh, you can just open a new window to solve this problem as well.

8.2.3   *Using the iPhone Simulator*

As we've mentioned several times, Apple provides an iPhone Simulator for the Macintosh. All of the screenshots of the iPhone in this book come from the iPhone Simulator, captured with Apple's Grab utility.

The iPhone Simulator comes as part of the iPhone SDK, but even if you're not planning to do any SDK programming, it might be worth getting the current version of the SDK (though the download is *quite* large, at over 1 GB at the time of this writing). We'll talk more about what all is in the SDK package (and how to get it) in chapter 10.

The iPhone Simulator doesn't come with any additional development tools, like the other browsers we've been talking about, but it provides the most precise simulation of an iPhone that you'll find, other than using the iPhone itself (and though we both have iPhones sitting right on our desks, hooked up to our computers, we still find it faster to pop up the Simulator, even to test out a simple web page, let alone SDK programs that require downloading).

Besides looking at pages on your iPhone Simulator, you can also rotate the iPhone left or right using the arrows keys. Perhaps most importantly, you can simulate two-finger gestures by holding down the option key when you scroll over the Simulator. Option-shift will additionally lock these fingers in relation to each other, allowing you to generate a two-fingered scroll event.

Seeing exactly what things look like on an iPhone should help you quickly resolve many purely visual problems with your iPhone web code. With that said, we're going to leave servers and browsers behind, and instead move on to more in-depth debugging tools.

8.3   *Debugging with your iPhone*

Desktop programs are generally the best way to debug your iPhone web apps. That's because you can have a fully featured desktop browser sitting next to an iPhone Simulator. With the first, you can easily look at source and use any number of development tools, and with the second you can look at how something appears on the iPhone.

Conversely, debugging on the iPhone itself can be more troublesome. Not only don't you have those development tools, but you can't even look at the source code! You'll be falling back to using `alert()`s for debugging purposes, and you'll be heading back to your desktop anyway to read and modify code. So, especially if you have access to the iPhone Simulator, our best advice for debugging directly with your iPhone is: *don't do it.*

Despite that admonition, you might have to if you don't have access to a Mac or if the iPhone Simulator and the actual iPhone aren't showing the same results (though we've only seen a difference thus far on websites that used too precise an agent lookup when trying to discover if you were using an iPhone). In these situations, there are a few tips and tricks likely to help you out. We'll be referring to two freeware programs within this section, both of which are listed in table 8.4.

Of these two programs, the first is intended specifically for JavaScript debugging, while the second—which we'll meet as part of a larger discussion of bookmarklets shortly—supports more general HTML debugging. We will not be discussing a third tool, Firebug on the iPhone, because it stopped working when iPhone OS 2.0 was released, with no indication that it's going to be repaired in the future.

Table 8.4 Publicly available programs can make debugging on the iPhone easier.

Program	Location
iPhone Debug	http://code.google.com/p/iphonedebug/
iPhone Web Developer	http://www.manifestinteractive.com/iphone/#_Webdev

8.3.1   Using iPhone Debug

Jon Brisbin's iPhone Debug is a JavaScript debugging program. He explained the need for the new tool by saying:

> *The iPhone Debug Console is meant to give greater visibility and interactivity on your iPhone/iPod Touch while doing development. I grew frustrated having to go through the "include console.log statement then reload" method of debugging. I wanted something similar to Firebug's fantastic console and debugger.*
>
> *In trying to find something that would fit my needs, I came across Joe Hewitt's iPhone/Firebug integration, but I wanted something more robust and that worked without firebug and requiring "console.log" in the desktop browser.*

> (http://code.google.com/p/iphonedebug/)

The complete installation instructions are listed at the iPhone Debug website. They're complex enough that they're likely to change through additional revisions, so we haven't repeated them here.

Generally, iPhone Debug provides you with similar functionality to the desktop Firebug, centering around a desktop console that you can use to receive data about a page you're viewing on your iPhone.

8.3.2   Using bookmarklets

iPhone Debug is great for troubleshooting JavaScript code on your iPhone, but you may also want to debug plain HTML—possibly when you don't have a desktop computer available. For this situation, there's one more tool that you might find useful: bookmarklets.

The word *bookmarklet* comes from combining the words *bookmark* and *applet*. They're little bits of JavaScript code that are encoded as URLs. Thus, when you see a mini-application that you like, you save it as a bookmark, and then you can activate it at any time just by selecting the appropriate link from your bookmark list.

For the iPhone, bookmarklets can give you all the client-side functionality that you want but don't have access to: things like viewing source, and using client-side tools like those found in Firefox and Safari.

Listing 8.1 shows code that will view the source of a page, written by Erwin Harte based on original code by Abe Fettig.

Listing 8.1   The code for a show-source bookmarklet

```
var sourceWindow = window.open("about:blank");
var newDoc = sourceWindow.document;
```

```
newDoc.open();
newDoc.write("<html><head><title>Source of " + document.location.href +
    "</title><meta name=\"viewport\" id=\"viewport\"
    content=\"initial-scale=1.0;" + "user-scalable=0; maximum-scale=0.6667;
    width=480\"/><script>function do_onload()" +
    "{setTimeout(function(){window.scrollTo(0,1);},100);}
    if(navigator.userAgent.indexOf" + "(\"iPhone\")!=-1)
    window.onload=do_onload;</script></head><body></body></html>");
    newDoc.close(); var pre =
    newDoc.body.appendChild(newDoc.createElement("pre"));
    pre.appendChild(newDoc.createTextNode(
    document.documentElement.innerHTML));
```

The code itself is basic JavaScript, and we won't go too far into the details. When clicked, this bookmarklet jumps to a brand new window that contains the entire text of the web page as a `<pre>` element. There's also a tiny bit of magic to scroll the chrome on an iPhone, solely for aesthetic purposes. In order to turn this code into a bookmarklet, you just need to `urlencode()` it so that the JavaScript is properly formatted as a URL.

You can create whatever bookmarklets you want, to add functionality to your iPhone. But there's already a large collection of them available at the iPhone Web Developer site listed in table 8.4. To use that site, browse to it on your desktop Safari, adding the bookmarklets that you like to your bookmarks menu. Then you sync your Safari bookmarks to your iPhone through iTunes, and you'll have instant access to the Web Developer bookmarks that you wanted.

Having now looked at several ways in which you can make your iPhone web pages work *correctly*, we're going to finish up with a look at how they can work *better*.

8.4 Profiling for the iPhone

Profiling—or performance analysis—is important for any type of computer program. But we've gotten sloppy about profiling web pages in the last several years as bandwidth has gotten cheaper and more abundant. With the iPhone, we now need to sit up and start paying attention again. This is because of some of the unique features of the iPhone.

We've already touched several times upon the fact that an iPhone in the wild depends upon the EDGE or 3G network for downloading. This means that we have to create smaller, leaner pages. But we also always need to remember that one of the iPhone's unique features is its energy consciousness, and as a result we shouldn't be creating web pages that take more juice than they need.

Solving the bandwidth problem is an easy one, because there are lots of tools available to help out, such as Safari's Network Timeline, which shows where your web page is getting slowed down. Generally, when analyzing bandwidth you should try to remember the low-bandwidth lessons of yesteryear, including the following:

- Minimize the size of graphics by using more lossy JPEGs.
- Keep your iPhone stylesheets small.

- Don't use stylesheets or other page inclusions that aren't required for your iPhone pages.
- Use Ajax when you can to reload parts of pages rather than whole new pages.

Besides showing you how long files are taking to load, the Network Timeline might also remind you of files that you hadn't even realized were being loaded, and thus are slowing you down for no reason.

Solving the energy problem is more difficult only because it's not an issue you usually have to think about when creating pages for desktop use. The top thing you need to watch is JavaScript. You don't want to include timeouts or event handlers that are constantly going to go out to the network or engage in other high-cost activities. Generally, you should think carefully before putting any sort of infinite loop on a page that's going to be viewed by an iPhone. It might be OK for an animation or some other client-side activity that will quietly shut off when the phone goes to sleep, but you should make sure you're polite about anything more than that. You don't want users avoiding your iPhone website because you drained their batteries.

The general lesson for iPhone profiling is this: pay attention and write the sort of carefully considered code that you were probably thinking more about in the 1990s, before bandwidth and CPU became cheap.

Lessons for SDK developers

In all honesty, we don't have much for you SDK developers this time. Most of our discussion in this chapter was about web servers and clients, and clearly those are going to have little crossover with SDK development. The only point of particular relevance is the section on iPhone profiling.

Thinking about energy consciousness is going to be even more important when we get into SDK development. There's a limit to how much damage a web programmer can do, because the iPhone ultimately controls access through the Safari interface. On the other hand, as an SDK programmer you're going to have access to a lot more fundamental code. Apple has done what it can to keep native programs from gobbling up an iPhone's battery, but you're going to need to do your part too.

8.5   *Summary*

Over the past six chapters, we've covered the libraries and tools that you can use to write iPhone web pages, but we've also done our best to show standard iPhone architectures in the process. Generally, it all comes down to understanding the iPhone's key features. Knowing them and what they mean is as important to writing iPhone code as the actual features and functions we've been discussing.

An *always-on internet* is the thing that allows us to write web apps, and, as we've just seen, *power consciousness* is an important consideration in profiling. The iPhone's unique *input* and *output* have been relevant to almost everything we've talked about in writing web pages, while its *orientation awareness* has occasionally been both an issue and an opportunity. The iPhone's *location awareness* is the only topic we haven't been able to look at when thinking about web applications, because that information isn't available yet to the web developer.

In this chapter, we're offered our best advice on how to make the web programming that you're doing easier. But the internet is a huge place increasingly full of great open source software and freeware. It's entirely possible that you've already found tools that you like better than the ones we've suggested.

Now that we've finished with our overview of the web, we invite you to take some time to get used to how the iPhone works in its web incarnation. Web programs will always remain among the easiest and most accessible programs that you can create for the iPhone, thanks to the simplicity of the web-based languages that have been developed over the last 15 years. As we discussed in chapter 2, even if you move on to the SDK, we believe that web programming will continue to have its place.

But we also hope that you'll eventually be champing at the bit to do more: that you'll want to get into the guts of the iPhone and learn how to create native applications that can better utilize the iPhone's unique features and that can run with or without access to a network. For that reason, we invite you to move on to part 3 of this book, which explores the flip side of iPhone programming: the SDK. To help you get there, we've written a special chapter intended to help bootstrap web developers into Objective-C coders; that's up next.

SDK *programming*
for web developers

This chapter covers

- Rigorous programming languages
- Object-oriented programming languages
- The MVC architectural pattern

We've spent the last six chapters talking about how to program great web pages and applications using familiar languages such as HTML, JavaScript, and PHP; brand-new libraries such as the WebKit, iUI, and Canvas; and helpful tools such as Dashcode, Firebug, and Safari. As we discussed in chapter 2, though, web development isn't the be-all and end-all of iPhone programming. There are some programs that will just be better suited for native programming on the iPhone. Apple provides a development platform for doing this sort of programming called the SDK (software development kit). The SDK includes a set of tools, frameworks, and templates that we'll meet fully in the next chapter. It also depends on a particular programming language: Objective-C.

If you've never worked with Objective-C, don't panic. This chapter is intended to build on your experiences with web development (which we assume includes some sort of dynamic programming language, such as PHP, Ruby on Rails, Python, or Perl) so that you'll be prepared to work with Objective-C when you encounter it in the next chapter.

We'll do this by focusing on three major topics. First we'll talk about C, which is a more complex and rigorous programming language than many of the somewhat free-form web-based languages. It's also the core of Objective-C. Then we'll talk about object-oriented programming, which is the style of programming used by Objective-C. Finally we'll hit on MVC, an architectural model used by many different programming languages, including Objective-C. If you're already familiar with some of these concepts, just skip the section where the concept is described.

Before we start this whirlwind tour, we'll offer one caveat: none of these short overviews can possibly do justice to the topics. There are complete books on each of these topics, and if you feel like you need more information, you should pick one up. This chapter will prepare you so that you will not only understand the code in part 3 of this book, but will also be ready to dive right in yourself by tweaking and ultimately building on the copious examples that we'll provide.

9.1 An introduction to C's concepts

The syntax of C will look a lot like whatever language you're familiar with. However, it may vary from your web language of choice in how it deals with some big-picture areas, such as declarations, memory management, file structure, and compilation. We've summarized all these ideas in table 9.1, but we're going to talk about each of them in turn at more length.

Our goal here *isn't* to teach you how to program in C. If you want more information on that, the definitive reference is *The C Programming Language, Second Edition*, by Brian W. Kernighan and Dennis M. Ritchie (Prentice Hall, 1988). Instead, our goal is

Table 9.1　The rigorous style of C requires you to think about a few new programming topics.

C concept	Summary
Declaration and typing	You must declare variable types. You must declare function argument types and return types. You may need to repeat these declarations in a header file.
Memory management	You may sometimes need to explicitly manage the memory usage of your variables.
Pointers	Some variables are represented as pointers to spaces in memory.
File structure	Programs are divided between source (.c) and header (.h) files.
Directives	Precompiler commands are marked with the # sign. This includes the `#include` directive, which incorporates header files into source files.
Compiling	Your code is turned into a machine-readable format when you compile it, not at runtime.

to explain the programming concepts that you may not have encountered in your web-based programming language.

The reasoning behind this section is ultimately that Objective-C is built right on top of C. Therefore, we'll show you how each of these concepts is used in C (though we're going to save the guts of Objective-C for the next chapter).

9.1.1   Declarations and typing

C is generally a more rigorous programming language than some of the casual languages found on the web. That means there's a bit more time spent saying what you're going to do before you do it.

The purpose of this is not only to make it easier for a computer to understand and efficiently run your program (which was more important in the early 1970s, when C was first invented, than it is today), but also to make it easier to catch errors (which is still important today, alas). If you tell the computer what you're going to do, then it can give you a warning if that isn't quite what happens.

First, we see this rigor in the *typing* of variables. Before you're allowed to use a variable in C, you must say how it's going to be used. For example, the following says that the variable n will be used as an integer:

```
int n;
```

Second, we see it in functions, where you must declare not only what types of variables you'll be passing a function, but also what type of variable you're going to return. This is all done as part of the line of code that kicks off the function. For example, the following function takes two floating-point numbers as arguments and returns a floating-point number:

```
float divide(float numerator, float divisor) {
```

These variable and function *declarations* often get done a second time as part of a header file, which is a topic that we'll return to momentarily.

A close relative to type declaration is the idea of type casting. This occurs when you take a variable and temporarily treat it ("cast it") as a different type of variable. You do so by marking the cast in parentheses before the variable that's being used. It usually looks something like this:

```
float a = 6.00;
int b;
b = (int) a;
```

Here, the float value of a is turned into an integer so that it can be saved to the integer variable b. Casting can sometimes lead to unexpected results, so you should be careful when using it and you shouldn't do it often.

OBJECTIVE-C DECLARATIONS AND TYPING

Declarations and typing work largely the same way in Objective-C. The only notable difference is that you'll more frequently use special Objective-C classes as types than some of the fundamentals like int and float.

Casting comes up the most in Objective-C when you use some of the older frameworks, such as Core Foundation. Older frameworks tend to have classes for fundamental types that aren't the ones you'd usually use but that are equivalent. For example, you can freely move between the `CFStringRef` type (from the Core Foundation framework) and the `NSString *` type (from Cocoa's Foundation framework), but to avoid compiler warnings and improve clarity, you should cast when doing so.

9.1.2   *Memory management and pointers*

You didn't have to worry at all about how memory was used to store your active data in most web-based languages. Conversely, in C if you're dynamically changing your data during your program's runtime, you often do. Memory management is usually done in C through the function's `malloc()` and `free()` methods.

When you specifically allocate memory, you also have to de-allocate it when you're done. If you don't, your program will "leak," which means that it will gradually increase its memory footprint over time due to memory that's been "lost."

When you allocate memory, you end up working with memory addresses that lead to your data, rather than the data itself. This is also by default the case with some sorts of data, such as strings. To address this, C introduces the concept of a *pointer*, wherein a variable refers to a memory address rather than to the data itself. When this is the case, you can *dereference* the memory address, and thus access your data, with the * character.

For example, the following would define a pointer to an integer:

```
int *bignumber;
```

Sometimes you need to do the opposite and get a memory address back from a regular variable. This is done with the & symbol:

```
int variable = 72;
int *variablepointer = &variable;
```

This symbol is often used to pass error messages back from a function.

OBJECTIVE-C MEMORY MANAGEMENT AND POINTERS

On the iPhone, memory management is vitally important because of the device's limited memory. If you're sloppy with your memory usage, you'll start receiving `didReceivedMemoryWarning` messages and eventually your program could get shut down. Objective-C uses the same general concepts of memory allocation and memory deallocation that we've already discussed, but it has some specific rules for when you have to worry about freeing up memory yourself. Because these rules are based on functionality of the iPhone OS, we cover them in the next chapter in section 10.4.2.

Although Objective-C objects are generally built using pointers to memory, you don't have to worry about dereferencing them because the details are hidden by the SDK's classes. However, when you initially declare objects, you'll always do so with a *; you'll constantly be creating pointers to objects.

In addition, you may occasionally run into a library that isn't built around object-oriented classes. In that situation you'll need to make full use of pointers. As you'll see, this is the case with SQLite, which is discussed in chapter 16.

9.1.3   *File structure and directives*

When you look at the file structure of a complex C program, you'll see that it includes a variety of files with .c and .h suffixes. The .c files include all the source code: the various functions that you're accustomed to using in a program, split up in a (hopefully) rational way.

The .h (or header) files, meanwhile, are the tools that allow you to easily integrate the source code from one .c file into the rest of your program. They contain all the declarations for variables that you want to make available outside of specific functions. In addition, they contain *function prototypes*. These are declarations for your functions that effectively describe a protocol for using them:

```
float divide(float numerator, float divisor);
```

Just as header declarations make a variable available outside its own function, function prototypes make a function available outside its own file.

To use a header file, you need the capability to include one file inside another. For example, if you want to access some of the global variables or some of the functions of file2.c in file1.c, you do so by incorporating file2.c's header file. You do this by inserting an include command into file1.c:

```
#include "file2.h"
```

The appropriate file is then inserted as part of the C preprocessor's work. This is what's known as a *compiler directive*, or just a *macro*.

OBJECTIVE-C FILE STRUCTURES AND DIRECTIVES

Objective-C replaces .c files with .m or .mm files and replaces the `#include` directive with `#import`, but the overall ideas are the same.

9.1.4   *Compiling*

The final major difference between C and most web-based programming languages is that you must *compile* it. This means that the human-readable source code is turned into machine-readable object code. The same thing happens to your web programs, but whereas they compile at runtime, C instead compiles in advance, providing for more efficient program startup at the cost of portability. C compilation can be done by a command-line program (like "cc" or "gcc") or by some fancy integrated development environment (IDE).

Because C programs tend to include many files, they need special instructions to tell the compiler how to put all the code together. This is most frequently done with a *makefile*, though integrated environments might have their own ways to list what should be used, possibly shielding the user entirely from worrying about this.

OBJECTIVE-C COMPILING

All of these details will be taken care of for you by Xcode, Apple's development environment. You therefore don't have to worry about makefiles or how the compiling works. The only catch is that you must remember to always add files (such as databases or images) to your project through Xcode so that they get added in correctly.

9.1.5   *Other elements*

C is full of other features that may or may not have turned up in your programming language of choice. Among them are symbolic constants (which are permanent declarations, typically used to increase readability), special sorts of loops (such as `while` and `do-while`), older-style branching statements (such as `goto` labels), and some more complex structures (such as unions).

We can't cover all of these topics with any justice here, so we've held ourselves to the big-picture stuff. If you see something unfamiliar in Objective-C code that you've been handed, and it looks like it's a foundational C structure of some sort, we again point you to Kernighan and Ritchie's definitive book on the topic.

With C now covered in the depth that we can give it, we're ready to move on to the next major element that will define your Objective-C programming experience: object-oriented programming.

9.2   *An introduction to object-oriented programming*

C is fundamentally a procedural language, as are early web-based languages like Perl and PHP (though that's changing for both through current releases). You make calls to functions that do the work of your program. Object-oriented programming (OOP) moves away from this old paradigm. Specifically, there are no longer separate functions and variables; instead, data and commands are bound into a more cohesive whole.

As in our previous section, we've created a summary chart of the major elements of object-oriented programming, which you can find in table 9.2.

If you need more information than what we've written here, *Design Patterns: Elements of Reusable Object-Oriented Software* (Addison-Wesley Professional, 1994) by Erich Gamma, Richard Helm, Ralph Johnson, and John Vlissides is an influential book that generally talks about OOP, then specifically covers some design patterns usable in it. But you should find all the basics here, starting with a look at the fundamentals of object-oriented programming.

Table 9.2   Object-oriented programming introduces a number of new concepts.

OOP concept	Summary
Class	A collection of variables and functions that work together
Framework	A collection of class libraries
Inheritance	The way in which a subclass gets variables and functions from its parent
Message	A call sent to an object, telling it to execute a function
Method	A function inside a class, executed by a message
Object	A specific instance of a class
Subclass	A descendent of a class, with some features in common and some variance

9.2.1   *Objects and classes*

The central concept in OOP is (as you might guess) the *object*. Think of it as a super-variable, or (if you prefer) as a concrete, real-world thing—which is what's actually being modeled in the OOP paradigm.

An object combines values (which Objective-C calls *instance variables* and *properties*) and functions (which Objective-C calls *methods*). These variables and methods are intrinsically tied together. The variables describe the object while the methods give ways to act on it (such as creating, modifying, or destroying the object).

Objects are specific *instances* of *classes*. The class is where the variable and method descriptions actually appear. An individual object then takes all of that information, and starts setting its own version of the variables as it sees fit.

Classes gain power because they support *inheritance*. That means that you can subclass an existing class. Your subclass starts off with all the default variables and methods for its parent class, but you can now supplement or even *override* them.

It's frequent for a subclass to override a parent method by first calling the parent method, then doing a few things to vary how it works. For example, if you had a class that represented an eating utensil, it might include a simple method to transfer solid food from your plate to your mouth. If you created a subclass for a spoon, it could include a new method that worked on liquids.

Once you've created a hierarchy of classes and subclasses, you can store them away in a *class library*. When you put several of those together you have a *software framework*, and that's what we'll be working with in the SDK, as Apple has provided numerous frameworks to make your programming of the iPhone easier.

9.2.2   *Messaging*

If the object is the OOP equivalent of the variable, then the *message* is the OOP equivalent of the function. To get something done in an object-oriented program, you send a message to a specific object that asks it to execute a specific method. The object then does so internally, using its own variables and reporting its results out to the calling object.

One of the most frequent types of messages that you'll see in OOP is a call to a method that looks at or changes an object's variables. A *getter* is an *accessor* that looks at data, while a *setter* is a *mutator* that changes data (though Apple calls both accessors in some of its documentation).

It's important to use accessors and mutators because they support the core OOP ideal of *encapsulation*. The actual variables in objects are hidden away from the rest of the world, freeing up that global namespace, which otherwise could become quite cluttered. If one object uses a `foo` variable, that no longer impacts the use of a `foo` variable by another object. Instead, each variable can only be accessed by the methods of its own class.

Some OOP languages support two different types of messages. You might see calls to class *methods* (where a special class object does general stuff like create an object) or to instance *methods* (where a specific instance object acts in some way). As you'll see, this is the case with Objective-C.

Figure 9.1 combines many of the ideas that we've talked about so far into a single diagram using a vegetative example.

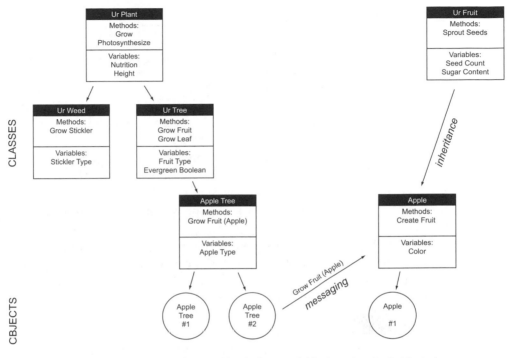

Figure 9.1   Inheritance and messaging combine to form an intricate network of objects in object-oriented programming.

When we look at the classes that are built into the iPhone OS, we'll see that there are even more levels of inheritance and overall a much larger web of classes and messages.

That ends our brief overview of object-oriented programming, but there's one more high-level abstraction that you should be familiar with before you dive into the iPhone SDK: the MVC architectural pattern.

9.3   *The Model-View-Controller (MVC) pattern*

Programming languages innately have their own philosophies and models that underlie how they work. Encapsulation and inheritance are two of the philosophies that are critical to OOP. A philosophy that's critical to good Objective-C programming is the Model-View-Controller (MVC) architectural pattern.

This method of software design goes back to 1979, when Trygve Reenskaug—then working on Smalltalk at Xerox PARC—described it. It's widely available today in OOP frameworks, including the iPhone's Cocoa Touch, and as libraries for other programming languages.

The MVC pattern breaks a program into three parts. The *model* is the data at the heart of a program. Meanwhile, the view and the controller together comprise the presentation layer of your application. The *view* is essentially the user interface side of things, while the *controller* sits between the view and the model, accepting user input and modifying the other elements appropriately. Figure 9.2 shows what this model looks like.

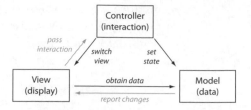

Figure 9.2   The MVC model covers how user input and other changes affect your program's design.

In figure 9.2 you can see the core ideal of the controller accepting input and making changes, but note that there will also be direct interaction between the model and the view.

Dynamic web design offers a simple example of an MVC pattern. The model represents your data, usually stored in a database or XML. The view is your HTML page itself. Finally, the controller is your JavaScript, PHP, Perl, or Ruby on Rails code, which accepts the input, kicking out HTML code on one side and modifying the database on the other.

Within Objective-C, you'll see an even more explicit interpretation of MVC. There are objects specifically called views and view controllers. If you're following good Objective-C programming practice, you'll make sure your controllers are accepting input from your view and doing the work themselves.

9.4   *Summary*

We think it's entirely possible for someone without object-oriented experience (and even without C experience) to make the transition from creating iPhone-based web pages to creating iPhone-based native apps. We've already seen it happen at iPhone development camps. We also think there are good reasons for doing so. As we said back in chapter 2, the SDK and web development can each do different things well, and it's always best to use the right tool for the job at hand.

We won't promise that it'll be easy. The whole idea of objects replacing procedural calls is a pretty big switch in how you do things when programming. When you meet actual Objective-C code, you'll also see that even with the simplest program you're going to have several files to work with that each serve very different purposes.

Fortunately you're going to have four terrific advantages on your side. First, of course, you'll have this book to help guide you. Second, you'll have access to the SDK's programming tools: Xcode and Interface Builder. The first will constantly offer you whatever documentation you need for the objects, methods, and properties you're using, while the second will provide you with a simple, graphical way to create objects. Third, you're going to have Objective-C itself on your side. Although its code looks a bit different from most programming languages you've worked with, its code is simple, elegant, and overall quite easy to read. Fourth, you're going to be able to use the iPhone OS's enormous library of frameworks, making your code even shorter and simpler thanks to many years of development on Apple's part, just as we already saw with Apple's Dashcode and web development.

You're going to be amazed at what you can do in just a few lines of code.

Part 3

Learning SDK fundamentals

Web development is just one way to program for the iPhone. In part 3 of this book we cover the fundamentals of another sort of iPhone development: SDK programming.

The SDK is Apple's Software Development Kit, an object-oriented set of frameworks that help you write great iPhone programs. Over the next several chapters we'll introduce all of its basics, starting with how to use Objective-C and the iPhone OS (chapter 10) and how the SDK's two most important programs, Xcode (chapter 11) and Interface Builder (chapter 12), work. From there we'll move on to some of the most important concepts in SDK programming, including view controllers (chapters 13 and 15) and events (chapter 14).

The SDK is a large topic, and these chapters represent the lightest introduction to it: we're going to delve more in depth into many elements of the SDK toolkit in part 4 of this book.

Learning Objective-C
and the iPhone OS

Over the next several chapters we're going to dig into the *other* side of iPhone development, where you'll be programming native applications using Apple's own toolkit. As we discussed back in chapter 2, there are a number of reasons that the SDK is better than web development, just as the opposite is the case, depending on your particular needs.

In this chapter, we assume you have a good understanding of a rigorous programming language (like C), that you know the basic concepts behind object-oriented programming (OOP), and that you understand what the MVC architectural model is. If you aren't familiar with any of these topics, just jump back to the previous chapter, where we give each of these topics an overview.

With that said, we're now ready to move into the world of SDK development. We'll download the SDK first thing so that we can see what it consists of, but then we're going to take a step back to examine the programming language and frameworks that you'll be using when you program with the SDK.

10.1   Getting ready for the SDK

The iPhone SDK (Software Development Kit) is a suite of programs available in one gargantuan (over 1GB) download from Apple. It'll give you the tools you need to program (Xcode), debug (Instruments), and test (iPhone Simulator) your iPhone code. Note that you must have an Apple Macintosh running Mac OS X 10.5.3 or higher to use the SDK.

10.1.1   Installing the SDK

To obtain the SDK, download it from Apple's iPhone Dev Center, which at time of this writing is accessible at http://developer.apple.com/iphone/. You'll need to register as an iPhone Developer in order to get here, though it's a fairly painless process. Note that this is also the site you can use to access Apple documents, as we've mentioned earlier.

The Apple docs and the SDK

We've already highlighted the fact that the Apple Developer Connection (ADC) provides access to numerous programming documents. For your SDK needs, you'll want to visit http://developer.apple.com/iphone/, which contains a few introductory papers, of which we think the best are "iPhone OS Overview" and "Learning Objective-C: A Primer," plus the complete class and protocol references for the SDK.

As we'll discuss in the next chapter, you can also access all of these docs from inside Xcode. We usually find Xcode a better interface because it allows you to click through from your source code to your local documents. Nonetheless, the website is a great source of information when you don't have Xcode handy.

As with the web chapters of this book, we've been constantly aware of Apple's documents while writing this part of the book, and we've done our best to ensure that what we include complements Apple's information. We'll continue to provide you with the introductions to the topics and to point you toward the references when there's need for in-depth information.

Once you've downloaded the SDK, you'll find that it leaves a disk image sitting on your hard drive. You just need to double-click it and then click on iphone SDK in the folder that pops up, as shown in figure 10.1.

This will bring you through the entire install process, which will probably take 20–40 minutes. You'll also get a few licensing agreements that you need to sign off on, including the iPhone Licensing Agreement, which lists some restrictions on what you'll be able to build for the iPhone.

Figure 10.1   Clicking iPhone SDK will start your installation.

iPhone SDK licensing restrictions

Although they're making the iPhone SDK widely available for public programming, Apple has placed some restrictions on what you can do with it. We expect these restrictions will change as the SDK program evolves, but what follows are some of the limitations at the time of this writing.

Among the most notable technical restrictions: you can't use the code to create plug-ins, nor can you use it to download non-SDK code. It was the latter that apparently spoiled Sun's original plans to port Java over to the iPhone. You also can use only Apple's published APIs. In addition, there are numerous privacy-related restrictions, the most important of which is that you can't log the user's location without permission. Finally, Apple includes some specific application restrictions. You can't create a program that does real-time route guidance, you can't write programs that include pornography or other objectionable content, and you can't include voice-over IP functionality.

In order for your program to run on iPhones, you're going to need an Apple certificate, and Apple maintains the right to refuse those certs if they don't like what you're doing. So, if you're planning on writing anything that might be questionable, you should probably check whether Apple is likely to approve it first.

When the SDK finishes installing, you'll find it in the /Developer area of your disk. Most of the programs appear in /Developer/Applications, which we suggest you make accessible using the Add to Sidebar feature in your Finder. The iPhone Simulator is located separately at /Developer/Platforms/iPhoneSimulator.platform/Developer/Applications. Since this is off on its own, you might want to add it to your Dock.

Warning: installation dangers

The SDK development tools will replace any existing Apple development tools that you have. You'll still be able to do regular Apple development, but you'll now be working with a slightly more bleeding-edge development environment.

You've now got everything that you need to program for the iPhone, but you won't actually be able to *release* iPhone programs. That takes a special certificate from Apple. See appendix C for complete information on this process, which is critical for moving your iPhone programs from the iPhone Simulator onto a real iPhone. For now, though, we'll assume that you're using the iPhone Simulator, and will warn you when you can't. The iPhone Simulator turns out to be just one of several programs that you installed, each of which can be useful in SDK programming.

10.1.2 *The anatomy of the SDK*

Xcode, Instruments, and Dashcode were all available as part of the development library of Mac OS X even before the iPhone came along. Many of these programs are expanded and revised for use on the iPhone, so we've opted to briefly summarize them all, in decreasing order of importance to an SDK developer:

- Xcode is the core of the SDK's integrated development environment. It's where you'll set up projects, write code in a text editor, compile code, and generally manage your applications. It supports code written in Objective-C (a superset of C that we'll cover in more depth shortly) and can also parse C++ code. You'll learn the specifics of how to use it in chapter 11.

- Interface Builder is a tool that lets you put together the graphical elements of your program, including windows and menus, via a quick, reliable method. It's tightly integrated with Xcode, and you'll always be using it, even when you don't call up the program. We'll introduce you to Interface Builder in chapter 12.

- iPhone Simulator allows you to view an iPhone screen on your desktop. We've already seen that it's a great help for debugging web pages. It's an even bigger help when working on native apps, because you don't have to get your code signed by Apple to test it out here.

- Instruments is a program that allows you to dynamically debug, profile, and trace your program. Whereas we had to point you to a slew of browsers, add-ons, and remote web sites to do this sort of work for web apps, for your native apps that's all incorporated into this one package. Space precludes us from talking much about this program.

- Dashcode we list here only for the sake of completeness since it's part of the /Developer area. It's a graphical development environment that is used to create web-based programs incorporating HTML, CSS, and JavaScript. You won't use it in SDK development, but we described its usefulness for web programmers back in chapter 7.

Figure 10.2 shows off the three most important Developer tools.

Besides the visible tools that you've downloaded into /Developer, you've also downloaded the entire set of iPhone OS frameworks, a huge collection of header files

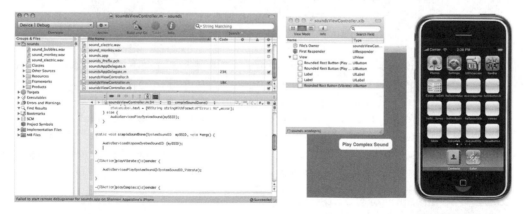

Figure 10.2   The SDK includes Xcode (left), Interface Builder (center), and the iPhone Simulator (right).

and source code—all written in Objective-C—which is going to greatly simplify your programming experience. Rather than jumping straight into your first program, we instead want to touch on these foundational topics. Let's begin by looking at Objective-C, the SDK's programming language, then by examining some of the basics of the iPhone OS, which contains that set of iPhone frameworks.

> **Jumping ahead**
>
> If you'd prefer to immediately dive into your first iPhone program, which will of course be Hello, World!, then simply head on to the next chapter. You can then pop back here to see what it all means.

10.2 *Introducing Objective-C*

All of the SDK's programming is done in Objective-C, a programming language created primarily by Brad Cox and Tom Love in the early 1980s. It's a full superset of C, allowing you to write any traditional C code. It adds powerful object-oriented capabilities as well. These extensions come by way of the design philosophies of Smalltalk, one of the earliest object-oriented languages. Because of its origin beyond the standard boundaries of C, Objective-C's messaging code may look a little strange to you at first, but once you get the hang of it, you'll discover that it's elegant and easy-to-read, providing some nice improvements over traditional ANSI C code.

Although this overview will give you enough to get started with Objective-C, it can't provide all the details, particularly for more complex functionality like properties and categories. If you need more information than we've been able to provide, take a look at Apple's own references on the topic, particularly "Object-Oriented Programming with Objective-C" and "The Objective-C 2.0 Programming Language," both of which can be found in Apple's iPhone developer library.

10.2.1 *The big picture*

Let's start with a look at Objective-C's big picture. It's an object-oriented language, which means it's full of classes and objects, instance variables, and methods. If you need a refresher on any of these topics, check section 9.2 in chapter 9.

As implemented by Apple and used throughout the iPhone OS's frameworks, Objective-C is built *entirely* around objects. Windows, views, buttons, sliders, and controllers will all be exchanging information with each other, responding to events and passing actions in order to make your program run.

A header (.h) file and a source code (.m) file together represent each object in Objective-C. Sometimes you'll access standard classes of objects that come built into the iPhone OS frameworks, but often you'll instead subclass objects so that you can create new behaviors. When you do this, you'll add a new header and source code file to your project that together represent the new subclass that you've invented.

Although we won't dwell on it much, note that C++ code can be mixed in with Objective-C code. We leave the specifics of that for the experienced object-oriented programmer (and, as usual, there's more detail on Apple's website). You can also freely insert older C syntax; as we'll discuss shortly, this will become a necessity when you're working with older libraries.

With all that said, we're ready to dive into Objective-C's unique syntax. Table 10.1 summarizes the six major elements of syntax.

Table 10.1 Objective-C code can look quite different from ANSI C; it depends on just a handful of syntactic changes.

Syntax element	Summary
Categories	Categories can be used to add to classes without subclassing.
Classes	Classes define object types in matched .h and .m files.
Messages	Messages send commands to objects in [bracketed] code.
Properties	Properties allow for the easy definition of accessors and mutators.
Protocols	Protocols define methods that a class promises to respond to.
@	@ directives are used by the compiler for a variety of purposes.

We'll offer a more technical summary at the end of this section, showing all the syntax of these elements. But first, we'll discuss these syntactic elements at length, in approximate order of importance.

10.2.2 *The message*

Objective-C's most important extension to the C programming language is the message. A message is sent when one object asks another to perform a specific action; it's Objective-C's equivalent to the procedural functional call. Messages are also the place in which Objective-C's syntax varies the most from ANSI C standards—which means that once you understand them, you'll be able to read most Objective-C code.

A simple message call looks like this:

```
[receiver message];
```

Here's a real-life example that we'll meet in the next chapter:

```
[window makeKeyAndVisible];
```

That message sends the `window` object the `makeKeyAndVisible` command, which tells it to appear and start accepting user input.

There are three ways in which this message could be slightly more complex. First, it could accept arguments; second, it could be nested; and third, it could be a call to one of a few different recipients.

MESSAGES WITH ARGUMENTS

Many messages will include just a simple command, as in our previous example. But sometimes you'll want to send one or more arguments along with a message to provide more information on what you want done. When you send a single argument, you do so by adding a colon and the argument after the message, like so:

```
[receiver message:argument];
```

Here's another real-world example:

```
[textView setText:@"These are the times ..."];
```

When you want to send multiple arguments, each additional argument is sent following a label, as shown here:

```
[receiver message:arg1 label2:arg2 label3:arg3];
```

For example:

```
[myButton setTitle:@"Goodbye" forState:UIControlStateNormal];
```

This is the way in which Objective-C's messages vary the most from C's functions. You're really going to come to love it. You no longer need to remember the ordering of the arguments because each gets its own title, clearly marking it. The result is much more readable.

NESTED MESSAGES

One of the most powerful elements of Objective-C's messaging system is the fact that you can nest messages. This allows you to replace either the recipient or the argument of a message (or both) with another message. Then, the return of that nested message automatically fills in the appropriate space of the message it's nested inside.

Object creation frequently replaces the receiver in this manner:

```
[[UITextView alloc] initWithFrame:textFieldFrame];
```

The object created by sending the `alloc` message to the `UITextView` class object is then initialized. (We'll get to class objects in just a moment.)

When you're passing a color as an argument, you almost always do so by nesting a call to the `UIColor` class object:

```
[textView setTextColor:[UIColor colorWithWhite:newColor alpha:1.0]];
```

Message nesting is a core Objective-C coding style, and thus you'll see it frequently. It also shows why Objective-C's bracketed messaging style is cool. With good use of code indentation, it can make complex concepts very readable.

MESSAGE RECIPIENTS

As we've seen over the last couple of examples, there are two different types of objects in Objective-C. Class objects innately exist and each represents one of the classes in your framework. They can be sent certain types of requests, such as a request to create a new object, by sending a message to the class name:

```
[class message];
```

For example:

```
UIButton *myButton =
    [UIButton buttonWithType:UIButtonTypeRoundedRect];
```

Instance objects are what you're more likely to think of when you hear the term "object." You create them yourself, and then the majority of your programming time is spent manipulating them. Except for those examples of creating new objects, all of our real-life examples so far have involved instance objects.

In addition to calling an object by name, you can also refer to an object by one of two special keywords: `self` and `super`. The first always refers to the object itself, while the second always refers to the class's parent.

We'll often see `self` used internal to a class's source code file:

```
[self setText:@"That try mens' souls. "];
```

We'll often see `super` used as part of an overridden method, where the child calls the parent's method before it executes its own behavior:

```
[super initWithFrame:frame]
```

All your message calls should follow one of these four patterns when naming its receiver. They can call something by its class name (for a class method), by its instance name (for an instance method), by the `self` keyword, or by the `super` keyword.

Now that you know how to send messages between objects, you'd probably like to know how to create those classes that your objects are instantiated from in the first place. That's the topic of our next section.

10.2.3 *Class definition*

As we've already noted, each class tends to be represented by a matched pair of files: a header file and a source code file. To define a class, each of these files must contain a special compiler directive, which is always marked in Objective-C with an @ symbol.

First, you define the interface for the class, which is a simple declaration of its public variables and methods. You do this in the header (.h) file. Next, you define the implementation for the class, which is the actual content of all of its methods; this is done in a source (.m) file.

Figure 10.3 shows this bifurcation graphically; we'll look at it in more depth in the next few sections.

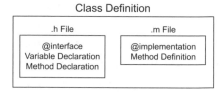

Figure 10.3 Headers and source code files each contain distinctive parts of your Objective-C classes.

THE INTERFACE

Interfaces begin with an `@interface` directive and finish with an `@end` directive. They contain instance variable declarations in curly brackets, then method declarations. Listing 10.1 shows an example of their usage. It's the first of several examples that we're going to offer in this section that will depict a fake class, `AppleTree`.

Listing 10.1   The `@interface` directive

```
::: AppleTree.h :::

@interface AppleTree : UrTree        ❶
{
    NSString *appleType;        ❷
}
- (id)growFruit:(NSString *)appleColor        ❸
@end    ❹
```

We began our interface command with the `@interface` directive ❶ and ended it with the `@end` directive ❹. Note that our `@interface` directive included not only our class name, but also the name of its superclass, following a colon. It could also include a list of protocols, a topic we'll return to later in this section.

The variable declaration ❷ is entirely normal. `NSString` is a type that we'll meet when we look at the iPhone OS later in this chapter. Note that you don't have to declare all of your variables in your `@interface`, but just those instance variables that you want to be accessible outside their methods. You'll declare variables that are used within only individual methods inside those methods, as you'd expect.

Our method declaration ❸ contains a typed description of a method with one argument, matching the syntax we've seen for messages already. It also contains one other new element: we've started it with a -. That means that this is an instance method, which is a method that can only be used by an instance object. Its opposite number, which is marked with a +, is the class method, which is used by a class object.

The `id` type used as the return of `growFruit` is another Objective-C innovation. Objective-C allows for dynamic typing, where type is decided at runtime. To support this, it includes the weak type of `id`, which can be a pointer to any object.

Before we finish our discussion of method declarations, we'd like to mention that, as with variables, you only have to declare those methods that can be called externally. Methods that remain internal to a class can remain hidden if you so desire.

THE IMPLEMENTATION

Once you've declared a class with an `@interface`, you can then define it with the `@implementation` directive. Listing 10.2 shows a brief example of what the implementation might look like for our AppleTree class, including a single example method.

Listing 10.2   The `@implementation` directive

```
::: AppleTree.m :::

#import "AppleTree.h"        ❶
#import "Apple.h"

@implementation AppleTree    ❷

- (id)growFruit:(NSString *)appleColor
{
   Apple *fruit = [Apple appleWithColor:appleColor];    ❸
   return fruit;
}
@end        ❹
```

Our code starts out with the #import directive ❶. This is Objective-C's variant for the #include macro. It includes the file unless it's already been included, and is the preferred alternative when using Objective-C. In this case we've included AppleTree.h, which should contain the interface we described in listing 10.1. Without including it, we'd need to redefine all of our instance variables and include our super class in the @implementation statement. Thus, the #import helps us avoid redundant code. We've also included the Apple.h file so that we can create an Apple.

As with our interface, the implementation code begins with a directive ❷ and ends with an end ❹. In between, we describe what our method does ❸, which includes sending a message to the Apple class object.

WHAT WE'RE MISSING

We've now got two parts of a puzzle: how to create new classes of objects and how to send messages among instantiated objects. What we're missing is how to instantiate an object from a class.

Generally object instantiation will follow the same pattern. First, you allocate memory for the object, and then you initiate any variables and perform any other setup. The precise manner in which this is done can vary from class to class. It's usually a framework that will decide how object creation works—which for our purposes means the iPhone OS. As you'll see later in this chapter, the iPhone OS specifies two methods for object instantiation: the alloc-init method and the class factory method. We'll meet each of these soon, when we talk about the iPhone OS, but first let's finish up with the core syntax of Objective-C.

10.2.4 *Properties*

What we've covered so far should be sufficient for you to understand (and write) most simple Objective-C code. There's one other major feature in Objective-C that deserves some extended discussion because of its unique syntax: the property.

THE PURPOSE OF PROPERTIES

Because instance variables are encapsulated, you usually have to write tons of getter and setter methods when doing OOP. This can get tedious, and you must also be careful about consistency so that you don't have dozens of different syntaxes for your accessors and mutators.

Objective-C offers you a solution to these problems: you can declare an instance variable as a property. When you do so, you standardize the variable's accessor and mutator methods by automatically declaring a getter and a setter. The setter is called `setVariable` and the getter is called `variable`.

For example, if we return to the apples that we've been talking about in our major examples, if we defined our `NSString *appleType` variable as a property, the following declarations would automatically occur:

- `(void)setAppleType:(NSString *)newValue;`
- `(NSString *)appleType;`

You'll never see these declarations, but they're there.

SETTING A PROPERTY

You declare an instance variable as a property by using the `@property` directive as part of your `@interface` statement. Listing 10.3 shows how to do so, in the full context of our example so far.

> **Listing 10.3   The `@property` directive**

```
::: AppleTree.h :::

@interface AppleTree : UrTree
```

```
{
    NSString *appleType;        ❶
}
@property NSString *appleType;        ❷

- (id)growFruit:(NSString *)appleColor
@end

::: AppleTree.m :::

#import "AppleTree.h"
#import "Apple.h"

@implementation AppleTree

@synthesize appleType;        ❸

- (id)growFruit:(NSString *)appleColor
{
    Apple *fruit = [Apple appleWithColor:appleColor];
    return fruit;

}
@end
```

Our header file shows that any property must start with the declaration of an instance variable ❶. The `@property` directive ❷ then repeats that declaration. If you wish, you can stop here. You've now implicitly declared your accessor and mutator methods, and you can go and write those methods on your own if you see fit.

Objective-C will also write these methods for you if you just ask it to! This is done with the `@synthesize` declaration in the `@implementation` statement ❸. This will create accessor methods that read and set the variable by the simple methods that you'd expect. The setter method is by default of type `assign`, but you can choose a different method using property attributes, which we'll talk about down the road.

USING THE ACCESSORS

If you're not doing anything fancy, you can immediately use your class's default getter and setter methods, as shown in the following three examples:

```
NSString *choosenType = [AppleTree appleType];
[AppleTree setAppleType:@"Washington Red"];
[AppleTree setAppleType:myAppleType];
```

Besides providing you with automatically created accessors and mutators, properties also give you access to a bit of syntactic sugar, which can make using them that much easier.

THE DOT SYNTAX

Objective-C offers a dot syntax that makes it easy to use an object's accessor and mutator methods (whether you synthesized them or created them yourself). The following are the dot syntax equivalents to the messages that we sent earlier:

```
NSString *ChoosenType = AppleTree.appleType;
AppleTree.appleType = @"Washington Red";
AppleTree.appleType = myAppleType;
```

The dot syntax can also be nested, just like you can nest messages. In the following example, the `treeType` property returns a tree object that has an `AppleType` property:

```
Apple1.treeType.AppleType
```

With that in hand, you should now be able to write simpler and more intuitive code.

PROPERTY COMPLEXITIES

There are several complexities of properties that we've opted not to delve into here.

First, property declarations can include attributes. They let you change getter and setter names, change setter assignment methods, set non-atomic accessors (which are accessors which can be interrupted by the CPU scheduler while in usage), and determine whether the property is read-only or read-write. These can all be set as part of the `@property` line.

Second, there's another directive called `@dynamic`, which lets you add accessor and mutator methods at runtime.

Third, it's possible to override default values that you've synthesized through normal method creation as part of your `@implementation`.

There's a variety of information on properties in Apple's Objective-C reference, and if you need to delve into any of these complexities, you should refer to that.

10.2.5 *Other compiler directives*

We're almost done with our overview of Objective-C, but we've got one other frequently used bit of syntax that we want to alert you to. As we've seen, the `@` symbol denotes a compile directive. It's a core part of class definition and it's required for properties. You'll also see it in a few other places in Objective-C code.

> **Warning: Common coding error**
>
> We have found that forgetting to mark a string with an `@` is our most common error in iPhone programming, so keep an eye out for this one!

Sometimes an `@` is used to create variables of certain types. This is most frequently used to create a variable of type `NSString *`. We saw this in a few of our messaging examples. You just include the `@` symbol, followed by the string value you want to set:

```
NSString *mySample = @"What does this have to do with apples?";
```

In chapter 14 you'll also encounter the `@selector` directive, which is used to create a variable of type `SEL`. This is a method selector, which is what you use when you want to pass the name of a method as an argument, as will occur when we get to events and actions. A standard usage looks like this:

```
SEL mySelector = @selector(growFruit:);
```

There are many other directives that you can use in Objective-C. Our purpose here is merely to highlight those you're most likely to see in this book and most likely to use in introductory SDK programming.

10.2.6 *Categories and protocols*

There are two final elements of Objective-C that we think it's important to at least touch on: the category and the protocol. We're going to broadly define what they do, but we won't delve too deeply into their details. To learn more, refer to Apple's Objective-C documentation.

THE CATEGORY

Categories are used if you want to add behavior to a class without subclassing. As usual, you do so by creating a new pair of files containing @interface and @implementation code. This time you no longer need to worry about the super class name, but must include a category name in parentheses, as follows:

```
@interface AppleTree (MyAppleChanges)
@implementation AppleTree (MyAppleChanges)
```

As a result, the categorized methods and variables that you describe for the classes will be added to the core class definition in your program.

We won't be using categories in this book.

THE PROTOCOL

A protocol is effectively an interface that's not tied to a class. It declares a set of methods, listing their arguments and their returns. Classes can then state that they're using the protocol in their own @interface statements. For example, if we had a Growing protocol that was used by plants and animals alike, we could define its usage as follows:

```
@interface AppleTree : UrTree <Growing>
```

The AppleTree class would thus be promising that it'd respond to all the methods defined in the Growing protocol.

We won't be creating any new protocols in this book. However, we will be making use of existing ones because within Apple's iPhone OS, they're tied integrally to the MVC model. Views hand off protocol descriptions of how they should be used to view controllers via datasource and delegate properties—both topics that we'll introduce when we talk about the iPhone OS in just a moment.

With that, we feel like the shine has gone off our apples, so we're going to be returning to real-life examples when we move on to the iPhone OS. But first, having provided an overview of a whole new programming language in an impossibly short number of pages, we're going to summarize what we've learned.

10.2.7 *Wrapping up Objective-C*

Table 10.2 summarizes the syntax specifics of the Objective-C elements that we've been discussing. This table can serve as a quick reference whenever you want to revisit how Objective-C code works differently from traditional C.

And with that, we've completed our look at the syntax and structure of the Objective-C programming language. However, that's only half of the foundation you'll need in order to use the SDK. You also need to be familiar with the specific methods and programming styles provided by the iPhone OS's extensive set of frameworks.

Table 10.2   Objective-C uses many typical object-oriented coding elements, but its syntax is somewhat unique.

Object-oriented element	Syntax
Object messaging	`[recipient message];`

Class creation	`::: .h file :::`
	`@interface class: super`
	`     (declarations)`
	`@end`
	`::: .m file :::`
	`@implementation class`
	`     (definitions)`
	`@end`

Method declaration	`- (return type)instancemethod:arguments` `+ (return type)classmethod:arguments`
Property declaration	`@property (declaration)`
Property synthesis	`@synthesize (property);`
Property accessor	`[object property];`
Property mutator	`[object setProperty:value];`
Property dot syntax	`object.property`
Category declaration	`@interface class: super (category)` `@implementation class: super (category)`
Protocol declaration	`@interface class: super <protocol>`

10.3   Introducing the iPhone OS

In the previous section, we started out not with a discussion of how to define objects, but with a look at how to send messages to them. That was our intent. Apple's SDK will provide you with a vast library of objects arranged into several frameworks. As a result, you're going to spend a lot more time sending messages to objects that are ready-made for your use than creating new ones.

Let's begin our look at the iPhone OS by exploring several of these objects and how they're arranged.

10.3.1   The anatomy of the iPhone OS

The iPhone OS's frameworks are divided into four major layers, as shown in figure 10.4.

Each of these layers contains a variety of frameworks that you can access when writing iPhone SDK programs. Generally, you should prefer the higher-level layers when you're coding (those shown toward the top in the diagram).

Figure 10.4   Apple provides you with four layers of frameworks to use when writing iPhone SDK programs.

Cocoa Touch is the framework that you'll become most familiar with. It contains the `UIKit` framework—which is what we'll spend most of our time on in this book—and the address book UI framework. The `UIKit` includes window support, event support, and user-interface management, and allows you to create both text and web pages. It further acts as your interface to the accelerometers, the camera, the photo library, and device-specific information.

Media is where you can get access to the major audio and video protocols built into the iPhone. Its four graphical technologies are OpenGL ES, EAGL (which connects OpenGL to your native window objects), Quartz (which is Apple's vector-based drawing engine), and Core Animation (which is also built on Quartz). Other frameworks of note include Core Audio, Open Audio Library, and Media Player.

Core Services offers the frameworks used in all applications. Many of them are data related, such as the internal Address Book framework. Core Services also contains the critical Foundation framework, which includes the core definitions of Apple's object-oriented data types, such as its arrays and sets.

Core OS includes the kernel-level software. You can access threading, files, networking, other I/O, and memory.

C vs. Objective-C

Most of your iPhone programming work will be done using the UIKit (UI) or Foundation (NS) frameworks. These libraries are collectively called Cocoa Touch; they're built on Apple's modern Cocoa framework, which is almost entirely object-oriented, and in our opinion, much easier to use than older libraries. The vast majority of code in this book will be built solely using Cocoa Touch.

However, you'll sometimes have to fall back on libraries that are instead based on simple C functionality. Examples include Apple's Quartz 2D and Address Book frameworks, as well as third-party libraries like SQLite. Expect object creation, memory management, and even variable creation to work differently for these non-Cocoa libraries.

When you fall back on non-Cocoa libraries, you'll sometimes have to use Apple's Core Foundation framework, which lies below Cocoa. Your first encounter with Core Foundation will be when we discuss the Address Book framework in chapter 16; we'll provide more details on how to use Core Foundation at that point.

Although Core Foundation and Cocoa are distinct classes of frameworks, many of their common variable types are "toll-free bridged," which means that they can be used interchangeably as long as you cast them. Thus, for example, `CFStringRef` and `NSString *` are toll-free bridged, as we'll see when we talk about the Address Book. The Apple class references will usually point out this toll-free bridging for you.

10.3.2 The hierarchy of the iPhone's objects

Within these frameworks you'll be able to access an immense wealth of classes that are arranged in a huge hierarchy. You'll see many of these used throughout this book, and you'll find a listing of even more in appendix A. Figure 10.5 shows many of the classes that we'll use over the next several chapters, arranged in hierarchy. They're just a fraction of what's available.

As shown in figure 10.5, the objects you're most likely to use fall into two broad categories.

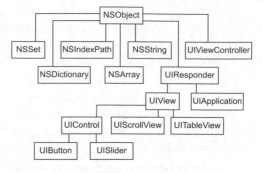

Figure 10.5 This hierarchy shows just a small selection of the classes available in the iPhone OS.

THE NS CLASSES

The NS classes come from Core Services' Foundation framework (the Cocoa equivalent of the Core Foundation framework), which contains a huge number of fundamental data types and other objects.

You should use the fundamental Cocoa classes like `NSString` and `NSArray` whenever you can, rather than C fundamentals like `string *` or a plain array. This is because they tend to play nicely with each other and with the `UIKit` frameworks, and therefore you're less likely to encounter bizarre errors. Although not shown, `NSNumber` is another class that you should be aware of. It should be your main numerical object when you're doing any sort of complex work with a number. It can be used to hold many sorts of numerical values, from floats to integers and more.

The objects that can hold collections of values like `NSArray` (a numerical array) and `NSDictionary` (an associative array) are picky about your sticking to their NS brethren. You'll need to wrap C variables inside Cocoa classes whenever you hand off objects to these arrays. Finally, though `NSString` can take many sorts of objects when you're formatting a string, you should be aware that Cocoa objects may require a different formatting string than their C equivalents.

There are two situations when you'll find that these NS classes can be a deficit. First, if you're using the Core Foundation framework you'll often have to take advantage of toll-free bridging by casting variables, as we'll see starting in chapter 16, when we look at the Address Book. Second, if you're using external APIs, you may need to convert some classes into their C equivalents. Chapter 16's look at the SQLite API explores this possibility, with `NSString` objects often being converted to their UTF-8 equivalent.

The most important of Cocoa's Foundation objects is the `NSObject`, which contains a lot of default behavior, include the iPhone's methods for object creation and memory management, all of which you'll learn about later in this chapter.

THE UI CLASSES

The second broad category of classes contains the UI classes. These come from Cocoa Touch's `UIKit` framework. It includes all of the graphical objects that you'll be using as well as all the functionality for the iPhone OS's event model, much of which appears in `UIResponder`. That's another topic we'll return to soon.

10.3.2 *Windows and views*

As the UI classes demonstrate, the iPhone OS is deeply rooted in the idea of a graphical user interface. Therefore, let's finish our introduction to the iPhone OS by looking at some of the main graphical abstractions embedded in the UIKit. There are three major abstractions: windows, views, and view controllers.

A *window* is something that spans the entire screen of the iPhone. There's only one of them for your application, and it's the overall container for everything that your application does.

A *view* is the actual content holder in your application. You may have several of them, each covering different parts of the window or doing different things at different times. They're all derived from the UIView class. However, don't just think of a view as a blank container. In actuality, almost any object that you use from the UIKit will be a subclass of UIView that features a lot of behavior all of its own. Among the major subclasses of UIView are UIControls, which give you buttons, sliders, and other items that users may manipulate your program with, and UIScrollableViews, which give users access to more text than can appear at once.

A *view controller* does what its name suggests. It acts as the controller element of the MVC model and in the process manages a screenful of text, which is sometimes called an *application view.* As such, it takes care of events and updating for your view.

In this book, we've divided view controllers into two types. *Basic view controllers* are those which just manage a screenful of text (such as the table view controller), while *advanced view controllers* are those that let a user move around among several pages of text (such as the navigation bar controller and the tab bar controller). Figure 10.6 shows how these three types of objects interrelate.

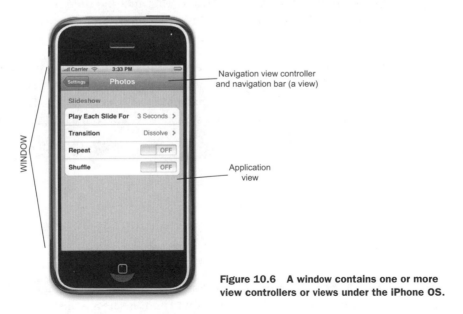

Figure 10.6 A window contains one or more view controllers or views under the iPhone OS.

Windows, views, and view controllers are ultimately part of a *view hierarchy*. This is a tree of objects that begins with the window at its root. A simple program might just have a window with a view under it. Most programs will start with a window, have a view controller under that, perhaps supported by additional view controllers, each of which controls views that might have their own subviews. We'll depict this concept more clearly in chapter 13 when we start looking at the basic view controllers that make this sort of hierarchy possible.

10.4   *The iPhone OS's methods*

As you've seen, the iPhone OS has a complex and deep structure of classes. Two of the most important are `NSObject` and `UIResponder`, which contain many of the methods and properties that you'll use throughout your programming. Thanks to inheritance, these important functions (and others) can be used by many different iPhone OS objects.

We'll cover some of these foundational methods here to provide a single reference for their usage, but we'll be sure to point them out again when we encounter them for the first time in future chapters.

10.4.1   *Object creation*

We talked earlier about how to define classes, but as we said at the time the specifics of *how* instance objects are created from classes depend on the implementation of your framework. In the iPhone OS it's the `NSObject` that defines how object creation works.

You're going to meet a few different interfaces that are used to support object creation, but they all ultimately fall back to a two-step procedure that uses the `alloc` class method and the `init` instance method. The `alloc` method allocates the memory for your object, and then returns the object itself. The `init` method then sets some initial variables in that method. They usually occur through a single, nested message:

```
id newObject = [[objectClass alloc] init];
```

The `alloc` method from `NSObject` should always do the right thing for you. However, when you write a new subclass you'll almost always want to write a new `init` method, because that's where you define the variables that make your class what it is. Listing 10.4 shows a default setup for an `init`, which would appear as part of your `@implementation`.

Listing 10.4   A sample `init` method for preparing an object

```
- (id)init
{
    if (self = [super init]) {          ❶
// Instance variables go here          ❷
    }
    return self;          ❸
}
```

Listing 10.4 shows all the usual requirements of an `init` method. First, it calls its parent ❶ to engage in its class's usual initialization. Then, it sets any instance variables that should be set ❷. Last, it returns the object, usually with `return self;` ❸.

The bare `init` is just one of a few major ways that you can use to create objects in the iPhone OS.

THE ARGUMENTATIVE ALTERNATIVE

Sometimes you'll want to send an argument with an `init`. You can do so with an initialization function that you name using the format `initWithArgument:`. Other than the fact that you're sending it an argument, it works exactly like a bare `init`. Here's another example drawn from actual code we'll see in upcoming chapters:

```
[[UITextView alloc] initWithFrame:textFieldFrame];
```

Initialization methods with arguments allow you to create nonstandard objects set up in ways that you choose. They're quite common in the `UIKit`.

One initialization method with an argument deserves a bit of extra mention. `initWithCoder:` is a special initialization method that's called whenever you create an object with Interface Builder—and thus is important if you want to do setup for such objects. We'll return to Interface Builder in chapter 12.

THE FACTORY METHOD ALTERNATIVE

A final sort of `init` supported through the iPhone OS is the factory method. This is a one-step message that takes care of both the memory allocation and initialization for you. All factory methods are named with the format `objecttypeWithArgument:`. Here's another real example:

```
[UIButton buttonWithType:UIButtonTypeRoundedRect];
```

Class factory methods make messaging a little clearer, but they also have the advantage of taking care of some memory management for you, which is the topic of our next major category of iPhone OS methods.

OBJECT CREATION WRAP-UP

We've summarized the four major ways that the iPhone OS supports the creation of objects in table 10.3.

As witnessed by our examples, we'll use all of these methods as we move through the upcoming chapters.

Table 10.3   iPhone OS supports several methods that you can use to create objects; different methods will be supported by different classes.

Method	Code	Summary
Simple	`[[object alloc] init];`	Plain initialization
Argument	`[[object alloc] initWithArgument:argument];`	An initialization where one or more arguments is passed to the method
Coder	`[[object alloc] initWithCoder:decoder];`	An initialization with an argument used for Interface Builder objects
Factory	`[object objecttypeWithArgument:argument];`	A one-step initialization process with an argument

10.4.2 *Memory management*

Because of power considerations, the iPhone OS doesn't support garbage collection. That means that every object that's created must eventually have its memory released by hand—at least if you don't want to introduce a memory leak into your program.

The fundamental rule of memory management in the iPhone OS is this: if you allocated the memory for an object, you must release it. This is done via the release message (which is once again inherited from NSObject):

```
[object release];
```

You just send that message when you're all done using an object, and you've done your proper duty as a programmer.

You'll note that we said you only must release the memory if you *allocated* the memory for it. If you look back to the class factory methods we talked about in the previous section, you'll see that we didn't actually allocate the memory for those (because we didn't send any alloc message), which means we're not responsible for releasing it. Instead, the class object that actually did the creation has to clean up its memory.

How does the OS know when we've finished working with the object it created for us? That's done through the wonders of autorelease.

THE AUTORELEASE ALTERNATIVE

If you're responsible for the creation of an object and you're going to pass it off to some other class for usage, you should autorelease the object before you send it off. This is done with the autorelease method:

```
[object autorelease];
```

You'll typically send the autorelease message just before you return the object at the end of a method. Once an object has been autoreleased, it's watched over by a special NSAutoreleasePool. The object is kept alive for the scope of the method that it's been passed to, and then the NSAutoreleasePool cleans it up.

RETAINING AND COUNTING

So what if you want to hold onto an object that has been passed to you, and that is going to get autoreleased? In that case, you send it a retain message:

```
[object retain];
```

When you do this, you're saying you want the object to stay around but now you've become responsible for its memory as well: you must send a release message at some point to balance your retain.

At this point, we should probably back up and explain the underlying way that the iPhone OS actually manages these memory objects. It does so by maintaining a count of object usage. By default it's set to 1. Each retain message increases that count by 1, and each release message reduces that count by 1. When the count drops to 0, the memory for the object is freed up.

Therefore, all memory management can be thought of as pairs of messages. If you balance every alloc and every retain with a release, your object will eventually be freed up when you're done with it.

MEMORY MANAGEMENT WRAP-UP

Table 10.4 provides a quick summary of the methods we've looked at to manage the memory used by your objects.

Table 10.4　**The memory management methods help you to keep track of the memory you're using and clean it up when you're done.**

Method	Summary
alloc	A part of the object-creation routine, but this is what actually allocates the memory for an object's usage.
autorelease	A request to reduce an object's memory count by 1 when it goes out of scope; this is maintained by an NSAutorelease pool.
release	Reduces the object's memory count by 1.
retain	Increases the object's memory count by 1.

For more information on memory management, including a look at the copy method and how this all interacts with properties, take a look at Apple's Objective-C references, but what we've discussed here should be enough for you to write good Objective-C code.

10.4.3 *Event response*

The next-to-last category of methods that we'll examine for the iPhone OS is event response. Unlike with object creation and memory management, we'll only tackle this issue briefly, because it's much better documented in chapter 14. The topic is important enough that we want to offer a quick overview of it now.

There are three main ways that events can appear on the iPhone: through bare events (or actions), through delegated events, or through notification.

Whereas the methods of our earlier topics all derived from NSObject, iPhone event response instead comes from the UIResponder object, while iPhone notification comes from the NSNotificationCenter. You won't have to worry about accessing responder methods and properties as UIResponder is the parent of most UIKit objects, but the NSNotificationCenter will require special access.

EVENTS AND ACTIONS

Most user input on the iPhone results in an *event* being placed into a *responder chain*. This is a linked set of objects that, for the most part, goes backward up through the view hierarchy. Any input is captured by the *first responder*, which tends to be the object that the user is directly interacting with. If that object can't resolve the input, it sends it up to its superview (e.g., a label might send it up to its full-screen view), then to its superview, all the way up the chain (e.g., up through the views, then up through the view controllers). If input gets all the way up the view hierarchy to the window object, it's next sent on to the application itself, which tends to pass it off to an *application delegate* as a last resort.

Any of these objects could choose to handle an event, which stops its movement up the responder chain. Following the standard MVC model, you'll often be building event response into `UIViewControllers` objects, which are pretty far up the responder chain.

For any `UIControl` objects, such as buttons, sliders, and toggles, events are often turned into *actions*. Whereas events report touches to the screen, actions instead report manipulations of the controls and are thus easier to read. Actions follow a slightly different hierarchy of response.

DELEGATES AND DATA SOURCES

There's another way that events can be sent to an object other than first responder: through a delegate. This is an object (usually a view controller) that says that it's going to take care of events for another object (usually a view). It's a close kin to a data source, which is an object (again, usually a view controller) that promises to do the data setup and control for another object (again, usually a view).

Delegation and data sourcing are each controlled by a protocol, which is a set of methods that the delegate or data source agrees to respond to. For example, a table's delegate might have to respond to a method that alerts it when a row in the table has been selected. Similarly, a table's data source might describe what all the rows of the table look like.

Delegates and data sources fit cleanly into the MVC model used by Objective-C, as they allow a view to hand off its work to its controller without having to worry about where each of those objects is in the responder chain.

NOTIFICATIONS

Standard event response and delegation represent two ways that objects can be alerted to standard events, such as fingers touching the screen. There's also a third method that can be used to program many different sorts of activities, such as an iPhone's orientation changing or a network connection closing: the notification.

Objects register to receive a certain type of notification with the `NSNotification-Center`, and afterward may process those notifications accordingly. Again, we'll discuss this topic in chapter 14.

10.4.4 *Life-cycle management*

By now you know how to create objects using the iPhone OS and how to release their memory when you're done with them. In that discussion we've neglected one other topic: how to recognize when objects are being created and destroyed—starting with your application itself.

Table 10.5 summarizes some of the important messages that will be sent as part of the life cycle of your program. To respond to them, you just fill in the contents of the appropriate methods in either an object or its delegate—which of course will require writing a subclass, and indeed is one of the prime reasons to do so.

Note that we've included our old friend `init` here, since it forms a natural part of the object life cycle. You should look at the individual Apple class references, particularly `UIApplicationDelegate`, for other methods you might want to respond to when writing your program.

Table 10.5 Several important methods let you respond to the life cycle of your application or its individual objects.

Method	Object	Summary
`applicationDidFinish-Launching:`	`UIApplicationDelegate`	The application has loaded up; you should create initial windows and otherwise start your program.
`applicationDidReceive-MemoryWarning:`	`UIApplicationDelegate`	The application received a low-memory warning; you should free up memory.
`applicationWill-Terminate:`	`UIApplicationDelegate`	The application is about to end; you should free up memory and save state.
`init`	`NSObject`	The object is being created; you should initiliaze it here.
`dealloc`	`NSObject`	The object is freeing up its memory; you should release any objects that haven't been autoreleased.

With that, we've completed our look at the big picture methods of the iPhone OS. You've not yet seen them in real usage, so bookmark these pages as we'll be referring back to them when we begin actual programming in just a couple of pages.

10.5 Summary

As you begin work with the SDK, you're not just approaching a new programming language for the iPhone, but a totally new way to create iPhone programs. This means you have to learn an entirely new programming suite.

The SDK is our toolbox. Its most important elements are Xcode, the integrated development environment, and Interface Builder, the graphical object creator.

Objective-C is our programming language. It's an object-oriented version of C that has some pretty unique syntax thanks to its elegant Smalltalk inspiration. Once you get used to it, you'll find it simple and easy to read.

The iPhone OS is a layered set of frameworks, which contains everything you need to make your iPhone programming easy. Much of the rest of this book will talk about how to make use of the right frameworks at the right time.

With all of that in your back pocket, you're ready to start programming using the SDK, a task that will begin on the next page when we dive into the Xcode program that you downloaded at the start of this chapter.

Using Xcode

This chapter covers

- Learning how Xcode works
- Writing a simple Hello, World! program
- Creating new classes

Now that you've learned a bit about the puzzle pieces needed to build an SDK program, you're ready to start programming. The main purpose of this chapter is to show you how Xcode, the SDK's main development environment, works. Via a traditional Hello, World! program, we'll look at the parts of a standard SDK program. We'll also examine how to create new classes of objects, and with that under our belt, we'll finish up by looking at some of Xcode's most interesting features.

11.1 Introducing Xcode

Apple programming begins with Xcode, an integrated development environment (IDE) that you can call up from the Developer directory. To write iPhone programs, you must have downloaded the iPhone SDK, as discussed in the previous chapter. Once you've done that, choosing File > New Project will get you started. You'll immediately be asked to select a template for your new project.

The template you choose will fill your project with default frameworks, default files, default objects, and even default code. As you'll see, it'll be a great help in jump-starting your own coding. For your first program, we want to go with the simplest template we can find: Window-Based Application.

Once you've selected a template, you'll also need to name your project and choose where to save it, but after you're done with that, you're ready to start coding. Before we get there, however, let's take a closer look at how the Xcode environment works.

11.1.1 The anatomy of Xcode

When called up, Xcode displays just one window. Figure 11.1 shows what it looks like for our first example project, helloworldxc.

As you can see in figure 11.1, Xcode's main project window includes three parts. Off to the left is a sidebar that contains a listing of all the files that are being used in your project, organized by type. Whenever you need to add frameworks, images, databases, or other files to your projects, you'll do so here. The left pane also contains some other useful elements, in particular an Errors and Warnings item that you can click open to quickly see any problems in your compilation.

The top-right pane contains an ungrouped list of files used by your project. When you click on one of those, its contents will appear in the bottom-right pane. As you can see, even the simplest program will include over a half-dozen files. Table 11.1 summarizes what each is.

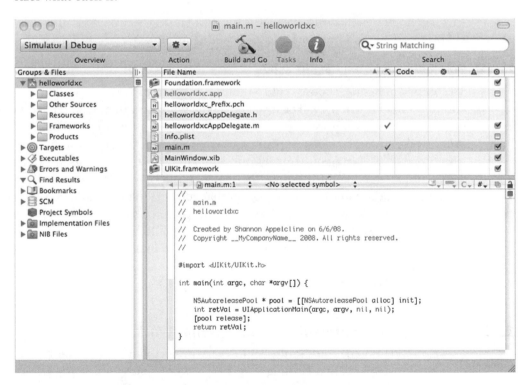

Figure 11.1 Xcode's main project window shows you all your files and also allows you to quickly view them.

Table 11.1   Several types of files will show up in your Xcode projects.

File	Summary
project.app	A compiled application.
*.framework	A standard framework included as part of your project. By default, every project should include Foundation, giving you access to NS objects, UIKit, giving you access to UI objects, and CoreGraphics, giving you access to various graphics functions. We'll talk about adding additional frameworks later on.
*.h	A header file, usually containing the @interface for a class.
*.m	A source code file, usually containing the @implementation for a class.
*.mm	A source code file with C++ code. Not used in this book.
project_Prefix.pch	A file containing special prefix headers, which are imported into every one of your source code files. It's here that the two main frameworks are imported.
Info.plist	An XML property list. It contains a number of instructions for your program compilation, the most important of which is probably the reference to the .xib file used in your program.
MainWindow.xib	An Interface Builder file, more broadly called a "nib file." This is your connection to the graphical design program that may be used to easily create objects for your project. We'll discuss it in depth in the next chapter.

In this chapter we'll focus exclusively on header and source code files. In the next chapter we'll extend our work to also include the .xib Interface Builder file.

11.1.2 *Compiling and executing in Xcode*

To compile in Xcode, choose Build > Build and Run from the menus. Your program will compile and link. Then it will be installed on the iPhone Simulator, and the iPhone Simulator will start it up. If you try this out using the project that we just created using the Window-Based Application template, you'll see the whole process, resulting in an empty white screen displaying on your iPhone Simulator. Note that programs only exist on your Simulator (or in the iPhone); they can't be run on your Macintosh directly.

If you want to later rerun a program that you've already compiled, you can do so in one of three ways. You can just click the program's button, which should now appear in your Simulator. Or, you can choose Run > Run from within Xcode. Finally, you can choose Build and Go in Xcode, which only builds if required, then executes your program.

That's it! With a rudimentary understanding of Xcode now in hand, you're ready to write your first SDK program.

11.2 *Creating a first project in Xcode: Hello, World!*

As we already noted, you should begin every project by running File > New Project, choosing a template, and naming your file. For our first sample project, we selected the Window-Based Application template and the name helloworldxc.

Before you start writing new code, you need a basic understanding of what's there already, so we'll examine contents of the three most important files that our basic template created: main.m, helloworldxcAppDelegate.h, and helloworldxcAppDelegate.m.

11.2.1 Understanding main.m

The first file created by Xcode is main.m, which contains your main function, as shown in listing 11.1.

> **Listing 11.1   main.m, which comes with standard code preinstalled for you**

```
#import <UIKit/UIKit.h>        ❶

int main(int argc, char *argv[]) {

    NSAutoreleasePool * pool = [[NSAutoreleasePool alloc] init];    ❷
    int retVal = UIApplicationMain(argc, argv, nil, nil);          ❸
    [pool release];
return retVal;
}
```

The creation of this main routine is automatic, and you generally shouldn't have to fool with it *at all*. However, it's worth understanding what's going on. You start off with an #import directive ❶, which you'll recall is Objective-C's substitute for #include. More specifically, you've included the UIKit framework, the most important framework in Objective-C. This actually isn't needed, because it's also in the Prefix.pch file, but at least at the time of this writing, it's part of the default main.m file.

You next create an NSAutoreleasePool ❷. You'll recall that we mentioned this in our discussion of memory management in the previous chapter. It's what supports the NSObject's autorelease method. Also note that you release the pool itself after you've run your application's main routine, following the standard rule that if you allocate the memory for an object, you must also release it.

The UIApplicationMain line ❸ is what creates your application and kicks off your event cycle. The function's arguments look like this:

```
int UIApplicationMain (
    int argc,
    char *argv[],
    NSString *principalClassName,
    NSString *delegateClassName
);
```

As with the rest of the main.m file, you should never have to change this, but we're nevertheless going to briefly touch on what the latter two arguments mean—though they'll usually be set to their defaults, thanks to the nil arguments.

The principalClassName defines the application's main class, UIApplication, by default. This class does a lot of the action- and event-controlling for your program, topics that we're going to return to in chapter 14.

The UIApplication object is created as part of this startup routine, but you'll note that no link to the object is provided. If you need to access it (and you will), you can use a UIApplication class method to do so:

```
[UIApplication sharedApplication];
```

This will return the application object. It will typically be sent as part of a nested message to a UIApplication method, as you'll see in future chapters. For now, the application does two things of note: it calls up your default .xib file and it interfaces with your application delegate.

The delegateClassName defines the application object's delegate, an idea we introduced in chapter 10. As we noted there, this is the object that responds to some of the application's messages, as defined by the UIApplicationDelegate protocol. Among other things, the application delegate must respond to life-cycle messages, most importantly the applicationDidFinishLaunching: message which is what runs your program's actual content, as we'll talk more about momentarily.

In Xcode's templates, your delegate class files will always have the name projectAppDelegate. Your program finds them, thanks to a delegate property that's built into Interface Builder.

You *could* change the arguments sent to UIApplicationMain and you *could* add other commands to the main.m file, but generally you don't want to. The defaults should work fine for any program you're likely to program in the near future. So, let's put main.m away for now and turn to the file where any programming actually starts: your application delegate.

11.2.2 *Understanding the application delegate*

As you've already seen, the application delegate is responsible for answering many of the application's messages. You can refer to the previous chapter for a list of some of the more important ones or to Apple's UIApplicationDelegate protocol reference for a complete listing.

More specifically, an application delegate should do the following:

- At launch time, it must create an application's windows and display them to the user.
- It must initialize your data.
- It must respond to "quit" requests.
- It must handle low-memory warnings.

Of these topics, it's the first that's of importance to you now. Your application delegate files, helloworldxcAppDelegate.h and helloworldxcAppDelegate.m, get your program started.

THE HEADER FILE

Now that you've moved past main.m, you're actually using classes, which is the sort of coding that makes up the vast majority of Objective-C code. Listing 11.2 shows the contents of your first class's header file, helloworldxcAppDelegate.h.

Listing 11.2 The Application Delegate header

```
@interface helloworldxcAppDelegate : NSObject <UIApplicationDelegate> {   ❶
    UIWindow *window;        ❷
}

@property (nonatomic, retain) IBOutlet UIWindow *window;     ❸
```

Again, there's nothing you're going to change here, but we want to examine the contents, both to reiterate some of the lessons you learned in the previous chapter and to give you a good foundation for work you're going to do in the future.

First, you'll see an interface line ❶ that subclasses your delegate off NSObject (which is appropriate, since the app delegate is a nondisplaying class) and includes a promise to follow the UIApplicationDelegate protocol.

Next, you have the declaration of an instance variable, window ❷.

Finally, you declare that window as a property ❸. You'll note this statement includes some of those property attributes that we mentioned, here nonatomic and retain. This line also includes an IBOutlet statement, which tells you that the object was actually created in Interface Builder. We'll examine this concept in more depth in the next chapter, but for now you just need to know that you have a window object already prepared for your use.

Although you won't modify the header file in this example, you will in the future, and you'll generally be repeating the patterns you see here: creating more instance variables, including IBOutlets, and defining more properties. You may also declare methods in this header file, something that this first header file doesn't contain.

THE SOURCE CODE FILE

Listing 11.3 displays the application delegate's source code file, helloworldxcAppDelegate.m, and it's here that you're going to end up placing your new code.

Listing 11.3   The Application Delegate object that contains your startup code

```
#import "helloworldxcAppDelegate.h"          ❶

@implementation helloworldxcAppDelegate      ❷

@synthesize window;          ❸

- (void)applicationDidFinishLaunching:(UIApplication *)application {     ❹
    [window makeKeyAndVisible];      ❺
}

- (void)dealloc {
    [window release];
    [super dealloc];
}

@end
```

The source begins with an inclusion of the class's header file ❶ and an @implementation statement ❷. Your window property is also synthesized ❸.

It's the content of the applicationDidFinishingLaunching method ❹ that's really of interest to you. As you'll recall, that's one of the iPhone OS life-cycle messages that we touched on in chapter 10. Whenever an iPhone application gets entirely loaded into memory, it'll send an applicationDidFinishingLaunching: message to your application delegate, running that method. You'll note there's already some code to display that Interface Builder–created window ❺.

For this basic project, you'll add all your new code to this same routine—such as an object that says Hello, World!

11.2.3 *Writing Hello, World!*

We've been promising for a while that you're going to be amazed by how simple it is to write things using the SDK. Granted, your Hello, World! program may not be as easy as a single `printf` statement, but nonetheless it's pretty simple considering that you're dealing with a complex, windowed UI environment.

As promised, you'll be writing everything inside the `applicationDidFinishingLaunching` method, as shown in listing 11.4.

Listing 11.4 The iPhone presents... Hello, World!

```
- (void)applicationDidFinishLaunching:(UIApplication *)application {

    [window setBackgroundColor:[UIColor redColor]];            ❶

    CGRect textFieldFrame = CGRectMake(50, 50, 150,40);        ❷
    UILabel *label = [[UILabel alloc] initWithFrame:textFieldFrame];   ❸

    label.textColor = [UIColor whiteColor];
    label.backgroundColor = [UIColor redColor];
    label.shadowColor = [UIColor blackColor];                  ❹
    label.font = [UIFont systemFontOfSize:24];

    label.text = @"Hello, World!";

    [window addSubview:label];                                 ❺
    [window makeKeyAndVisible];                                ❻

    [label release];                                           ❼

}
```

Since this is your first look at real live Objective-C code, we're going to examine everything in some depth.

ABOUT THE WINDOW

You start off by sending a message to your window object, telling it to set your background to red ❶. You'll recall from our discussion of the header file that Interface Builder was what created the window. The `IBOutlet` that was defined in the header is what allows you to do manipulations of this sort.

Note that this line also makes use of a nested message, which we promised you'd see with some frequency. Here, you make a call to the `UIColor` class object and ask it to send you the red color. You then pass that on to your window.

In this book, we're going to hit a lot of UIKit classes without explaining them in depth. That's because the simpler objects all have standard interfaces; the only complexity is in which particular messages they accept. If you ever feel as if you need more information about a class, you should look at appendix A, which contains short descriptions of many objects, or in the complete class references available online at developer.apple.com (or in Xcode).

ABOUT FRAMES

You're next going to define where your text label is placed. You start that process off by using `CGRectMake` to define a rectangle ❷. Much as with Canvas, the SDK uses a grid with the origin (0,0) set at the top left. Your rectangle's starting point is thus 50 to the right and 50 down (50,50) from the origin. The rest of this line of code sets the rectangle to be 150 pixels wide and 40 tall, which is enough room for your text.

You're going to be using this rectangle as a "frame," which is one of the methods you can use to define a view's location.

Choosing a view location

Where your view goes is one of the most important parts of your view's definition. Many classes use an `initWithFrame:` init method, inherited from `UIView`, which defines location as part of the object's setup.

The frame is simply a rectangle that you've defined with a method like `CGRectMake`. Another common way to create a rectangular frame is to set it to take up your full screen:

```
[[UIScreen mainScreen] bounds];
```

Sometimes you'll opt not to use the `initWithFrame:` method to create an object. `UIButton` is an example of a UIKit class that instead suggests you use a class factory method that lets you define a button shape.

In a situation like that, you must set your view's location by hand. Fortunately, this is easy to do because `UIView` also offers a number of properties that you can set to determine where your view goes, even after it's been initialized.

`UIView`'s frame property can be passed a rectangle, just like the `initWithFrame:` method. Alternatively, you can use its `center` property to designate where the middle of the object goes and the bounds property to designate its size internal to its own coordinate system.

All three of these properties are further explained in the `UIView` class reference.

Note that `CGRectMake` is a function, not a method. It takes arguments using the old, unlabeled style of C, rather than Objective-C's more intuitive manner of using labeled arguments. Once you get outside of Cocoa Touch, you'll find that many frameworks use this older paradigm. For now, all you need to know is what it does and that you don't need to worry about releasing its memory. If you require more information, read the sidebar "Using Core Foundation" in chapter 16.

ABOUT THE LABEL

The label is a simple class that allows you to print text on the screen. We included figure 11.2 so you can see what your label (and the rest of your program) looks like.

As you'd expect, your label work begins with the actual creation of a label object ❸. Note that you follow the standard methodology of nested object creation that we introduced in the previous chapter. First you use a class method to allocate the object, and then you use an instance method to initialize it.

Afterward you send a number of messages to your object ❹, this time using the dot shorthand. We offer this as a variation from the way you set the window's background color. If you prefer, you can use the dot shorthand of `window.backgroundColor` there, too. The two ways to access properties are totally equivalent.

The most important of your messages sets the label's text. You also set a font size and some colors. You even can give the text an attractive black shadow, to demonstrate how easy it is to do cool stuff using the iPhone OS's objects.

Every object that you use from a framework is going to be full of properties, methods, and notifications that you can take advantage of. The best place to look all these up in is the class references.

Figure 11.2   Hello, World! is easy to program on the iPhone using the SDK.

FINISHING UP OUR WORLD

The final steps in your program are all pretty simple and standard.

First, you connect your label and your window by using the window's `addSubview` method ❺. This is a standard (and important!) method for adding views or view controllers to your window. You'll see it again and again.

Second, you create your window on the screen, using the line of code that was here when we started ❻. Making the window "key" means that it's now the prime recipient of user input (for what that's worth in this simple example), while making it "visible" means that the user can see it.

Third, you remember the standard rule that you must release anything you allocated? Here, that's just the label ❼.

And that's a simple Hello, World! program, completely programmed and working, with some neat graphical nuances.

Although it was sufficient for our purposes, Hello, World! didn't make much use of the class creation that's possible in an object-oriented language. Sure, we depended on some existing classes—including `UIColor`, `UILabel`, and `UIWindow`—but all of our new code went into a single function, and we didn't create any classes of our own. We'll address this issue in our next example, when we start working with new classes.

11.3   *Creating a new class in Xcode*

New programs will usually be full of new classes. Here are three major reasons why you might create new classes:

- You might create a totally new class, with different functionality from anything else. If it's a user interface class, it'll probably be a subclass of UIView. If it's a nondisplaying class, it'll probably be a subclass of NSObject.
- You might create a new class that works similarly to an old class but with some standardized differences. This new class would generally be a subclass of the old class.
- You might create a new class that has specific event responses built in. This class would also generally be a subclass of the old class.

Creating a new class and linking it in is easier than you think. In our next example you're going to create a project called newclass that will include the brand-new labeledwebview subclass. Again we'll build it using the Window-Based Application template.

11.3.1 *The new class how-to*

Once you've gotten your new project going, the process of creating a new class (see table 11.2) is simple, with Xcode (as usual) doing most of the work for you in file creation.

Table 11.2   A few steps in Xcode will quickly create a brand-new object.

Step	Description
1. Create your new file.	Choose File > New File. Choose the class to use as your parent from among the Cocoa Touch Classes options. Select your filename, preferably an intuitive name reflecting your object. Accept the default setup, including the creation of an .h file.
2. Modify your files.	If you weren't able to select your preferred class to subclass, change that now by modifying the parent class in the @interface line of yourclass.h.
3. Import your object.	Add an #import line for your class's header in whatever file will be using it.

For our sample program, we created the labeledwebview class as a subclass of UIView and then imported our new .h file into our application delegate:

```
#import "labeledwebview.h"
```

Afterward it's a simple matter of designing your class to do the right thing. For our purposes, we've decided to create an object that will display both a web page and the URL of that web page on the iPhone screen by linking together some existing classes.

There are three steps to the process, all of which we'll touch on in this section: you need to write your new header file, you need to write your new source code file, and you need to use the new class inside your program.

11.3.2 *The header file*

As usual, you've got the start of a header file already, thanks to Xcode. Listing 11.5 shows how you'll expand it to create your new class.

Listing 11.5   A header file for a new class

```
@interface labeledwebview : UIView {

    UILabel *URLLabel;                      ❶
    UIWebView *myWebView;
}

@property(nonatomic, retain) UILabel *URLLabel;      ❷
@property(nonatomic, retain) UIWebView *myWebView;

- (void)loadURL:(NSString *)url;        ❸
@end
```

This is the last time that we're going to look at a header file that has only basic information in it, but since it's your first time working with a new class, we figure it's still worthwhile. Within the header file, you again engage in some of those common declarations that you saw back in our Hello, World! program.

First, you declare some instance variables that you want to use throughout your class ❶. Second, you define those variables as properties ❷. Third, you declare a method ❸ that you want to make available outside the class.

Now you're ready for the actual code.

11.3.3 *The source code file*

The source code file contains the guts of your new class, as shown in listing 11.6.

Listing 11.6   A source code file for a new class

```
#import "labeledwebview.h"        ❶

@implementation labeledwebview

@synthesize URLLabel;             ❷
@synthesize myWebView;

- (id)initWithFrame:(CGRect)frame {
    if (self = [super initWithFrame:frame]) {

        URLLabel = [[UILabel alloc]
            initWithFrame:CGRectMake(20, 20, 280, 50)];
        myWebView = [[UIWebView alloc]
            initWithFrame:CGRectMake(20,60,280,400)];

        URLLabel.textColor = [UIColor whiteColor];
        URLLabel.shadowColor = [UIColor blackColor];    ❸
        URLLabel.adjustsFontSizeToFitWidth = YES;
        myWebView.scalesPageToFit = YES;

        [self addSubview:URLLabel];
        [self addSubview:myWebView];

    }
    return self;
}

- (void)setBackgroundColor:(UIColor *)color {        ❹
    [super setBackgroundColor:color];
```

```
    [URLLabel setBackgroundColor:color];
}
- (void)loadURL:(NSString *)url {
    [myWebView loadRequest:
        [NSURLRequest requestWithURL:[NSURL URLWithString:url]]];

    URLLabel.text = url;
}
- (void)dealloc {
    [myWebView release];
    [URLLabel release];
    [super dealloc];
}
@end
```

Figure 11.3 shows the end results of your class creation in actual use, but we're also going to explain the parts of the code that will get you there.

You always have to import your header file into its matched source code file ❶. You follow that up by synthesizing your two properties ❷, making them available for use.

You put together the pieces of your new class in the initWithFrame: method ❸. As usual, you called your parent's init. Then you create the two objects your new class will contain: a label and a web view. After setting some basic values for each, you make them subviews of your new labeledwebview class.

Don't worry about the fact that we're not spending much time on how the web view works; it's one of the UIKit objects that will get more attention down the road, when we talk about the SDK and the web in chapter 20.

Because you want your label's view to always match the background color, you override your view's setBackgroundColor: method ❹. After calling the parent's method, which sets the view's background color, your class adjusts the color of its label too.

Figure 11.3   A brand-new class makes it easy to display a URL and call it up on our screen; what you've created is the first step in building a web browser.

The real magic occurs in the brand-new loadURL: method ❺. First you load the URL in the web view. This requires a two-step process that goes through the NSURLRequest and NSURL class objects. (You can find more information in the Apple class references and in chapter 20.) That's all you have to do to generate a fully functional web page, which is pretty amazing. If you try to use it, you'll even find that it has much of the iPhone's unique output functionality: you can pinch, tap, and zoom just like in Safari. You finish the method by setting the label to match your URL.

Your new class ends with the standard dealloc method ❻, where you clean up the two objects that you allocated as part of your object creation.

So there you have less than a page of code that creates an object that would require a lot more work if you were programming it by hand. But there are just so many tools available to you in the SDK that knocking out something like this is, as you can see, simplicity itself. You could definitely improve this example. For example, you could link in to the `UIWebViewDelegate` protocol to update your label whenever the web view changed, but for now we're pleased with what we have as a second example of Xcoding.

11.3.4   *Linking it in*

Just creating a new class isn't enough: you also need to use it. Listing 11.7 shows the code that you put in the application delegate to use your new subclass.

Listing 11.7   A few app delegate calls

```
#import "labeledwebview.h"

- (void)applicationDidFinishLaunching:(UIApplication *)application {

    labeledwebview *myWeb = [[labeledwebview alloc]
        initWithFrame:[[UIScreen mainScreen] bounds]];
    [myWeb loadURL:@"http://www.manning.com/callen/"];
    [myWeb setBackgroundColor:[UIColor grayColor]];

    [window addSubview:myWeb];

    [window makeKeyAndVisible];
}
```

As you can see in listing 11.7, you create your new object just as you would any object that comes naturally in the iPhone's frameworks, and then you take advantage of those methods you coded into it.

You've now seen how to make use of Xcode to write a few simple programs, but before we finish up this chapter let's take a quick look at some of Xcode's other features, which you'll be doubtless taking advantage of.

11.4   *Other Xcode functionality*

What follows are some notes on how to undertake other common Xcode tasks, ending with an overview of a lot of Xcode's cooler bells and whistles that you may not be familiar with.

11.4.1   *Adding frameworks with Xcode*

To date, our programs have included three frameworks: CoreGraphics, Foundation, and UIKit. You may find someday that you want to add another framework, to get access to some other set of classes that will make your life easier. In particular, if you have problems accessing a method defined in the SDK, you can look in the appropriate class reference, and see the framework that's required near the top of the reference.

All you have to do to add a framework to Xcode is Ctrl-click on the Frameworks folder in your Xcode sidebar and choose Add > Existing Frameworks. Xcode will show you a long list of frameworks. When you choose one, it'll automatically be set up as a target when you compile.

Your default frameworks are selected by the template that you choose to use when you create our project. For our example, the Window-Based Application determines that you have access to the CoreGraphics, Foundation, and UIKit frameworks at the start. Other templates may give access to other frameworks with no additional work required on your part.

11.4.2 *Using alternate templates with Xcode*

When you're creating a new program in Xcode, you always have the option to select among several templates, each of which will give you a different basis for your code. Besides a Window-Based Application, you can create a project as a View-Based Application, a Tab Bar Application, a Navigation-Based Application, a Utility Application, or an OpenGL ES Application.

Most of these templates build in view controllers and give access to other functionality that we won't encounter for a couple of chapters. We'll give you a glance at them all now so you can see the possibilities that Xcode offers. Figure 11.4 shows how much different templates can vary from the Window-Based Application that we've been using.

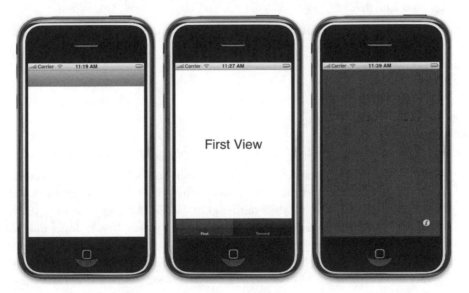

Figure 11.4 By using the appropriate templates, you can lay out a nav bar (left), a tab bar (middle), or a flipside function (right) before you start writing any code.

Here's a bit more information on each of the six templates:

- A *Window-Based Application,* as we've seen, is entirely minimalist. You'll need to create a default UIView for your application before you can do anything. That's what we've used so far.
- A *View-Based Application* has just a hair more functionality. It includes a basic view controller that will allow you to autorotate iPhone content. We'll use it in chapter 13 (and most of the time thereafter).

- A *Tab Bar Application* creates a tab bar along the bottom that allows you to switch between multiple views. The template does this by creating a tab bar controller and then defining what each of its views looks like. We'll make use of it in chapter 15.

- A *Navigation-Based Application* sets you up with a navigation controller, a nav bar along the top, and a table view in the middle of the page so that you can easily build hierarchical applications. We'll also use it in chapter 15.

- A *Utility Application* defines a flip-side controller that has two pages, the first with an info button that allows you to call up the second page. This is the last view controller that we'll explore in chapter 15.

- An *OpenGL ES Application* is another minimalistic application. Its main difference from the Window-Based Application is that it includes GL frameworks, sends the `glView` messages to get it started, and otherwise sets certain GL properties. We won't get to GL until chapter 19, and even then we'll only touch on it very lightly.

11.4.3 Xcode tips and tricks

Before we leave Xcode behind, let's explore a few of the great features that it includes to make your coding easier. You can investigate these features in any of the projects that we've written so far.

EDITING WINDOW

You'll see a file's code in an editing window whenever you single-click on an .h or .m file. If this window isn't big enough, you can instead double-click to get a brand-new window.

The editing window includes a number of nice features, among them:

- *Autocompletion*—Whenever you write code in the editing window, Xcode will try to autocomplete words for you. This will include framework methods, your own methods, and even variable names. For methods, it goes a step further and shows you the standard arguments you should pass. If you don't like what you see, just keep typing, but if you do, hit the Tab key, and the rest will be inserted. We've torn out our hair way too many times due to misbehaving code that turned out to be the results of a typo. Xcode's autocompletion can easily resolve that problem—in advance.

- *Class controls*—Ctrl-click on the class name in an `@implementation` line and you'll see a Refactor option. Select this option to not only change the name of your class in all files, but to modify the names of the files for that class as well. Also see *variable controls* for a similar feature.

- *Code folding*—As with many modern IDE environments, you can fold your code, making it easier to read by hiding the contents of functions and/or comments. You can easily fold a function by clicking in the narrow gray bar to the left of your code. The functionality is also available in the View menu or by Ctrl-clicking inside the editing window.

- *Doc lookup*—Option-double-click any standard structure, method, or property, and you'll see information on it in an Xcode Workspace Guide window (which we discuss more in a moment). We think this is the best feature of the IDE, because it makes programming with otherwise unknown frameworks very simple.
- *Superclass lookup*—At the top of your editing window is a menu labeled "C." You can use this window to look up the superclass of your current object. Doing so will lead you to the superclass's header file, which will reveal its properties and methods.
- *Variable controls*—Click on a variable and you'll see a gray underline materialize; shortly after that you'll see a triangle appear to the right, allowing you to pull down a menu. From there you can jump straight to the definition of the variable or Edit All in Scope, which allows you to change the name of a variable within your current scope. Also see *class controls* for a similar feature.

ORGANIZER

You call up this window by choosing Window > Organizer. You can store references to a number of projects here, linking them in from the "+" menu that appears at the bottom of the window. In addition to easily accessing your projects, you can compile them in a variety of configurations and even see debugging logs and crash logs related to them.

XCODE WORKSPACE GUIDE

Open this window by selecting Help > Xcode Workspace Guide. Here you can access local copies of the documents that are found at developer.apple.com. To keep your copies of the docs up to date, you should "subscribe" to individual doc sets (such as "iPhone OS 2.1"). This will download the newest copies of the docs whenever they become available.

11.5 Summary

Xcode is ultimately the basis of all SDK programming. It's where you'll write the actual code that allows you to create, manipulate, and destroy objects. As we've seen in this chapter, it's quite easy to use Xcode to do pretty sophisticated things. This is in part due to Objective-C's elegant methods of messaging and in part due to the iPhone OS's massively library of classes.

However, there's another way to do things. Basic objects can also be created graphically, using Interface Builder. It allows you to lay out objects using a graphical UI that makes their arrangement a lot easier. That's going to be the basis of our next chapter, when we delve into the other side of SDK programming.

Using Interface Builder

12

This chapter covers

- Learning how Interface Builder works
- Writing a simple program using the interface
- Linking Interface Builder and Xcode

In the last chapter, you built a `labeledwebview` class that included both a label and a web view. As is typically the case with programmatic design, you had to crunch numbers to make sure all your objects fit correctly on the screen.

What if you didn't have to do that? What if you could lay out objects using a graphical design program and then immediately start using them in Xcode? With the SDK, you can, thanks to Interface Builder.

As we write this, Apple doesn't offer any extensive iPhone Interface Builder documentation. The "Interface Builder User Guide" contains some good information, but it's still more desktop-centric than we'd like. If you need more information than we provide here, you might still want to read that document, because it does have some iPhone specifics.

Because we consider Interface Builder to be an alternative to Xcode (in appropriate situations), our exploration of it will mirror the structure of last chapter's look at Xcode. We'll give an overview of the program, and then we'll put together a simple first project using it. Afterward, we'll explore a more complex but fundamental technology—connecting Interface Builder to Xcode. Finally, we'll briefly touch upon other functionality.

With all that said, what exactly is Interface Builder, and how does it work?

12.1 An introduction to Interface Builder

Interface Builder is a graphical development environment integrally tied in to Xcode. Whenever you write an Xcode project, it includes an .xib file that contains Interface Builder definitions for where graphical objects are placed. Each of the different Xcode templates comes with different objects prebuilt this way. Some of them have multiple, linked .xib files, with one file representing each separate screen of information.

We'll get into the specifics of how the two programs link over the course of this chapter. For now, be aware that Xcode and Interface Builder are designed to go together.

12.1.1 The anatomy of Interface Builder

You usually access Interface Builder by double-clicking an .xib file in your project. Your default .xib file is generally called MainWindow.xib. Clicking it brings up the file in Interface Builder, showing how default objects have been designed.

> **Nib vs. xib**
>
> You'll see talk of both .xib files and .nib files in this and later chapters. They're pretty much the same thing: a .nib file is a compiled .xib file. They'll appear to you as .xib files in Xcode, but some methods call them nib files, as we'll see later in this chapter, and Apple documents refer to a nib document window in Interface Builder; we've done the same here.

THE INTERFACE BUILDER WINDOWS

When you call up Interface Builder, you initially see three windows: the nib document window, the main display window, and the Library window. The fourth important window—the inspector window—doesn't appear by default, but you'll call it up pretty quickly. These windows are shown in figure 12.1.

The *nib document window* displays top-level objects, which are objects without a parent. A default MainWindow.xib file includes four such objects. The window object is the one real object here; you can play with it in Interface Builder and also link it out to Xcode. As you'd expect, this is the window object that was created by default in the templates you've used so far.

The other three top-level objects are all *proxies*, which means they're placeholders for objects not contained in Interface Builder. Generally, you can only see objects in Interface Builder that were created there; if you need to access something else in Interface Builder, you do so by creating a proxy.

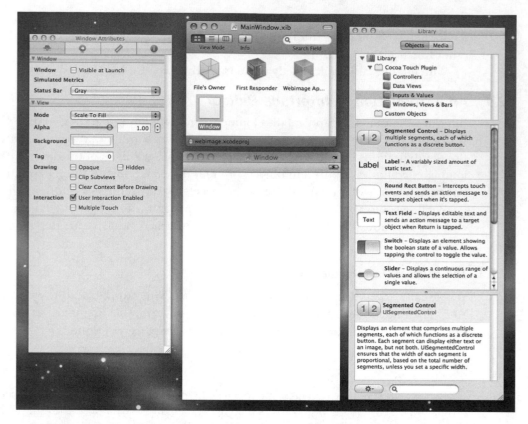

Figure 12.1 The nib document window (middle top) and the main display window (middle bottom) are two of the fundamental displays in Interface Builder. The Library (right) also comes up when you start Interface Builder. You need to call up the inspector (left) by hand.

The Webimage App Delegate is a proxy for the app delegate object. (This one is for a program you'll create shortly.) File's Owner refers to the object that manages the .xib file (usually either the application or a view controller), and First Responder refers to the top object in the responder chain (which we introduced in chapter 10 and will cover in depth in chapter 14). You'll meet these proxies again when we touch on IBOutlets.

The *main display window* shows what the .xib file currently looks like. Because we used the Window-Based Application template in Xcode, there's nothing here yet. If we'd used one of the other templates, you'd see tab bars or other prebuilt elements. In any case, this is where you arrange your user interface elements as you create them.

Together, the nib document and main display windows contain all the objects understood by Interface Builder (which will likely omit many objects you created in Xcode).

The *Library window* is where you can find all the UI elements you may want to add to your program. You can start exploring the library with a little mousing. Click the Library and Cocoa Touch Plugin toggles, and you'll see four main classes of UI elements:

- Controllers gives you different ways to manage your views.
- Data Views gives you different ways to display data.
- Inputs & Values gives you a variety of simple input mechanisms.
- Windows, Views & Bars gives you the core window and view objects, plus a variety of other elements.

The *inspector window* gives you access to a wide variety of information about an object and lets you change it; but as we said earlier, it doesn't open automatically. You can call up the inspector by choosing Tools > Inspector. Afterward, whenever you click an object, its data will appear in the inspector. By default, the inspector window has four tabs: Attributes, Connections, Size, and Identity.

We'll talk about everything on these tabs in depth when you start writing your first program, but for the meantime we want to introduce two additional core concepts: IBOutlets and IBActions.

IBOUTLETS AND IBACTIONS

In order for Interface Builder–created objects to be useful, Xcode must be able to access their properties and respond to actions sent to them. This is done with IBOutlets and IBActions.

You saw an IBOutlet in the last chapter, as part of the default header file for your first project. It looked like this:

```
@interface helloworldxcAppDelegate : NSObject <UIApplicationDelegate> {
    IBOutlet UIWindow *window;
}
```

An IBOutlet provides a link to an Interface Builder–created object. It's what you use to access that object's properties and methods.

You won't see an IBAction until we get to chapter 14, where we'll deal with events and actions, but it's similar. You declare a method in your @interface, including IBAction as its return:

```
- (IBAction)changeSlider:(id)sender;
```

An IBAction is a message that's executed when a specific action is applied to an Interface Builder–created object, such as when a slider moves or a button is clicked.

Figure 12.2 shows how these two types of actions link in to your Xcode files. As shown, all the declarations go into your header file. For IBActions, you also need to define the methods that should occur when the related actions happen. Those methods will go in your main source-code file.

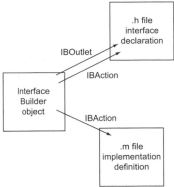

Figure 12.2 Through outlets and events, you can export Interface Builder information to different parts of your Xcode.

12.1.2 Simulating in Interface Builder

You can't compile your full program in Interface Builder, but you can choose File > Simulate Interface, which mocks up all your Interface Builder objects, but without any additional Xcode work. That can sometimes be useful, but more often you'll want to do a real compilation, which requires going back to Xcode and compiling there as normal.

12.2 Creating a first project in Interface Builder: pictures and the web

With the overview of Interface Builder out of the way, you're ready to create a simple first project. We want to keep expanding the cool things you're doing with the SDK, so in this example you'll add one more object to the set you've worked with so far. In addition to labels and web views, you'll also incorporate a picture. And because you're using Interface Builder, it will all be quicker and more efficient to put together.

To begin the project, create a window-based application and then double-click MainWindow.xib to call up the new project's Interface Builder objects. Right now there's nothing but a window, but you'll quickly change that.

12.2.1 Creating new objects

Imagine a program that uses an image as a background, sets up a web view on top of that, and has a label running above everything. We'll show you how easy it is to create those entirely usable objects in Interface Builder.

You'll find the Image View object under Data Views in the Library. Drag it over to your main window, and it'll quickly resize to suggest a full-screen layout. You should be able to arrange it to fit exactly over the screen, and then release your mouse button to let it go. One object created!

The Web View object is right under the Image View. Drag it over to the main window. If you move it toward the center of the view, you'll see dashed lines appear: they're intended to help you center your object. If you mouse over the middle of the screen, a dashed line will appear in each direction, forming a sort of crosshairs. When that happens, release the mouse button—you now have a web view in the middle of the screen. Two objects created!

Finally, select Label, which is under Inputs & Values. Drag it toward the top left of your screen, and let go. You're done! You now have three objects laid out in Interface Builder, as shown in figure 12.3.

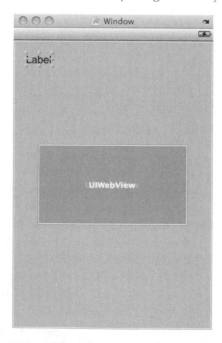

Figure 12.3 A few seconds of work results in three objects ready for use in Interface Builder.

You can now either manipulate these objects in Interface Builder or create IBOutlets to manipulate them in Xcode. We'll look at Interface Builder manipulations first, starting with the work you can do in the graphical interface.

12.2.2 Manipulating objects graphically

Interface Builder is fundamentally a graphic design program, so it makes sense that you can do some simple manipulation of your objects graphically. For example, if you want to change the name of your label, double-click it; you're given the option to fill in your own text. You can similarly move and resize most objects using the main window.

If you want to engage in more complex manipulations of your Interface Builder–created objects, you'll need to use the inspector window. If you don't already have it available, you can call it up by selecting Tools > Inspector.

12.2.3 Using the inspector window

As we noted in our overview of Interface Builder, the inspector window contains four tabs: Attributes, Connections, Size, and Identity. We'll look at each of these in turn, as we inspect the humble label.

THE ATTRIBUTES TAB

The Attributes tab contains all the basic information about your object. It will generally be your first stop whenever you want to modify an object that exists in Interface Builder. Figure 12.4 shows the Attributes tab for our label.

When we manipulated our label graphically, we changed the text to "My Apple Stock" for reasons that will become obvious shortly. You can see that this change has already been made in the label's attributes. You can set a lot of other properties via this single window, with no programming required.

Do you want your text to be a nice maraschino-cherry red? No problem: click the Text Color box. Doing so will lead you to a window that offers several ways to set colors. Choose the tab that allows selection by name, and you'll find maraschino cherry on the list. You can also set shadows, alignments, and a number of other text options from this panel.

Besides the label options, the Attributes tab contains several options that relate to the view—they're the UIView properties that most graphical objects inherit. You can change alpha transparency, background color, and a number of

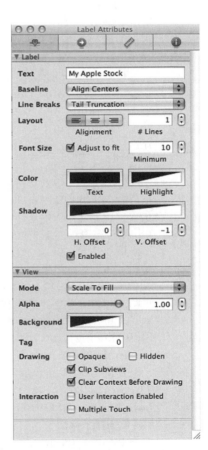

Figure 12.4 The Attributes tab shows all of an item's basic information.

other elements. For now, you can stop after having changed the color of the text and having generally seen what the Attributes tab can do.

The Attributes tab is available for all Interface Builder–generated objects, but it has different contents depending on the object in question. If you look at the attributes for the web-view and image-view objects you created, you'll see that you can set them in specific ways as well, but we'll save those for later. For now, we're concentrating on that label.

THE CONNECTIONS TAB

The second tab in the inspector window is the Connections tab. It shows an object's `IBOutlets` and `IBActions`.

The example label doesn't have any, which means it can't be accessed from Xcode. But this is fine; we're happy with how the label is set up in Interface Builder and don't need to adjust it during runtime.

We'll look at the Connections tab in depth when you use it in the next section.

THE SIZE TAB

You can use the Size tab to adjust the size and position of an object. Figure 12.5 shows the options you can change here.

This tab leads off with values for size and position. Not only can you change an object's starting point, but you can also define where that starting point is, relative to the object, using the grid at the top left. Width and height are available here too.

The Autosizing box controls how your object resizes its subviews when it resizes. For now, leave it be; it'll be of more importance when we talk about basic view controllers in chapter 13.

The Alignment section allows you to make multiple objects line up along an edge. Although you won't use them for your label, this is a frequent desire in layout. To make this work, you select all the objects you want to align (by Shift-clicking) and then choose the appropriate box in the Alignment section.

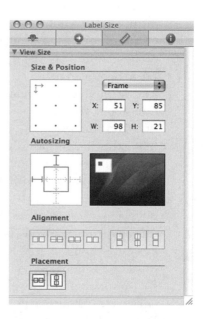

Figure 12.5   You can change an object's positioning and size from the Size tab.

Finally, the Placement section lets you align your current object relative to its parent. This works like the crosshairs you saw when you initially created your objects. If you click both placement buttons, your label would move to the center of the iPhone screen.

THE IDENTITY TAB

The final panel in the inspector window is the Identity tab. Like the Connections tab, it's not of much use for this label, but we'll cover its functionality for the sake of completeness. Figure 12.6 shows what it looks like.

For simple Interface Builder objects (like this example label), you only use the Interface Builder Identity section at the bottom of the Identity tab. This lets you name your object, which makes it easier to see what you're accessing in Interface Builder. It's strictly for your own use. For our purposes, we named the label "Hello Label".

The other three sections of the Identity tab are for more advanced purposes, and we'll give them more attention later in this chapter.

You use Class Identity if you want to link in an external object. You'll do this, for example, when you subclass view controllers and then want to make a link to that new controller in Interface Builder.

The Class Actions and Class Outlets sections show `IBAction` and `IBOutlet` declarations that you've made in your object's header file. For example, the app delegate object has a window `IBOutlet` (which you've seen several times), and your web-view object has a few system-defined actions. These are the things to which you can build connections.

For now, leave them be. They're not required for the label. But you have two more objects to work with in Interface Builder: the image view and the web view.

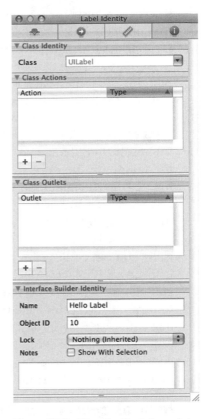

Figure 12.6 The Identity tab contains some deeper stuff that's mostly beyond the needs of this simple example program.

12.2.4 *Working with pictures*

We promised you that we were going to introduce a totally new object in this section: the image view. As with web views, we'll get more into the guts of images several chapters down the line; for now, we want to show how easy it is to work with an unfamiliar object type—like the image view—in Interface Builder.

ADDING THE IMAGE

To use an image in an SDK program, you need to load that image into your project. That means you drag the image file into Xcode's sidebar, alongside all your other files.

Once you've done that, you can go to your image view's Attributes tab in Interface Builder and type in the filename of your image file. In our case, it was apples.jpg. Most SDK programs use PNGs instead, but the JPG was much smaller, so we went with it. As soon as you enter this name, your picture should automatically pop up in Interface Builder's main window.

You then may wish to use the Attributes tab to change how the picture displays in the window (including automatically resizing it if you didn't build your image to be a specific size) or to adjust other elements. For example, we opted to change the image's alpha transparency to .5, to make it easier to see the text over the image.

If you want, you can now go out to Xcode and compile this program, which was built *entirely* in Interface Builder. You can see the results in figure 12.7.

It's clear that the program has a bit of a problem.

WHAT'S MISSING

The problem is that an unsightly white box is sitting in the middle of the display. That's the web view. If you inspect the Attributes tab for the web view, you'll see why we didn't do anything more with it: you can't set the starting URL from inside Interface Builder.

You *can* do other things in Interface Builder. Specifically, you can easily resize the window. We chose to set it to 280x391 pixels, which various Interface Builder guidelines suggested was the right size. We also opted to turn off the Scales Page to Fit option, which would make the web view act as if it had a viewport 980 pixels wide, like iPhone Safari. But to fill the web-view window, you have to access it from Xcode, which means building a new IBOutlet.

Figure 12.7　Combining graphics and text can be hard in some programming languages, but under the SDK it can be done entirely with Interface Builder.

12.3　*Building connections in Interface Builder*

As we've already discussed, an IBOutlet gives you the ability to access an Interface Builder–created object from within Xcode. This is critical when you want to reset properties in Xcode or when you want to set a property that's not available from within Interface Builder, as is the case with the web view's URL.

Creating this connection is a three-step process, as outlined in table 12.1.

Table 12.1　You can link together an Interface Builder object with a instance variable in Xcode through a few simple steps.

Step	Description
1. Declare your variable.	In Xcode, add an IBOutlet variable to the header file of the appropriate object. Save your Xcode files.
2. Connect your object.	In Interface Builder, drag a connection from your Interface Builder object to the appropriate Xcode top-level object, which should highlight. Select the appropriate variable name from the pop-up listing.
3. Code!	You can now access your Interface Builder object from Xcode, as if it were created there.

We'll now look at each of these steps in more depth.

12.3.1 *Declaring an IBOutlet*

You met IBOutlets in the previous chapter. They're declared like normal instance variables in your header file, but you precede them with the IBOutlet statement to show that the object was created in Interface Builder.

For the example web-view object, that means you need to update the header file of your app delegate class as shown in listing 12.1.

Listing 12.1 New IBOutlet to link to an Interface Builder object

```
@interface webimageAppDelegate : NSObject <UIApplicationDelegate> {
    IBOutlet UIWindow *window;
    IBOutlet UIWebView *myWebView;
}

@end
```

You've now finished the first step of connection building.

12.3.2 *Connecting an object*

At this point, you can build the physical connection from your object in Interface Builder to the IBOutlet in your Xcode. You start this process by bringing up the Connections tab for the object you want to connect: in this case, a web view.

Each web view built in Interface Builder comes with five potential connections built in. You can automatically define your web view's delegate in Interface Builder by creating a connection. You can also connect up a few actions—a topic we'll return to shortly. For now, you want to connect the object to Xcode.

You do that by clicking the New Referencing Outlet circle in the object and then dragging a line to the top-level object that holds your new IBOutlet statement. In this case, you drag a connection from the web view to the app-delegate proxy. If the top-level object has IBOutlets that are waiting to accept connections, it is highlighted as shown in figure 12.8.

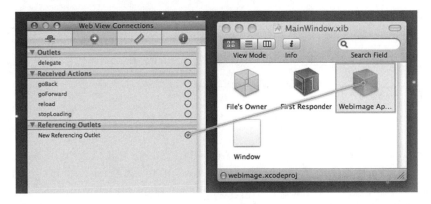

Figure 12.8 You need to drag and release to build a connection in Interface Builder.

When you mouse over to the correct top-level object, you should release the mouse button. A menu will pop up that lists all the available IBOutlets in that file. When you do this, you'll see the myWebView IBOutlet that you just built, plus a viewController, which you don't need to use. Select the appropriate outlet, click the mouse, and your object and your IBOutlet are connected.

At this point, you're done building a bridge between Interface Builder and Xcode. If you look at the Connections tab of the app delegate proxy, you'll see that it now acts as a referencing outlet for the web view.

If you want to look at both Connections tabs at the same time, you can do so by Ctrl-clicking each of the two objects to bring up stand-alone Connections panels, as shown in figure 12.9.

Through these Connections panels, you can see not only the reciprocal web view connection that you just built, but also the app delegate's existing connections: it acts as a delegate for the application (which is the .xib file's owner), and it acts as an outlet (which you already used) for a window.

Figure 12.9 You can Ctrl-click to access Connections panels if you want to see multiple connections at the same time.

Now, all you need to do is fall back on your existing Xcoding skills to make the web view do what you want.

12.3.3 *Coding with IBOutlets*

Heading back into Xcode, you only need to input a single line of new code to get the web view working, as shown in listing 12.2.

> **Listing 12.2   IBOutlet to access the object's usual properties**

```
- (void)applicationDidFinishLaunching:(UIApplication *)application {

    [myWebView loadRequest:[NSURLRequest requestWithURL:
        [NSURL URLWithString:
            @"http://quote-web.aol.com/?syms=AAPL&w=280&h=391..."]]];

// Full, dynamic URL not included, for readability
// Put a 280x391 sized page of your choice into the message

    [window makeKeyAndVisible];
}
```

Note that you don't have to allocate the web view, nor do you have to initialize it, nor do you have to add it as a subview of your window; all those details are taken care of by Interface Builder. But once you link to the object via an outlet, you can access it like any object you created yourself.

As we promised earlier, we'll take a more complete look at how web views work in chapter 20. But we wanted to include them here to demonstrate (again) how easy it is to incorporate an unfamiliar object into your code using Interface Builder.

In addition, a web view provides a nice example of client-server integration between web design and the SDK—a topic that we first touched on in chapter 2 and that turns out to be pretty simple to attain using the iPhone OS. By linking to a URL that sends dynamic content to your iPhone, you can make a sophisticated, always up-to-date program despite only designing the display engine on the SDK side of things.

Figure 12.10 shows what the final product looks like.

That brings us to the end of our Apple stock example. It presented some fundamental uses of Interface Builder that you'll encounter again and again. In particular, creating objects in Interface Builder and then hooking them up to Xcode will likely become a regular part of your SDK coding experience, so you should make sure you're entirely familiar with that process.

Before we leave this Interface Builder introduction, we'll touch on some additional functionality that will become more important in the chapters to come.

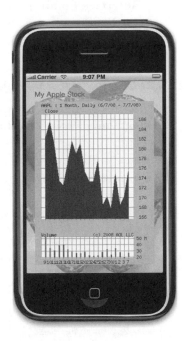

Figure 12.10   An image, a label, and a dynamic web view are put together in Interface Builder with only a single line of Xcode required (plus a couple of declarations).

12.4   Other Interface Builder functionality

When we finished our look at Xcode in chapter 11, we had a few slightly more advanced topics that we wanted to talk about: foundational stuff that nonetheless lay beyond the bounds of our first simple program. The same is true with Interface Builder.

12.4.1   Building other connections

As we've noted, IBOutlets are only one of a few major types of connections that you can build into Interface Builder. You've also seen delegate connections, which allow you to tie one object to another for purposes of delegation. You do this without ever setting the delegate property in Xcode: you link the one object to the other in Interface Builder.

The third major type of connection is the IBAction, which creates a connection that causes a method in your class file to be run whenever an action is sent to a UIControl. Building IBAction connections is almost exactly like building IBOutlet connections: you create a method declaration in the appropriate header file that uses IBAction as its return, and then you connect it to an appropriate action in

Interface Builder; each object comes with all of the potential actions built in. Finally, in Xcode, you write the method that specifies what happens when the control is modified.

We'll get into this in more depth in chapter 14.

12.4.2   *Creating external objects*

In the last chapter, you built your first subclass: the labeledwebview class. You'll be building lots more subclasses in your SDK work, and you'll often want to connect them into Interface Builder so you can connect outlets and actions to them. How do you access these new classes in Interface Builder when they don't appear in the Library window? It turns out that you need to use the Identity tab, which you've already met (in figure 12.6).

Table 12.2 outlines the two-step process.

Table 12.2   Creating a new proxy object to link to in Interface Builder takes a couple of steps.

Step	Description
1. Create a new object.	From the Controllers section of the Library, drag an appropriate object to the nib document window.
2. Change the class.	Open the Identity inspector tab, and change the class name to your new class.

We say that you start the process with an "appropriate object." For a totally new object, this will probably be the blank object, but if you're making a subclass of an existing object, you should start with the object you're subclassing from.

Once you type your new subclass name into your object's Class field, things will automatically be linked up. You'll use this technique in future chapters.

12.4.3   *Initializing Interface Builder objects*

Eventually, you'll realize that you want to do some initialization when an Interface Builder object is created. But if you try to build your setup into a standard init method, it won't work. As we've mentioned, Interface Builder objects use a special init method called initWithCoder:. You must create it by hand, as in listing 12.3.

Listing 12.3   initWithCoder: is required to initialize Interface Builder objects

```
- (id)initWithCoder:(NSCoder *)decoder {
    if (self = [super initWithCoder:decoder]) {
        // Setup code goes here
    }
    return self;
}
```

Other than its decoder argument (which you should be able to ignore), it should work like any other init method.

12.4.4 Accessing .xib files

Finally, we come to the .xib file. We've taken it pretty much for granted so far, but there are ways in which you can specify a different .xib file than MainWindow.xib, and even ways in which you can specify the use of multiple .xib files.

THE MAIN FILE

The main .xib file is defined in Info.plist, which you've seen is an XML file. You can look at its contents in Xcode, or you can read the XML from the command line. It's easy to find where the main .xib file (or rather, its compiled .nib twin) is defined:

```
<key>NSMainNibFile</key>
<string>MainWindow</string>
```

If you want to change the name of your main .xib file, do it here, using either Xcode or a command-line editor, but you shouldn't need to.

MULTIPLE FILES

As we've mentioned, an .xib file can only lay out the contents of a single program screen. Although this has been fine for our programs so far, it becomes a limitation when you want to create more-complex programs. Fortunately, it's easy to build multiple .xib files into a single program.

New .xib files are usually loaded through view controllers, which will be the topic of the next chapter. As we've discussed previously, a view controller tends to control a page full of objects, and it makes sense that they use .xib files to help manage that. To use a new .xib file for a new page in your program, all you need to do is associate the new .xib file with the appropriate view controller.

The easiest way to do that is through Interface Builder—provided that's where your view controller was created. A view controller's Attributes tab includes a space for you to enter an .xib filename.

Alternatively, if you want to create a view controller in Xcode, you can link in a new .xib file through its `init` method:

```
FlipsideViewController *viewController =
  [[FlipsideViewController alloc]
     initWithNibName:@"FlipsideView" bundle:nil];
```

If you're a little fuzzy on the concept of view controllers, don't worry, because we're about to dive into the topic wholeheartedly. For now, note this connection between view controllers and Interface Builder.

12.4.5 Creating new .xib files

Now that you understand how to load additional .xib files, you may wish to create new ones. You do so from within Interface Builder, where you can choose File > New to begin. Afterward, you'll be asked to choose a template: Application, Empty, View, or Window. You'll most often create new .xib files for view controllers, in which case you should select View.

To make your new .xib file part of your existing project, save the .xib file to the main project directory. You'll be asked if you want to add it to the project; answer Yes.

12.5 *Summary*

In the previous chapter, we showed you how to create some simple programs using Xcode alone. But Xcode is only half of the development environment that Apple provides. You also have access to Interface Builder, a powerful graphic design program that allows you to lay out objects by mousing and then to link those objects back to Xcode for use there.

The example in this chapter, which combines text, graphics, and web content, is both more complex than anything you wrote by hand in the previous chapter and a lot easier to put together. That's the power of Interface Builder, and it's something you'll take full advantage of as you make your way through the SDK over the course of this book.

Although you now have the two fundamental tools of the SDK well in hand, we've neglected two of the SDK building blocks you'll use to create projects: view controllers and events. In the next three chapters, we'll cover those topics, and in the process, we'll complete our look at the iPhone OS classes you'll use in almost any SDK program you write.

Creating basic view controllers

13

This chapter covers

- Understanding the importance of controllers
- Programming barc view controllers
- Utilizing table view controllers

In the last two chapters, we've offered a hands-on look at the two core tools used to program using the SDK: Xcode and Interface Builder. In the process, we haven't strayed far from the most fundamental building block of the SDK, the view, whether it be a `UILabel`, a `UIWebView`, or a `UIImageView`.

Ultimately, the view is only part of the story. As we mentioned when we looked at the iPhone OS, views are usually connected to view controllers, which manage events and otherwise take the controller role in the MVC model. We're now ready to begin a three-part exploration of what that all means.

In this chapter, we're going to look at basic view controllers that manage a single page of text. With that basis, we can look at events and actions in chapter 14, correctly integrating them into the MVC model. Finally, in chapter 15, we're going to return to the topic of view controllers to look at advanced classes that can be used to connect up several pages of text.

Over the course of our two view controller chapters (13 and 15), we're going to offer code samples that are a bit more skeletal than usual. That's because we want to provide you with the fundamental, reusable code that you'll need to use the controllers on your own. So, consider chapters 13 and 15 more of a reference—though a critical one. We'll be making real-world use of the controllers in the rest of this book, including when we look at events and actions in chapter 14.

13.1 *The view controller family*

When we first talked about view controllers in chapter 10, we mentioned that they come in several flavors. These run from the bare bones `UIViewController`, which is primarily useful for managing autorotation and for taking the appropriate role in the MVC model, to the more organized `UITableViewController`, on to a few different controllers that allow navigation across multiple pages.

All of these view controllers—and their related views—are listed in table 13.1.

Table 13.1   There are a variety of view controllers, giving you considerable control over how navigation occurs in your program

Object	Type	Summary
`UIViewController`	View controller	A default controller, which controls a view; also the basis for the flipside controller, which appears only as an Xcode template, not as a `UIKit` object.
`UIView`	View	Either your full screen or some part thereof. This is what a view controller controls, typically through some child of `UIView`, not this object itself.
`UITableViewController`	View controller	A controller that uses `UITableView` to organize data listings.
`UITableView`	View	A view that works with the `UITableViewController` to create a table UI. It contains `UITableCells`.
`UITabBarController`	View controller	A controller that works with a `UITabBar` to control multiple `UIViewControllers`.
`UITabBar`	View	A view that works with the `UITabBarController` to create the tab bar UI. It contains `UITabBarItems`.
`UINavigationController`	View controller	A controller used with a `UINavigationBar` to control multiple `UIViewControllers`.

Table 13.1 There are a variety of view controllers, giving you considerable control over how navigation occurs in your program *(continued)*

Object	Type	Summary
`UINavigationBar`	View	A view that works with `UINavigation-Controller` to create the navigation UI.
Flipside controller	View controller	A special template that supports a two-sided `UIViewController`.

`ABPeoplePickerNavigationController`
`ABNewPersonViewController`
`ABPersonViewController`
`ABUnknownPersonViewController`
`UIImagePickerController`
	View controller	Modal view controllers that allow interaction with sophisticated user interfaces for the Address Book and the iPhone photos roll.

As we've already noted, we'll be discussing these view controllers in two different chapters. Here we're going to look at the single-page view controllers: `UIViewController` and `UITableViewController`. In chapter 15, we're going to look at the multi-page view controllers: `UITabBarController`, `UINavigationController`, and the flipside controller. This is a clear functional split: the single-page controllers exist primarily to support the controller role of the MVC model, whereas the multipage controllers exist primarily to support navigation, and may even delegate MVC work to a simpler view controller lying below them. (As for the modal controllers, we'll get to them when we cover the appropriate topics in chapters 16 and 18.)

Though we've programmed without view controllers to date, they're an important part of SDK programming. You *could* write an SDK program without them, but every SDK program *should* include them, even if you use a bare-bones view controller to manage the rotation of the iPhone screen.

13.2 The bare view controller

The plain view controller is simple to embed inside your program. By why would you want to use a view controller? That's going to be one of the topics that we're going to cover here. Over the course of this section, we'll look at how view controllers fit into the view hierarchy, how you create them, how you expand them, and how you make active use of them. Let's get started with the most basic anatomical look at the view controller.

13.2.1 The anatomy of a view controller

A view controller is a `UIViewController` object that sits immediately above a view (of any sort). It, in turn, sits below some other object as part of the tree that ultimately goes back to an application's main window. This is shown in figure 13.1.

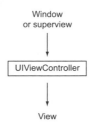

Figure 13.1 A bare view controller shows view-controlling at its simplest: it sits below one object and above another.

When we move on to advanced view controllers, in chapter 15, we'll see that the use of a bare view controller can grow more complex. Bare view controllers will often sit beneath advanced view controllers, to take care of the individual pages that the advanced view controller allows navigation among.

Looking at the iPhone OS's class hierarchy, we can see that the UIViewController is a direct descendent of NSObject. That means that it doesn't get any of the functionality of UIResponder or UIView, which you find in most other UIKit objects. It's also the parent object of all the other view controllers we'll be discussing. Practically, this means that the lessons learned here also apply to all the other controllers.

But learning about how a view controller works leaves out one vital component: how do you create it?

13.2.2 Creating a view controller

The easiest way to incorporate a plain view controller into your project is to select a different template when you create it. The View-Based Application template should probably be your default template for programming from here on out, because it comes with a view controller built in.

As usual, the template's work is primarily done through Interface Builder. Once you create a new project (which we've called "viewex" for the purpose of this example) you can verify this by looking up the view controller's IBOutlet command in the program's app delegate header file:

```
IBOutlet viewexViewController *viewController;
```

The app delegate's source code file further shows us that the view controller's view has already been hooked up to the main window:

```
[window addSubview:viewController.view];
```

This view is a standard UIView that's created as part of the template. Though a view controller only has one view, that view may have a variety of subviews, spreading out into a hierarchy. We're going to show you how to add a single object beneath the view in a moment, and we're going to make more complete use of it in the next chapter. But before we get there, we want to step back and look at how you could create a view controller by hand, if you needed to.

Creating another view controller is simple. First, in Interface Builder, drag a View Controller from the Library to your xib document window. Alternatively, in Xcode, you can alloc and init an object from the UIViewController class.

NOTE   Increasingly, we're going to assume that you're doing work through Interface Builder and using appropriate templates, but the same methods for object creation that we learned in the last couple of chapters remain available for all objects.

Second, note that the previous IBOutlet command shows that the controller isn't instantiated directly from the UIViewController class, but rather from its own subclass, which has its own set of files (`viewexViewController.{h|m}`), named after our example project's name. This is standard operating procedure.

Because we want a view controller to do event management, we'll often need to modify some of the controller's standard event methods, so we require our own subclass. To start, our view controller class files are mostly blank, but Xcode helpfully highlights a number of standard view controller methods that we might want to modify.

Once you've finished creating a bare view controller, you're mostly ready to go, but there's some slight opportunity to modify the view controller for your specific program, and that's what we're going to cover next.

13.2.3 *Building up a view controller interface*

In order to correctly use a view controller, you need to build your view objects as subviews of the view controller, rather than subviews of your main window or whatever else lies above it. This is easy in both Xcode and Interface Builder.

THE XCODE SOLUTION

The view controller class file gives you access to a pair of methods that can be used to set up your view controller's views. If the view controller's view is linked to an .xib file, you should use `viewDidLoad`, which will do additional work after the .xib is done loading; if it isn't created from inside Interface Builder (IB), you should instead use `loadView`.

Before you do any of this, your view controller will always start off with a standard `UIView` as its one subview. But by using these methods, you can instead create view controller's view as you see fit, even creating a whole hierarchy of subviews if you so desire.

Listing 13.1 shows how you could add a simple `UILabel` to your view controller using `viewDidLoad`. We've chosen a humongous font that gets automatically sized down so that later we can show off how rotation and resizing work.

Listing 13.1 You can add views to an IB-created view controller inside `viewDidLoad`

```
- (void)viewDidLoad {

    [super viewDidLoad];

    UILabel *myLabel = [[UILabel alloc]
        initWithFrame:[[UIScreen mainScreen] bounds]];
    myLabel.adjustsFontSizeToFitWidth = YES;
    myLabel.font = [UIFont fontWithName:@"Arial" size:60];
    myLabel.textAlignment = UITextAlignmentCenter;
    myLabel.text = @"View Controllers!";
    myLabel.backgroundColor = [UIColor grayColor];

    [self.view addSubview:myLabel];        ❶

    [myLabel release];
}
```

The `self.view` line is the only one of particular note ❶. It connects your label object as a subview of the view controller's `UIView`.

This example is also noteworthy because it's the first time you've definitively moved outside of your app delegate for object creation. You could have done this object creation over in the app delegate, but that's often sloppy programming. Now that you've got view controllers, you'll increasingly be doing most of your work in those class files. This not only better abstracts your object creation, but it also kicks off your support of the MVC model, because you've now got controllers instantiating the views they manage. Watch for a lot more of this in the future. We're also going to briefly return to the `viewDidLoad` and `loadView` methods when we talk about the bigger picture of the view controller life cycle, shortly.

THE INTERFACE BUILDER SOLUTION

In the last chapter, we noted that view controllers often have their own .xib files, allowing you to have one .xib file for each page of content. That's exactly what's going on in the program you created from the View-Based Application template. At creation, the template contains two .xib files, MainWindow.xib and viewexViewController.xib.

The MainWindow.xib file contains a view controller and a window. It also contains the all-important link to the second .xib file. If you click the view controller's Attribute tab, it'll helpfully show you that the controller's content is drawn from viewexView-Controller(.xib). This is shown in figure 13.2.

Now that you understand the hierarchy of .xib files that's been set up, how do you make use of them? In order to create an object as a subview of the view controller, you need to place it inside the .xib file that the view controller manages—in this case viewexViewController.xib. So, to add a `UILabel` to your view controller, you call up the viewexView-Controller.xib file and then drag a label to the main display window, which should represent the existing view. Afterward, you can muck with the label's specifics in the inspector window, as usual.

Practically, there's nothing more you need to do to set up your basic view controller, but there are still a few runtime fundamentals to consider.

Figure 13.2   To hook up a new .xib file to a view controller, enter its name in the view controller's attributes under NIB Name.

13.2.4 *Using your view controller*

If you've chosen to use a standard view controller, it should be because you're only managing one page of content, not a hierarchy of pages. In this situation, you don't need your view controller to do a lot, but your view controller is still important for three things, all related to event management:

- It should act as the hub for controlling its view and subviews, following the MVC model. To do this, it needs easy access to object names from its hierarchy.
- It should control the rotation of its view, which will also require resizing the view in rational ways. Similarly, it should report back on the orientation of the iPhone if queried.
- It should deal with life-cycle events related to its view.

We've split these main requirements up into six topics, which we'll cover in turn.

PUTTING THE MVC MODEL TO USE

Though we've talked about the Model-View-Controller (MVC) architectural pattern, you haven't yet put it to real use. To date, it's instead been a sort of abstract methodology for writing programs. But now that you're ready to use view controllers, you can start making use of MVC as a real-world ideal for programming.

As you'll recall, under MVC, the *model* is your back-end data and the *view* is your front-end user interface. The *controller* is what sits in between, accepting user input and modifying both of the other entities. The view controller should take the role of the controller in the MVC, as the name suggests. We're going to get into this more in the next chapter, but we can say confidently that event and action control *will* happen through the view controller.

We can say this confidently because we're pretty much going to be forced into using MVC. A view controller will automatically be set up to access and modify various elements of views that sit under it. For example, the view controller has a title property that is intended to be a human-readable name for the page it runs. In chapter 15, we'll learn that tab bars and navigation bars automatically pick up that information for their own use. In addition, we'll often see view controllers automatically linked up to delegate and datasource properties, so that they can respond to the appropriate protocols for their subviews.

So, when you start seeing view controllers telling other objects what to do, look at it from the MVC lens. You should also think about MVC, yourself, as you start to program more complex projects using view controllers.

FINDING RELATED ITEMS

If a view controller is going to act as a controller, it needs to have easy access to the objects that lay both above and below it in the view hierarchy. For this purpose, the view controller contains a number of properties that can be used to find other items that are connected to it. They're listed in table 13.2.

Table 13.2 When you start connecting a view controller up to other things, you can use its properties to quickly access references to those other objects.

Property	Summary
modalViewController	Reference to a temporary view controller, such as the Address Book and photo roll controllers that we discuss in chapter 16 and 18.
navigationController	Reference to a parent of the navigation controller type.
parentViewController	Reference to the immediate parent view controller, or nil if there is no view controller nesting.
tabBarController	Reference to a parent of the tab bar controller type.
tabBarItem	Reference to a tab bar item related to this particular view.
view	Reference to the controller's managed view. The view's subviews property may be used to dig further down in the hierarchy.

These properties will primarily be useful when we move on to advanced view controllers, because they're more likely to link multiple view controllers together. We're mentioning them here because they're related to the idea of MVC and because they're UIViewController properties that will be inherited by all other types of controllers.

For now, we're going to leave these MVC-related properties aside and get into some of the more practical things you can immediately do with a view controller, starting with managing view rotation.

ROTATING VIEWS

Telling your views to rotate is simple. In your view controller class file, you'll find a method called shouldAutorotateToInterfaceOrientation:. In order to make your application correctly rotate, all you need to do is set that function to return the Boolean YES, as shown in listing 13.2.

Listing 13.2 Enabling autorotation in a view controller

```
- (BOOL) shouldAutorotateToInterfaceOrientation:
     (UIInterfaceOrientation) interfaceOrientation {
   return YES;
}
```

At this point, if you compile your program, you'll find that when you rotate your iPhone, the label shifts accordingly. Even better, because you set its font size to vary based on the amount of space it has, it gets larger when placed horizontally. This is a simple application of modifying your content based on the iPhone's orientation.

There is one additional thing that you should consider when rotating your views: whether they will resize to account for the different dimensions of the new screen.

RESIZING VIEWS

When you change your iPhone's orientation from portrait to landscape, you're changing the amount of space for displaying content—the device goes from 320x480 to 480x320. As we saw, when you rotated your label, it automatically resized, but this doesn't happen without some work.

A UIView (not the controller!) contains two properties that affect how resizing occurs. The autoresizesSubviews property is a Boolean that determines whether autoresizing occurs or not. By default it's set to YES, which is why things worked correctly in the first view controller example. If you instead set it to NO, your view would stay the exact same size when a rotation occurs. In this case, your label would stay 320 pixels wide despite now being on a 480-pixel wide screen.

Once you've set autoresizesSubviews, which says that resizing *will* occur, your view will look at its autoresizingMask property to decide *how* it should work. The autoresizingMask property is a bitmask that you can set with the different constants listed in table 13.3.

If you wanted to modify how your label resized from within Xcode, you could do so by adding the following two lines to viewDidLoad:

```
myLabel.autoresizesSubviews = YES;
myLabel.autoresizingMask = UIViewAutoresizingFlexibleHeight |
  UIViewAutoresizingFlexibleWidth;
```

Table 13.3 `autoresizingMask` **properties allow you to control how your views resize.**

Constant	Summary
`UIViewAutoresizingNone`	No resizing
`UIViewAutoresizingFlexibleHeight`	Height resizing allowed
`UIViewAutoresizingFlexibleWidth`	Width resizing allowed
`UIViewAutoresizingFlexibleLeftMargin`	Width resizing allowed to left
`UIViewAutoresizingFlexibleRightMargin`	Width resizing allowed to right
`UIViewAutoresizingFlexibleBottomMargin`	Height resizing allowed to bottom
`UIViewAutoresizingFlexibleTopMargin`	Height resizing allowed to top

Note again that these resizing properties apply to a *view*, not to the view controller. You could apply them to any view that you've seen to date. There has been little need for them before you started rotating things.

Modifying the way resizing works is even easier from within Interface Builder. If you recall, the Resize tab of the inspector window contains an Autosizing section, as shown in figure 13.3.

You can click six different arrows that correspond to the six resizing constants other than None. Highlighting an individual arrow turns that type of resizing on. The graphic to the right of these arrows serves as a nice guide to how resizing will work.

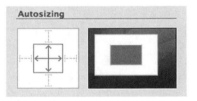

Figure 13.3 Interface Builder will graphically depict exactly what autoresizing looks like.

CHECKING ORIENTATION

Now that you've got an application that can rotate at will, you may occasionally want to know what orientation a user's iPhone is sitting in. This is done by querying the `interfaceOrientation` view controller property. It will be set to one of four constants, as shown in table 13.4.

You don't have to have a view controller to look this information up. A view controller's data is kept in tune with orientation values found in the `UIDevice` object—a

Table 13.4 The view controller's `interfaceOrientation` property tells you the current orientation of an iPhone.

Constant	Summary
`UIInterfaceOrientationPortrait`	iPhone is vertical, right side up
`UIInterfaceOrientationPortraitUpsideDown`	iPhone is vertical, upside down
`UIInterfaceOrientationLandscapeLeft`	iPhone is horizontal, tilted left
`UIInterfaceOrientationLandscapeRight`	iPhone is horizontal, tilted right

useful object that also contains other device information, like your system version. We'll talk about it a bit in chapter 17.

MONITORING THE LIFE CYCLE

We've covered the major topics of loading, rotating, and resizing views within a view controller. With that under our belt, we can now look at the life-cycle events that might relate to these topics.

We saw life-cycle events in chapter 10, where we examined methods that alerted us to the creation and destruction of the application itself, and some individual views. Given that one of the purposes of a controller is to manage events, it shouldn't be a surprise that the UIViewController has several life-cycle methods of its own, as shown in table 13.5.

Table 13.5 You can use the view controller's event handler methods to monitor and manipulate the creation and destruction of its views.

Method	Summary
loadView	Creates the view controller's view if it is not loaded from an .xib file
viewDidLoad	Alerts you that a view has finished loading; this is the place to put extra startup code if loading from an .xib file
viewWillAppear:	Runs just before the view loads
viewWillDisappear:	Runs just before a view disappears —because it's dismissed *or* covered
willRotateToInterfaceOrientation:duration:	Runs when rotation begins
didRotateToInterfaceOrientation:	Runs when rotation ends

We've already met loadView and viewDidLoad, which are run as part of the view controller's setup routine and which we used to add extra subviews. The viewWillAppear: message is sent afterward. The rest of the messages are sent at the appropriate times, as views disappear and rotation occurs.

Any of these methods could be overwritten to provide the specific functionality that you want when each message is sent.

OTHER VIEW METHODS AND PROPERTIES

The view controller object contains a number of additional methods that can be used to control exactly how rotation works, including controlling its animation and what header and footer bars slide in and out. These are beyond the scope of our introduction to view controllers, but information about them can be found in the UIViewController class reference.

That's our look at the bare view controller. You now know not only how to create your first view controller, but also how to use the fundamental methods and properties that you'll find in *every* view controller. But the other types of view controller also have special possibilities all their own. We're going to look at these, starting with the one other view controller that's intended to control a single page of data: the table view controller.

13.3 The table view controller

Like the plain view controller, the table view controller manages a single page. Unlike the plain view controller, it does so in a structured manner.

Our discussion of the table view controller is going to mirror the discussion we just completed of the bare view controller. We're going to examine its place in the view hierarchy, and then we're going to see how to create it, modify it, and use it at runtime.

Let's get started with this new view controller's anatomy.

13.3.1 The anatomy of a table view controller

The table view controller's setup is slightly more complex than that of the bare view controller. A `UITable-VieuController` controls a `UITableView`, which is an object that contains some number of `UITableView-Cell` objects arranged in a single column. This is shown in figure 13.4.

By default, the controller is both the delegate and the data source of the `UITableView`. As we've previously discussed, these properties help a view hand off events and actions to its controller. The responsibilities for each of these control types is defined by a specific protocol: `UITableViewDelegate` declares which messages the table view controller must respond to, and `UITableViewDataSource` details how it must provide the table view with content. You can look up these protocols in the same library that you've been using for class references.

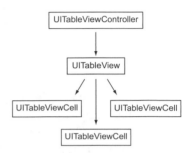

Figure 13.4 A table view controller controls a table view and its collection of cells.

Of all of the view controllers, the table view controller is the trickiest to create on its own, for reasons that we'll see momentarily.

13.3.2 Creating a table view controller

None of the Xcode templates support a plain table view. That's because a table view is usually linked up with a navigation controller, as we'll see in chapter 15. If you want to create a plain table view controller, you'll need to do so by hand, starting with the Window-Based Application template. Table 13.6 shows the entire process.

The project-creation, object-creation, and object-linking steps pretty much follow the lessons that you've already learned. You have to create the subclass for the table view controller because the class file is where you define what the table view contains; we'll cover this in more depth shortly.

Note that you use two of the more "advanced" Interface Builder techniques that you learned in chapter 12: first linking in a new class (by changing the Identity tab)

Table 13.6   Creating a table view controller is simple, but it involves several steps.

Step	Description
1. Create a new project.	Open a Window-Based Application.
2. Create a table view controller.	In Xcode, create a new file containing a subclass of `UITableView-Controller`; we've chosen `RootViewController` for our new class name. Import your new header into your app delegate. In Interface Builder, drag a Table View Controller to your nib display window. Change the class of your controller to your new Xcode subclass in the Identity tab of the inspector window.
3. Link your Interface Builder object.	In Xcode, create an `IBOutlet` for your Interface Builder object in the app delegate header file. In Interface Builder, link an outlet from your table view controller to the `IBOutlet` in the app delegate object using the Connections tab of the inspector window.
4. Connect your controller.	Link the controller's view to your main window.

and then creating a new connection from it to your app delegate (via the Connections tab). As a result, there ends up being two connections from Interface Builder to Xcode. On the one hand, the Interface Builder-created table view controller depends on your `RootViewController` files for its own methods; on the other hand, your app delegate file links to the controller (and eventually to the methods) via its outlet. This two-part connection to Interface Builder is very common, and you should make sure you understand it before moving on.

As usual, you could have elected to create this object solely in Xcode, by using an alloc-init command:

```
UITableViewController *myTable = [[RootViewController alloc]
    initWithStyle:UITableViewStylePlain];
```

Listing 13.3 finishes off the table-creation process, showing the simple code you'll use to link in the table's view in step 5 of the process.

Listing 13.3   After the warm up, creating a table view controller takes a few lines of code

```
- (void)applicationDidFinishLaunching:(UIApplication *)application {
    [window addSubview:myTable.view];
    [window makeKeyAndVisible];
}
```

Note that you link up your table view controller's view—not the controller itself—to your window. We've seen in the past that view controllers come with automatically created views. Here, the view is a table view.

If you want to see how that table view works, you can now go back into Interface Builder and click the table view to get its details. As shown in figure 13.5, it's already got connections created for its dataSource and delegate properties.

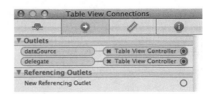

At this point, you could compile your program, but the result would be pretty boring; consisting only of an empty list. Next you need to fill that table with content.

Figure 13.5   A look at the connections automatically created for a controller's table view.

13.3.3   *Building up a table interface*

As the data source, the controller needs to provide the view with its content. This is why you created a subclass for your table view controller and why every one of your table view controllers should have its own subclass: each will need to fill in its data in a different way.

We've already mentioned that the UITableViewDataSource protocol declares the methods that your table view controller should be paying attention to in order to correctly act as the data source. The main work of filling in a table is done by the tableView:cellForRowAtIndexPath: method. When passed a row number, this method should return the UITableViewCell for that row of your table.

Before you can get to that method, though, you'll need to do some work. First, you must define the content that will fill your table. Then, you must define how large the table will be. Only afterward can you fill in the table using the tableView:cellForRow-AtIndexPath: method.

Besides these major table view elements, we'll also cover two optional variants that can change how a table looks: accessory views and sections.

CREATING THE CONTENT

There are numerous SDK objects that you could use to create a list of data your table should contain. In the future, we'll talk about SQLite databases and pulling RSS data off the internet. For now we're going to stay with the SDK's simpler objects. The most obvious are NSArray, which produces a static indexed array; NSMutableArray, which creates a dynamic indexed array; and NSDictionary, which defines an associative array.

For this example of table view content creation, we have elected to create an NSArray containing an NSDictionary that itself contains color names and UIColor values. As you can probably already guess, you're going to fill this skeletal table view example out with something like the color selector that you wrote back when we were learning iUI in chapter 5. The code required to create your content array is shown in listing 13.4.

Listing 13.4 An array of selective arrays is perfect for table creation

```
- (id)initWithCoder:(NSCoder *)decoder {          ❶

    self = [super initWithCoder:decoder];         ❷

    colorList = [NSArray arrayWithObjects:
        [NSDictionary dictionaryWithObjectsAndKeys:
            @"brownColor",@"titleValue",
            [UIColor brownColor],@"colorValue",nil],
        [NSDictionary dictionaryWithObjectsAndKeys:
            @"orangeColor",@"titleValue",
            [UIColor orangeColor],@"colorValue",nil],        ❸
        [NSDictionary dictionaryWithObjectsAndKeys:
            @"purpleColor",@"titleValue",
            [UIColor purpleColor],@"colorValue",nil],
        [NSDictionary dictionaryWithObjectsAndKeys:
            @"redColor",@"titleValue",
         [UIColor redColor],@"colorValue",nil],
        nil];
    [colorList retain];          ❹

    return self;          ❺
}
```

This sort of setup should be done as part of an initialization method. Note that you'll be using `initWithCoder:` ❶, which is the required init method if you're working with an Interface Builder object (and it's an alternative to the view controller's `viewDidLoad` method). Any `initWithCoder:` method should call its parent ❷ and return itself ❺, like a normal init.

The array and dictionary creations are pretty simple ❸. The Apple class references contain complete information on how to create and manipulate these objects, but, in short, you can create an `NSArray` as a listing of objects ending in a `nil`, and you can create an `NSDictionary` using pairs of values and keys, ending in a `nil`. Here, you're creating an array containing four dictionaries, each of which will fill one line of your table.

You also have to think a bit about memory management here. Because your array was created with a class factory method, it's going to get released when it goes out of scope. In order to use this array elsewhere in your class, you not only need to have defined it in your header file, but you also need to send it a `retain` message to keep it around ❹. You'll `release` it in your `dealloc` method, elsewhere in the class files.

BUILDING YOUR TABLE CELLS

Once you've got a data backend set up for your table, you need to edit three methods in your table view controller file: two that define the table and one that fills it, as shown in listing 13.5. We'll explain each of these in turn.

Listing 13.5 Three methods control how your table is created and runs

```
- (NSInteger)numberOfSectionsInTableView:(UITableView *)tableView {
    return 1;                                                        ❶
}

- (NSInteger)tableView:(UITableView *)tableView          ❷
        numberOfRowsInSection:(NSInteger)section {
```

```
        return colorList.count;
}
- (UITableViewCell *)tableView:(UITableView *)tableView
        cellForRowAtIndexPath:(NSIndexPath *)indexPath {        ❸

    static NSString *MyIdentifier = @"MyIdentifier";

    UITableViewCell *cell = [tableView
        dequeueReusableCellWithIdentifier:MyIdentifier];

    if (cell == nil) {
        cell = [[[UITableViewCell alloc] initWithFrame:CGRectZero
            reuseIdentifier:MyIdentifier] autorelease];
    }
    cell.textColor= [[colorList objectAtIndex:indexPath.row]        ❹
        objectForKey:@"colorValue"];
    cell.text = [[colorList objectAtIndex:indexPath.row]        ❺
        objectForKey:@"titleValue"];

    return cell;
}
```

(near `return colorList.count;`) ❷

All of these methods should appear by default in the table view controller subclass that you created, but you may need to make changes to some of them to accommodate the specifics of your table.

The first method is `numberOfSectionsInTableView:` ❶. Tables can optionally include multiple sections, each of which has its own index of rows, and each of which can have a header and a footer. For this first example, you're creating a table with one section, but we're going to look at multiple sections before we finish this chapter.

The second method, `tableView:numberOfRowsInSection:` ❷, reports the number of rows in this section. Here you'll return the size of the array that you created. Note that you ignore the `section` variable because you only have one.

The third method, `tableView:cellForRowAtIndexPath:` ❸, takes the table set up by the previous two methods and fills its cells one at a time. Though this chunk of code looks intimidating, most of it will be sitting there waiting for you the first time you work with a table. In particular, the creation of `UITableViewCell` will be built in. All you need to do is set the values of the cell before it's returned. Here you use your `NSDictionary` to set the text color ❹ and the text content ❺ of the cell. Also note that this is your first use of the `NSIndexPath` data class. It encapsulates information on rows and sections.

You may want to change more things than text content and color. Table 13.7 lists all of the cell features that you might want to muck with at this point.

Table 13.7 You can modify your table cells in a variety of ways.

Property	Summary
`font`	Sets the cell text's font using `UIFont`
`lineBreakMode`	Sets how the cell's text wraps using `UILineBreakMode`

Table 13.7   You can modify your table cells in a variety of ways. *(continued)*

Property	Summary
text	Sets the content of a cell to an NSString
textAlignment	Sets the alignment of cell's text using the UITextAlignment constant
textColor	Sets the color of the cell's text using UIColor
selectedTextColor	Sets the color of selected text using UIColor
Image	Sets the content of a cell to a UIImage
selectedImage	Sets the content of a selected cell to UIImage

Using all of these properties, you can make each table cell look entirely unique, depending on the needs of your program.

ADDING ACCESSORY VIEWS

Though we didn't in our color selector example, you can optionally set accessories on cells. Accessories are special elements that appear to the right of each list item. Most frequently, you'll set accessories using an accessoryType constant that has four possible values, as shown in table 13.8.

Table 13.8   A cell accessory gives additional information.

Constant	Summary
UITableViewCellAccessoryNone	No accessory
UITableViewCellAccessoryDisclosureIndicator	A normal chevron: >
UITableViewCellAccessoryDetailDisclosureButton	A chevron in a blue button: ⊙
UITableViewCellAccessoryCheckmark	A checkmark: ✔

An accessory can be set as a property of a cell:

```
cell.accessoryType = UITableViewCellAccessoryDetailDisclosureButton;
```

The normal chevron is usually used with a navigation controller, the blue chevron is typically used for configuration, and the checkmark indicates selection.

There is also an accessoryView property, which lets you undertake the more complex task of creating an entirely new view to the right of each list item. You create a view and then set the accessoryView to that view:

```
cell.accessoryView = [[myView alloc] init];
```

There's an example of this in chapter 16, where you'll be working with preference tables.

ADDING SECTIONS

Our example showed how to display a single section worth of cells, but it would be trivial to rewrite the functions to each offer different outputs for different sections within the table. Because of Objective-C's ease of accessing nested objects, you could prepare for this by nesting an array for each section inside a larger array:

```
masterColorList = [NSArray arrayWithObjects:colorList,otherColorList,nil];
```

Then you'd return the count from this uber-array for the `numberOfSections:` method:

```
return masterColorList.count;
```

You'd similarly return a subcount of one of the subarrays for the `tableView:number-OfRows:` method:

```
return [[masterColorList objectAtIndex:section] count];
```

Finally, you'd pull out content from the appropriate subarray when filling in your cells using the same type of nested messaging.

Once you're working with sections, you can also think about creating headers and footers for each section. Figure 13.6 shows what the revised application would look like so far, including two different sections, each of which has its own section header.

How do you create those section headers? As with all of the methods we've seen that fill in table views, the section header messages and properties show up in the `UITableViewDataSource` protocol reference.

In order to create section headers, you write a `tableView:titleForHeaderInSection:` method. As you'd expect, it renders a header for each individual section.

An example of its use is shown in listing 13.6, though you could probably have done something fancier instead, such as building the section names directly into your array.

Figure 13.6 Section headers can improve the usability of table views.

Listing 13.6 To set a section header, define the `return` in the appropriate method

```
- (NSString *)tableView:(UITableView *)tableView
    titleForHeaderInSection:(NSInteger)section {
  if (section == 0) {
    return @"SDK Colors";
  } else if (section == 1) {
    return @"RGB Colors";
  }
  return 0;
}
```

You can similarly set footers, and otherwise manipulate sections according to the protocol reference.

There's still more to the table view controller. Not only do you have to work with data when you're setting it up, you also have to do so when it's in active use, which usually occurs when your user selects individual cells.

13.3.4 *Using your table view controller*

We're not going to dwell too much on the more dynamic possibilities of the UITableViewController here. For the most part, you'll either use it to hold relatively static data (as we do here) or you'll use it to interact with a navigation controller (as we'll see in chapter 15). But before we finish up with table view controllers, we're going to look at one other fundamental: selection.

SELECTED CELLS

If you try out the sample "tableex" application that you've been building throughout section 13.3, you'll see that individual elements in a table view can be selected.

In table 13.7 we've already seen that there are some properties that apply explicitly to selected cells. For example, the following maintains the color of your text when it's selected, rather than changing it to white, as per the default:

```
cell.selectedTextColor =
    [[[masterColorList objectAtIndex:indexPath.section]
        objectAtIndex:indexPath.row] objectForKey:@"colorValue"];
```

As with the rest of the cell definitions, you'd put this line of code in tableView:cellForRowAtIndexPath:. You should also note that this is another example of using nested arrays to provide section- and row-specific information for a table list.

The tableView:didSelectRowAtIndexPath: method is the most important for dealing with selections. This method appears in the UITableViewDelegate protocol and tells you when a row has been selected. The message includes an index path, which, as we've already seen, contains both a row and a section number.

Listing 13.7 shows a simple example of how you might use this method to checkmark items in your list.

Listing 13.7 We can see when table cells are selected and act accordingly

```
- (void)tableView:(UITableView *)tableView
    didSelectRowAtIndexPath:(NSIndexPath *)indexPath {

    [[tableView cellForRowAtIndexPath:indexPath]
        setAccessoryType:UITableViewCellAccessoryCheckmark];

}
```

You are able to easily retrieve the selected cell by using the index path, and then you use that information to set the accessory value. We'll make more use of cell selection in chapter 15, when we talk about navigation controllers.

OTHER TABLE METHODS AND PROPERTIES

In the class reference, you'll also find information on how you can edit, delete, and insert cells, scroll your table view, and otherwise actively use it.

13.4 Summary

View controllers are the most important building blocks of the SDK that we had not yet seen. As was explained in this chapter, they sit atop views of all sorts and control how those views work. Even in this chapter's simple examples, we saw some real-world examples of this control, as our view controllers managed rotation, filled tables, and reacted to selections.

Now that we're getting into user interaction, we're ready to examine how it works in a bit more depth, and that's the focus of our next chapter. We'll examine the underpinnings of user interaction: events and actions.

14

Monitoring events and actions

This chapter covers

- The SDK's event modeling
- How events and actions differ
- Creating simple event- and action-driven apps

In the previous chapter you learned how to create the basic view controllers that fulfill the controller role of an MVC architectural model. You're now ready to start accepting user input, since you can now send users off to the correct object. Users can interact with your program in two ways: by using the low-level event model or by using event-driven actions. In this chapter, you'll learn the difference between the two types of interactions. Then we'll look at notifications, a third way that your program can learn about user actions.

Of these three models, it's the events that provide the lowest-level detail (and which ultimately underlie everything else), and so we'll begin with events.

14.1 An introduction to events

We briefly touched on the basics of event management in chapter 10, but as we said at the time, we wanted to put off a complete discussion until we could cover them in depth; we're now ready to tackle that job.

Part 1 of this book, dealing with web design, outlined how events tend to work on the iPhone. The fundamental unit of user input is the touch: a user puts his finger on the screen. This could be built into a multi-touch or a gesture, but the touch remains the building block on which everything else is constructed. It's thus the basic unit that we're going to be examining in this chapter.

14.1.1 *The responder chain*

When a touch occurs in an SDK program, you have to worry about something you didn't have to think about on the web: who responds to the event. That's because SDK programs are built of tens—perhaps hundreds—of different objects. Almost all of these objects are subclasses of the `UIResponder` class, which means they contain all the functionality required to respond to an event. So who gets to respond?

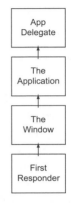

The answer is embedded in the concept of the *responder chain*. This is a hierarchy of different objects that are each given the opportunity, in turn, to answer an event message.

Figure 14.1 shows an example of how an event moves up the responder chain. It starts out at the *first responder* of the *key window*, which is typically the view where the event occurred—in other words, where the user touched the screen. As we've already noted, this first responder is probably a subclass of `UIResponder`—which is the class reference you'll want to look to for a lot of responder functionality.

Figure 14.1 Events on the iPhone are initially sent to the first responder, but then travel up the responder chain until someone accepts them.

Any object in the chain may accept an event and resolve it, but whenever that doesn't occur the event moves further up the list of responders. From a view, an event will go to its superview, then *its* superview, until it eventually reaches the `UIWindow` object, which is the superview of everything in your application. It's useful to note that from the `UIWindow` downward, the responder chain is the view hierarchy turned on its head, so when you're building your hierarchies, they'll be doing double duty.

Although figure 14.1 shows a direct connection from the first responder to the window, there could be any number of objects in this gap in a real-world program.

Often the normal flow of the responder chain will be interrupted by *delegation*. A specific object (usually a view) delegates another object (usually a view controller) to act for it. We already saw this put to use in our table view in chapter 13, but we now understand that a delegation occurs as part of the normal movement up the responder chain.

If an event gets all the way up through the responder chain to the window and it can't deal with an event, then it moves up to the `UIApplication` itself, which most frequently punts the event to its own delegate: the *application delegate*, an object that we've been using in every program to date.

First responders and keyboards

Before we leave this topic of responders behind, we'd like to mention that the first responder is a very important concept. Because this first responder is the object that can accept input, it'll sometimes take a special action to show its readiness for input. This is particularly true for text objects like `UITextField` and `UITextView`, which (if editable) will pop up a keyboard when they become the first responder. This has two immediate consequences.

If you want to pop up a keyboard for the text object, you can do so by turning it into the first responder:

```
[myText becomeFirstResponder];
```

Similarly, if you want to get rid of a keyboard, you must tell your text object to stop being the first responder:

```
[myText resignFirstResponder];
```

We'll discuss these ideas more when we encounter our first editable text object toward the end of this chapter.

Ultimately you, the programmer, will be the person who decides what in the responder chain will respond to events in your program. You should keep two factors in mind when you make this decision: how classes of events can be abstracted together at higher levels in your chain, and how you can build your event management using the concepts of MVC.

At the end of this section we'll address how you can subvert this responder chain by further regulating events, but for now let's build on its standard setup.

14.1.2 *Touches and events*

Now that you know a bit about how events find their way to the appropriate object, we can dig into how they're encoded by the SDK. First we want to offer a caveat: usually you won't need to worry about this level of detail because the standard UIKit objects will generally convert low-level events into higher-level actions for you, as we discuss in the second half of this chapter. With that said, let's look at the nuts and bolts of event encoding.

The SDK abstracts events by combining a number of touches (which are represented by `UITouch` objects) into an event (which is represented by a `UIEvent` object). An event typically begins when the first finger touches the screen and ends when the last finger leaves the screen. In addition, it should generally only include those touches that happened in the same view.

In this chapter we'll work mainly with `UITouch`es (which make it easy to parse single-touch events) and not with `UIEvent`s (which are more important for parsing multi-touch events). Let's lead off with a more in-depth look at each.

UITOUCH REFERENCE

A UITouch object is created when a finger is placed on the screen, moves on the screen, or is removed from the screen. A handful of properties and instance methods can give you additional information on the touch, as detailed in table 14.1.

Table 14.1   Additional properties and methods can tell you precisely what happened during a touch event.

Method or property	Type	Summary
phase	Property	Returns a touch phase constant, which indicates whether touch began, moved, ended, or was canceled
tapCount	Property	The number of times the screen was tapped
timestamp	Property	When the touch occurred or changed
view	Property	The view where the touch began
window	Property	The window where the touch began
locationInView:	Method	Gives the current location of the touch in the specified view
previousLocationInView:	Method	Gives the previous location of the touch in the specified view

Together the methods and properties shown in table 14.1 offer considerable information on a touch, including when and how it occurred.

Only the phase property requires additional explanation. It returns a constant that can be set to one of five values: UITouchPhaseBegan, UITouchPhaseMoved, UITouch-PhaseStationary, UITouchedPhaseEnded, or UITouchPhaseCancelled. You'll often want to have different event responses based on exactly which phase a touch occurred in, as you'll see in our event example.

UIEVENT REFERENCE

To make it easy to see how individual touches occur as part of more complex gestures, the iPhone SDK organizes UITouches into UIEvents. Figure 14.2 shows how these two sorts of objects interrelate.

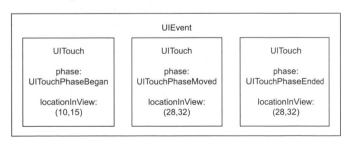

Figure 14.2   UIEvent objects contain a set of related UITouch objects.

Just as with the UITouch object, the UIEvent object contains a number of properties and methods that you can use to figure out more information about your event, as described in table 14.2.

Table 14.2   The encapsulating event object has a number of methods and properties that let you access its data.

Method or property	Type	Summary
timestamp	Property	The time of the event
allTouches	Method	All event touches associated with the receiver
touchesForView:	Method	All event touches associated with a view
touchesForWindow:	Method	All event touches associated with a window

The main use of a UIEvent method is to give you a list of related touches that you can break down by several means. If you want to get a list of every touch in an event, or if you want to specify just gestures on a certain part of the screen, then you can do that with UIEvent methods. This ends our discussion of event containers in this chapter.

Note that all of these methods compact their touches into an NSSet, which is an object defined in the Foundation framework. You can find a good reference for the NSSet at Apple's developer resources site.

THE RESPONDER METHODS

So, how do you actually access these touches and/or events? You do so through a series of four different UIResponder methods, which are summarized in table 14.3.

Each of these methods has two arguments: an NSSet of touches that occurred during the phase in question and a UIEvent that provides a link to the entire event's worth of touches. You can choose to access either one, as you prefer; as we've said, we're going to be playing with the bare touches. With that said, we're now ready to dive into an actual example that demonstrates how to capture touches in a real-life program.

Table 14.3   The UIResponder methods are the heart of capturing events.

Method	Summary
touchesBegan:withEvent:	Reports UITouchPhaseBegan events, for when fingers touch the screen
touchesMoved:withEvent:	Reports UITouchPhaseMoved events, for when fingers move across the screen
touchesEnded:withEvent:	Reports UITouchPhaseEnded events, for when fingers leave the screen
touchesCancelled:withEvent:	Reports UITouchPhaseCancelled events, for when the phone is put up to your head, or other events that might cause an external cancellation

14.2   *A touching example: the event reporter*

Our sample application for events is something we call the event reporter, which will offer a variety of responses depending on how and when the iPhone screen is touched. We have two goals with our sample program.

First, we want to show you a cool and simple application that you can write using events, one that should get you thinking about everything you can do.

Second, we want to show some of the low-level details of how events work in a visual form. Therefore, if you actually take the trouble to code and compile this program, you'll gain a better understanding of how the various phases work as well as how tapping works.

You'll kick off this development process by creating a project named eventreporter that uses the View-Based Application template. That means you'll start with a view controller already in place. We'll also use this example to show how an MVC program can be structured.

14.2.1 Setting things up in Interface Builder

By now you should be comfortable enough with Interface Builder that you can set up all of your basic objects using it. For this program we've decided that we want to create three new objects: two button-shaped objects that will float around the screen to mark the beginning and end of touches, plus a status bar to go at the bottom of the screen and describe a few other events when they occur.

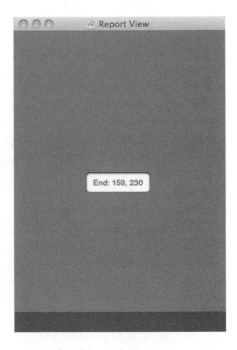

Figure 14.3 Two `UITextFields` (one of them hidden) and one `UILabel`, all set against an aluminum-colored background, complete the object creation we'll need for our eventreporter project.

Because you want all of our new objects to lie beneath the view controller in the view hierarchy, you call up the view controller's own .xib file, eventreporterViewController.xib. As usual, you'll add your new objects to the Main Display window that represents the view controller's view.

All of your work in Interface Builder is, of course, graphical, so we can't show the code of this programming process. However, we have included a quick summary of the actions you should take. The results are shown in figure 14.3.

- Set the background color of the `UIView` to an attractive aluminum color. You do this on the Attributes tab of the Inspector window, as you would with most of your work in this project.
- Create a `UILabel`, stretch it across the bottom of the screen, and set the color to be steel. Also, clear out its text so that it doesn't display anything at startup.
- Create two `UITextFields`. This class of objects is generally used to accept input, but we opted to use the objects for our pure display purposes because we liked their look. (Don't worry; we'll show how to use the full functionality of a `UIText-Field` toward the end of this chapter.)

- Center each UITextField at the center of the screen using Interface Builder's handy positioning icons. This location's coordinates will be 159, 230.
- For each UITextField, input text that lists its starting position; this will later be updated by the program as the text field moves. Deselect the user interaction–enabled option for each UITextField so that users can't manipulate them.

The process takes longer to explain than it takes to accomplish. You'll have a working interface in just a couple of minutes.

CREATING IBOUTLETS

Because you'll modify all three of these objects during the course of your program's runtime, you need to link them to variables inside Xcode. You'll want to link everything to your controller, since it'll be taking care of updates, as is appropriate under the MVC model.

The tricky thing here is that the view controller doesn't seem to appear in our eventreporterViewController.xib file—at least not by that name. Fortunately, there's a proxy for it. Since the view controller is what loads up the .xib, it appears as the file's owner in the nib document window. You can therefore connect objects to the view controller by linking them to the file's owner proxy. This is a common situation, since view controllers frequently load additional .xib files for you.

Listing 14.1 shows your view controller's header file, eventreportViewController.h, following the addition of these IBOutlets. The listing also contains a declaration of a method that you'll use later in this project.

> **Listing 14.1   An IB-linked header**

```
@interface eventreporterViewController : UIViewController {

    IBOutlet UITextField *startField;
    IBOutlet UITextField *endField;
    IBOutlet UILabel *bottomLabel;
}

- (void)manageTouches:(NSSet *)touches;

@end
```

To finish up this process, you connect your Interface Builder objects to the IBOutlets, using the procedures described in chapter 12.

14.2.2 *Preparing a view for touches*

Touch events can only be captured by UIView objects. Unfortunately, as of this writing, there's no way to automatically delegate those touches to a view controller. Therefore, in order to manage touch events using the MVC model, you'll typically need to subclass a UIView, capture the events there, and then send messages to the view controller.

In your project you'll create a new object class, reportView, which is a subclass of UIView. You then link that new class into the view controller's existing view through Interface Builder. You open up eventreporterViewController.xib, go to the Identity tab for the view object that you've been using, and change its name from UIView to reportView, just as you did in chapter 13 when you created a table view controller subview.

Any new methods that you write into reportView, including methods that capture touch events, will be now reflected in your view. To clarify this setup, figure 14.4 shows the view hierarchy that you've built for your eventreporter project.

Figure 14.4 Working primarily in Interface Builder, we've connected up six objects that we'll be using to report iPhone events.

With a brand-new UIView subclass in hand, you can now write methods into it to capture touch events and forward them on to its controller. This code, which appears in reportView.m, is shown in listing 14.2.

Listing 14.2 A collection of methods report touches in UIViews

```
- (void) touchesBegan:(NSSet *)touches withEvent:(UIEvent *)event {
    [self.nextResponder manageTouches:touches];
}

- (void) touchesEnded:(NSSet *)touches withEvent:(UIEvent *)event {
    [self.nextResponder manageTouches:touches];
}

- (void) touchesMoved:(NSSet *)touches withEvent:(UIEvent *)event {
    [self.nextResponder manageTouches:touches];
}
```

The code in listing 14.2 is pretty simple. You're just filling in standard methods so that your program will have the responses you want when those messages are sent. The overall structure of these methods reminds us of several important facts about events.

First, as promised, there are a variety of responder methods. Each of them reports *only* the events for their specific phase. So, for example, the touchesBegan:with-Event: method would only have UITouchPhaseBegan touches in it. In forwarding on these touches we could have kept the different phases distinct, but we've instead decided to throw everything together and sort it out on our own on the other side.

Second, we'll comment one final time that these methods send us two pieces of information: a set of touches and an event. They are partially redundant, and which one you work with will probably depend on the work you're doing. If you're not doing complex multi-touch events, then the NSSet of touches will probably be sufficient.

Third, note that you're sending the touches to the view controller by way of the nextResponder method. As you'll recall, the responder chain is the opposite of the view hierarchy at its lower levels, which means in this case the nextResponder of reportView is the UIViewController. We would have preferred to have the UIView-Controller just naturally respond to the touches messages, but we made use of our responder chain in the next best way. As of this writing, the compiler warns that next-Responder may not know about the manageTouches method, but it will; this warning can be ignored.

We'll see some other ways to use the `nextResponder` method toward the end of our discussion of events.

AN ASIDE ON THE TEXT FIELDS AND LABEL

If you were to actually code in this example, you'd discover that this program correctly responds to touch events even when the touches occurred atop one of the text fields or the label at the bottom of the page. How does your program manage that when you only built event response into the `reportView`?

The answer is this: it uses the responder chain. The text fields and the label don't respond to the event methods themselves. As a result, the events get passed up the responder chain to the `reportView`, which *does* leap on those events, using the code we've just seen.

14.2.3 *Controlling your events*

Intercepting touches and forwarding them up to the view controller may be the toughest part of this code. Once the events get to the view controller, they run through a simple method called `manageTouches:`, as shown in listing 14.3.

> **Listing 14.3   `manageTouches`, which accepts inputs and changes views**

```
::eventreporterViewController.m::

- (void)manageTouches:(NSSet *)touches {

   for (UITouch *touch in touches) {                        ❶
      if (touch.phase == UITouchPhaseBegan) {               ❷
         CGPoint touchPos = [touch locationInView:self.view];   ❸
         startField.center = touchPos;
         startField.text = [NSString stringWithFormat:            ❹
            @"Begin: %3.0f,%3.0f",touchPos.x,touchPos.y];
      } else if (touch.phase == UITouchPhaseMoved) {        ❷
         bottomLabel.text = @"Touch is moving ...";
      } else if (touch.phase == UITouchPhaseEnded) {        ❷
         if (touch.tapCount > 1) {                          ❺
            bottomLabel.text = [NSString stringWithFormat:
               @"Taps: %2i",touch.tapCount];
         } else {
            bottomLabel.text = [NSString string];
         }
         CGPoint touchPos = [touch locationInView:self.view];
         endField.center = touchPos;
         endField.text = [NSString stringWithFormat:
            @"End: %3.0f,%3.0f",touchPos.x,touchPos.y];
      }
   }
}
```

Touches are sent as an `NSSet`, which can be broken apart in a number of ways, as described in the `NSSet` class reference. Here, you'll use a simple `for … in` construction ❶ that lets you look at each touch in turn.

Once you get a touch, the first thing you do is determine what phase it arrived in. Originally you could have determined this information based on which method a touch arrived through, but since we combined everything you have to fall back on the phase property. Fortunately, it's easy to use. You just match it up to one of three constants ❷, and that determines which individual actions your program undertakes.

Having different responses based on the phase in which a touch arrive is common—which is in fact why the event methods are split up in the first place. Our example demonstrates this with some distinct responses: you move your start field when touches begin, you move your end field when touches end, and you update the bottom label in both the moved and ended phases.

In your UITouchPhaseBegan response, you delve further into your touch's data by using the locationInView: method to figure out the precise coordinates where a touch occurred ❸. You're then able to use that data to reposition your text field and to report the coordinates in the text field ❹. You later do the same thing in the UITouchPhaseEnded response.

Finally, you take a look at the tapCount in the UITouchPhaseEnded response ❺. This is generally the best place to look at taps since the iPhone now knows that the user's finger has actually come off the screen. As you can see, it's easy to both run a command based on the number of taps and to report that information.

Figure 14.5 shows what your event responder looks like in action. You should imagine a finger that set down on the space where the begin text field is sitting and that is currently moving across the screen.

And with that, your event reporter is complete. Besides illustrating how a program can respond to touches, we have highlighted how the MVC model can be used in a real application.

Your project contained four views: a reportView, a UILabel, and two UITextFields. It was tempting to process events in the reportView itself, especially since you had to create a subclass anyway, but instead you pushed the events up to the view controller, and in doing so revealed *why* you'd want to do MVC modeling.

Since it takes on the controller role, you gave the view controller access to all of its individual objects, and therefore you didn't have to try to remember what object knew about what other object. Tying things into the view controller, rather than scattering them randomly across your code, made your project that much more readable and reusable, which is what most architectural and design patterns are about.

Figure 14.5 Your event responder uses a few graphical elements to report events as they occur.

14.3 *Other event functionality*

Before we complete our discussion of events entirely, we'd like to cover a few more topics of interest. We're going to explore how to regulate the report of events in a variety of ways and then describe some deficiencies in the event model.

14.3.1 *Regulating events*

As we mentioned earlier, there are some ways that you can modify how events are reported (and whether they are at all). As you'll see, three different objects give you access to this sort of regulation: `UIResponder`, `UIView`, and `UIApplication`. We've listed all the notable options we're going to discuss in table 14.4.

Table 14.4 Properties in various objects allow for additional control of when events are monitored.

Method or property	Type	Summary
nextResponder	UIResponder method	Returns the next responder in the chain by default but can be modified
hitTest:withEvent:	UIView method	Returns the deepest subview containing a point by default but can be modified
exclusiveTouch	UIView property	A Boolean set to NO by default; controls whether other views in the same window are blocked from receiving events
multipleTouchEnabled	UIView property	A Boolean set to NO by default; controls whether multi-touches after the first are thrown out
beginIgnoringInteractionEvents	UIApplication method	Turns off touch event handling
endIgnoringInteractionEvents	UIApplication method	Turns on touch event handling
isIgnoringInteractionEvents	UIApplication method	Tells whether the application is ignoring touch events

Since `UIView` is a subclass of `UIResponder`, you'll generally have access to the methods from both classes in most `UIKit` objects. You'll need to do some additional work to access the `UIApplication` methods.

UIRESPONDER REGULATION

You've already seen that `UIResponder` is the source of the methods that let you capture events; as shown here, it's also the home of the methods that control how the responder chain works.

Most of the responder chain–related methods won't be directly used by your code, instead typically appearing deep in frameworks. `becomeFirstResponder` and resignFirstResponder (which control who the first responder is) and canBecome-FirstResponder, canResignFirstResponder, and isFirstResponder (which return Booleans related to the information in question) all typically fall into this category.

It's the last UIResponder method, nextResponder, which may be of use in your programs. As defined by UIResponder, nextResponder just returns the next responder, per the normal responder chain. We used it in our example to pass our touches up.

If you want to change the normal order of the responder chain, you can do so by creating your own nextResponder function in a subclass. This new function will override its parent method, and thus will allow your program to take a different path up your own responder chain.

UIVIEW REGULATION

When you move into the UIView class methods, you can take the opposite approach by overriding hitTest:withEvent:. This method is passed a CGPoint and an event, and by default it returns the deepest subview that contains the point. By writing a new method, you can cause your responder chain to start at a different point.

The two UIView properties that we noted both work as you'd expect. exclusive-Touch declares that the view in question is the only one that can receive events (which is an alternative way that we could have managed our eventreporter example, where we didn't want anything but the reportView to accept events). Meanwhile, multiple-TouchEnabled starts reporting of multi-touch events, which are otherwise ignored.

UIAPPLICATION REGULATION

Finally we come to the UIApplication methods. These lie outside of our normal hierarchy of objects, and thus we can't get to them from our view objects. Instead we need to call them directly from the UIApplication object as shown here:

```
[[UIApplication sharedApplication] beginIgnoringInteractionEvents];
```

As you may recall from chapter 11, sharedApplication is a UIApplication class method that provides us with a reference to the application object. Typically, we've used its return as the receiver for the beginIgnoringInteractionEvents message.

Each of the three methods listed under UIApplication works as you'd expect once you know the secret to accessing them.

14.3.2 *Other event methods and properties*

We've spent a lot of time on events, but at the same time we've only scratched the surface. We have mixed feelings on the subject.

On the one hand, events give you low-level access to the unique user input allowed by the iPhone. Since much of this book is about how the iPhone is unique, we'd like to delve into it much further.

On the other hand, you won't be using events that much. That's because you usually won't need this sort of low-level control over your user input. Instead, you'll use the iPhone's many control objects (and thus actions) in order to accept almost all user input.

As a result, this chapter has offered you a compromise: a solid look at how events work that should suffice for those times when you do need to descend to touch management, but not all of the intricacies. The thing that we've most clearly left out is how to work with multi-touch events. For that, we point you, as usual, to the Apple iPhone developer website. There's a good tutorial on multi-touch events available as part of the iPhone OS Programming Guide that you should read if you're one of that smaller percentage of developers—such as programmers creating games and novelties—who might need access to multi-touches and more complex gestures.

14.4   An introduction to actions

So if you're not usually going to be programming directly with events, how will you access user input? The answer is by using actions. You'll typically depend on preexisting text views, buttons, and other widgets to run your programs. When using these objects, you don't have to worry about raw events at all. Instead, you can build your programs around control events and actions that are generated by UIControls.

14.4.1   The UIControl object

When we were working with events, we found that the UIResponder class held many of the methods critical for event control. Similarly, we can access a lot of the methods important to SDK controls through the UIControl class.

UIControl is a child of UIView (and thus UIResponder). It is the parent of important user interface controls such as UIButton, UISwitch, UIPageControl, UISegmented-Control, UISlider, and UITextField. It is *not* used for some other control-looking objects such as UISearchBar, so you should check the Apple class references before trying to use its functionality. Also note that the higher-level UIControl class can't be used on its own; it just defines the common methods used by its children.

The UIControl class contains several properties that control its basic setup, such as enabled (which determines whether it's on), highlighted (which determines its visual state), and selected (which sets Boolean state for appropriate sorts of controls, such as switches). You can also directly access a control's touch events with beginTrackingWith-Touch:withEvent:, continueTrackingWithTouch:withEvent:, and endTracking-WithTouch:withEvent:, methods that work in a similar way to the event response functions that we played with in UIResponder. But we won't be using these methods at all, because they don't represent the simple advantages that you'll see when using control objects. For that, we turn to UIControl's action-target mechanism.

14.4.2   Control events and actions

The UIControl object introduces a whole new event-handling infrastructure that takes touch events of the sort that we might have directly handled in the previous section and (eventually) converts them into simple actions, without you having to worry about the specifics of how a user accessed your control. The complete sequence of events is outlined in figure 14.6.

When a touch event arrives at a UIControl object (via normal dispatching along the responder chain), the control does something unique. Inside the standard

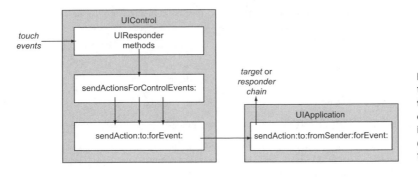

Figure 14.6 `UIControl` objects take standard touch events and turn them into actions that are dispatched by `UIApplication`.

`UIResponder` methods that we used in the previous section (such as `touches-Began:withEvent:`), a `UIControl` object turns standard touch events into special control events.

These control events broadly describe how the user has interacted with the controls rather than just recording gestures. For example, they might report that a button has been pushed or a slider moved. They're divided into three categories: touch events, editing events, and a slider event. The touch events describe how a user's finger interacted with the control; the editing events describe changes to a `UITextField`; and the `ValueChanged` event describes changes to a `UISlider`.

These control events are all enumerated in a bitmask that's defined in the `UIControl` object. An almost complete listing of them—including some composite control events—can be found in table 14.5. We've left out only a few `Reserved` values.

Table 14.5 `UIControl` objects recognize a number of special events.

Value	Summary
`UIControlEventTouchDown`	A finger touch.
`UIControlEventTouchDownRepeat`	A repeated finger touch (with `tapCount > 1`).
`UIControlEventTouchDragInside`	A finger movement ending inside the control.
`UIControlEventTouchDragOutside`	A finger movement ending just outside the control.
`UIControlEventTouchDragEnter`	A finger movement that enters the control.
`UIControlEventTouchDragExit`	A finger movement that exits the control.
`UIControlEventTouchUpInside`	A finger removed from the screen inside the control.
`UIControlEventTouchUpOutside`	A finger removed from the screen outside the control.
`UIControlEventTouchCancel`	A system event canceled a touch.
`UIControlEventValueChanged`	A slider (or other similar) object changed its value.
`UIControlEventEditingDidBegin`	Editing began in a `UITextField`.
`UIControlEventEditingChanged`	Editing changed in a `UITextField`.

Table 14.5 `UIControl` objects recognize a number of special events. *(continued)*

Value	Summary
`UIControlEventEditingDidEnd`	Editing ended in a `UITextField` due to a touch outside the object.
`UIControlEventEditingDidEndOnExit`	Editing ended in a `UITextField` due to a touch.
`UIControlEventAllTouchEvents`	Composite for all the touch-related events.
`UIControlEventAllEditingEvents`	Composite for the editing-related events.
`UIControlEventAllEvents`	Composite for all events.

Once a standard event has been turned into a control event, a sequence of additional methods is called, as shown in figure 14.6. First, the `UIControl` object calls `send-ActionsForControlEvents:`. That in turn breaks down the events it's been sent and calls `sendAction:to:forEvent:`, once per event. Here the control event is turned into an action, which is a specific method that's going to be run in a specific target object. Finally the `UIApplication` method `sendAction:to:fromSender:forEvent:` is called by the control, again once per event.

This is another situation where the application object does big-picture controlling of messaging. It's the application that sends the action to the target object. But there's one catch: if the target that the action is being sent to has been listed as `nil`, the action is sent to the first responder instead, and from there moves up the responder chain.

That whole process can be slightly exhausting, and fortunately you shouldn't normally need to know its details. For your purposes, you should be aware that a `UIControl` object turns a touch event first into a control event and then into an action with a specific recipient. Even better, it's only the last part of that conversion, from control event into targeted action, that you need to code.

14.4.3 *The addTarget:action:forControlEvents: method*

A `UIControl` object maintains an internal dispatch table that correlates control events with target-action pairs. In other words, this table says which method should be run by which object when a specified event occurs. You can add entries to this table with the `UIControl` object's `addTarget:action:forControlEvents:` method. The following example shows how it works:

```
[controlObject addTarget:recipientObject action:@selector(method)
    forControlEvents:UIControlEvents];
```

The first argument, `addTarget:`, says who the message will be sent to. It's frequently set to `self`, which usually refers to a view controller that instantiated the control object.

The second argument, `action:`, is the trickiest. First, note that it uses the `@selector` syntax that we mentioned in chapter 10. The selector should identify the name of the method that's going to be run in the target object. Second, be aware that you can either send the action argument without a colon (`method`) or with a colon (`method:`). In the latter case, the ID of the `controlObject` will be sent as an argument. Be sure that your receiving method is correctly defined to accept an argument if you include that colon in your selector.

The third argument, `forControlEvents:`, is a bitmasked list of possible control events, taken from table 14.5.

With all these puzzle pieces in place, you're now ready to write some code that will make actual use of actions (and this method). As a simple example, you're going to expand the functionality to your event reporter by adding a "reset" button.

14.5 Adding a button to an application

The simplest use of an action is probably just adding a button to an application and then responding to the press of that button. As we'll see, this turns out to be a *lot* easier than digging through individual touches.

We've opted to show you how to work with a button in two ways: first by using the `addTarget:action:forControlEvents:` method that we were just introduced to and then by using Interface Builder's `IBAction` declaration.

Both of these examples begin with your existing eventreporter program. You'll add a simple `UIButton` to it using Interface Builder. Place the button down atop the label at the bottom of our page and use the `attributes` tag to label it "Reset." With it in place and defined, it's now ready to be linked into your program by one of two different ways.

Both examples will call a method you've written called `resetPage:`, which just restores the three changeable objects in your eventreporter to their default states. It's shown in listing 14.4 of `eventreporterViewController.m`, and as you can see is entirely elementary.

Listing 14.4   A simple button action

```
- (void)resetPage:(id)sender {

    startField.text = @"Begin: 159,230";
    startField.center = CGPointMake(159,230);

    endField.text = @"Begin: 159,230";
    endField.center = CGPointMake(159,230);

    bottomLabel.text = [NSString string];
}
```

We can now take a look at the two ways you can call this method.

14.5.1 Using addTarget:action:forControlEvents:

On the one hand, you might wish to add actions to your button programmatically. This could be the case if you created your button from within Xcode or if you created your button within Interface Builder but want to change its behavior during runtime.

Your first step, then, is bringing your button into Xcode. If you created your button in Interface Builder, as we suggested earlier, you just need to create an `IBOutlet` for your button, which should be old hat by now. If for some reason you didn't create your button in Interface Builder, you could do so in Xcode. This probably means using the factory class method buttonWithType:, which lets you create either a rounded rectangle button or one of a few special buttons, like the info button. By either means, you should now have a button object available in Xcode.

Your second step is to send the addTarget:action:forControlEvents: message as part of your application's startup. Assuming that you're having your view controller manage the button's action, this message should be sent from the view controller's loadView method (if your controller was created in Xcode) or in its viewDidLoad method (if you created the controller in Interface Builder).

Listing 14.5 shows what the viewDidLoad method of your view controller looks like when applied to a button called myButton.

Listing 14.5 Adding an action to a control

```
- (void)viewDidLoad {

  [myButton addTarget:self action:@selector(resetPage:)
    forControlEvents:UIControlEventTouchUpInside];
  [super viewDidLoad];
}
```

This real-life example of addTarget:action:forControlEvents: looks much like the sample we showed in the previous section. You're sending a message to your button that tells it to send the view controller a resetPage: message when a user takes his or her finger off the screen while touching the button.

That single line of code is all that's required; from there on out, your button will connect to your resetPage: method whenever it's pushed (and released).

14.5.2 Using an IBAction

The other way that you can link up actions to methods is to do *everything* inside of Interface Builder. This is the preferred choice if you've already created your object in Interface Builder (as we've suggested) and you're not planning to change its behavior in runtime.

When you're using this procedure, you won't need to make your button into an IBOutlet. It'll be effectively invisible from Xcode, which is fine, because all you care about is what happens when the button is pushed. You also won't use the somewhat complex addTarget:action:forControlEvents: method that we just ran through; instead, you'll connect things up via intuitive Interface Builder means.

For the purposes of this example, you should start with a clean slate: with a button freshly crafted inside Interface Builder and no connections yet built.

To link an object in Interface Builder to an action in Xcode, you must declare the method you're using as having a return of IBAction. This means adding the following declaration to the header file of your view controller:

```
- (IBAction)resetPage:(id)sender;
```

The implementation of the method should share the same return.

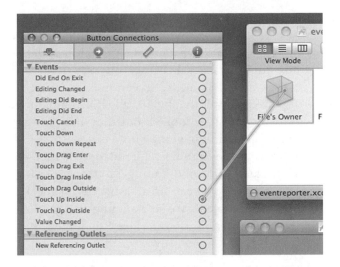

Figure 14.7 With an `IBAction`, there's no code, just a link.

Afterward you can go into Interface Builder and create a connection, as shown in figure 14.7.

As shown, when you're connecting a control, Interface Builder gives you access to the whole palette of possible control events. You select the one (or ones) that you want to connect up to `IBActions`, and then you drag over to the top-level object containing your `IBAction`. In this case, that's once again the file's owner object, which represents your view controller. As usual, a menu will pop up, this time showing you possible `IBActions` to which you could link your control event.

To our eyes, the results are almost magical. With that single graphical link, you've replaced the `addTarget:action:forControlEvents:` call and in fact any code of any type. The button now links to the targeted action "automagically."

What we've described so far covers the broad strokes of actions; everything else lies in the details. If we spent less time on actions than events, it's not because actions are less important than events, but because they're a lot simpler.

From here on, your challenge in using controls will simply be in figuring out how individual controls work. See appendix A for an overview of classes and the Apple Class References for specifics. However, there are a few controls that we'd like to give more attention to because they vary a bit from the norm.

14.6 Other action functionality

In this section we'll look at two controls that report back different signals than the simple button-up or button-down control events. The first is the `UITextField`, the prime control for entering text, and the second is the relatively simple (but unique) `UISlider`. In the process we'll also explore the other text-based entry formats, since they share some unique issues with `UITextField`.

14.6.1 The UITextField

There are four ways to display pure text in the SDK: the `UILabel`, the `UISearchBar`, the `UITextView`, and the `UITextField`. Each has a slightly different purpose. The `UILabel` and the `UISearchBar` are intended for short snippets of text; the `UITextView` is intended for multiple lines. Each of those text objects except the `UILabel` is editable, but only the `UITextField` is a `UIControl` subclass, with its own control events already defined.

If the `UITextField` sounds familiar, that's because you used it in your eventreporter example. If you go back and look at the screenshots, you'll see that the begin and end buttons were displayed in ovals that looked a lot like input boxes. As we mentioned at the time, we liked the way they looked, but they also gave us a good excuse to familiarize you with the object without getting into its details.

Usually, a `UITextField` *would* accept user input. It's intended to be used mainly for accepting short user input. The trickiest thing about using a `UITextField` is getting it to relinquish control of your iPhone after you call up a keyboard. Listing 14.6 shows the two steps needed to resolve this problem. We're assuming that you're working with a `myText UITextField` object created inside Interface Builder and instantiated inside a view controller.

Listing 14.6   A few commands required to get a `UITextField` working

```
- (void)viewDidLoad {

  myText.returnKeyType = UIReturnKeyDone;        ❶

  [super viewDidLoad];
}

- (BOOL)textFieldShouldReturn:(UITextField *)textField {        ❷
  [textField resignFirstResponder];
  return YES;                                      ❸
}
```

Your setup of an Interface Builder object begins, pretty typically, inside its controller's `viewDidLoad` method. Here you turn the text field's keyboard's Return key into a bright blue "Done" key ❶, to make it clear that's how you get out. You accomplish this by using part of the `UITextInputTraits` protocol, which defines a couple of common features for objects that use keyboards.

To do anything else, you need to declare a delegate for the `UITextField` that will follow the `UITextFieldDelegate` protocol. This can be done either by setting the text field's `delegate` property in Xcode or by drawing a delegate link in Interface Builder. (This sample code presumes you've taken the easier solution of doing so in Interface Builder.) Once you've done that, you can modify the `textFieldShouldReturn:` delegate method. We're assuming that the view controller has been set as the delegate, which would be typical, and which allows you to do this work ❷ in the same view controller class file.

Finally, you just enter two standard lines of code into this delegate method. They tell the text field to let go of first-responder status (which, as we've previously noted, is what's necessary to make a keyboard go away) and return a YES Boolean ❸.

With this code in place, a user will actually be able to get in *and* out of a `UITextField`. To use the text field afterward you just need to monitor the text field's special control events (especially `UIControlEventEditingDidEnd`) and also look at its `text` property.

In a moment we're going to provide a sample of how that works, but first let's examine a few other text objects that aren't controls but that you might use to accept text entry.

UILABEL

The `UILabel` is not user-editable.

UISEARCHBAR

The `UISearchBar` looks an awful lot like a `UITextField` with some nuances, such as a button to clear the field and a bookmark button. Despite the similarities in style, the `UISearchBar` is not a `UIControl` object, but instead follows an entirely different methodology.

To use a `UISearchBar`, set its `delegate` to be the object of your choice, likely your view controller. Then respond to the half-dozen messages that are described in `UISearchBarDelegate`. The most important of these is likely the `searchBarSearch-ButtonClicked:` method. Be sure to include `resignFirstResponder` in order to clear the keyboard, and then you can take actions based on the results. There's an example of a `UISearchBar` in chapter 16, section 16.5.3.

UITEXTVIEW

A `UITextView` works like a `UITextField`, except that it allows users to enter many lines of text. The biggest gotcha here is that you can't use the Return key as your Done button, because users will likely want to hit returns in their text. Instead, you must have a Done button somewhere near the top of your screen, where it can be seen when the keyboard is up. When that button is clicked, you can tell the text view to `resignFirst-Responder`. Beyond that, you must set the `UITextView`'s `delegate` property; then you can watch for delegate messages, most importantly `textViewDidEndEditing:`. There's an example of a text view in usage in chapter 16, section 16.3.4.

With our quick digression into this variety of text objects out of the way, we can now return to the other `UIControl` object that we wanted to discuss: `UISlider`.

14.6.2 *The UISlider*

The slider is a simple object, but we've singled it out because it's the one other class that has its own control event, `UIControlEventValueChanged`. If you target this event, you'll find that it gets called whenever the slider moves, but the control event won't tell you what the new value is. To get that information, your action method must query the slider's properties.

There are three properties of particular note: `value` shows a slider's current value, `minimumValue` shows the bottom of its scale, and `maximumValue` shows the top of its scale. You can use the `value` without modification if you've set your slider to return a reasonable number (as described in the class reference), or if you prefer you can use all three properties together to determine the percentage that the slider is moved over—which is exactly what you're going to do in one final control example.

14.6.3 *A TextField/Slider mashup*

Since we've got two `UIControl` objects that we want to examine more closely, it makes sense to quickly mash up an example that takes advantage of both of them. You'll do this in the View-Based RGB Application, which sets the background color of a view based on the word you type into a `UITextField` and the selected position of a `UISlider`.

As usual, you'll create all of these objects in Interface Builder. Then you'll need to go hog-wild linking objects to your view controller. In all you should create five links: an outlet each for your text field and slider; an action link for the important text field and the slider events; and a delegate link for the text field. Figure 14.8 shows what the view controller's Connections tab looks like after these have all been done.

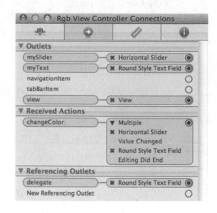

Figure 14.8   A heavily connected view controller will be a pretty normal sight as you gain experience in creating objects in Interface Builder.

As shown, the actions from both of the controls link into a single method, called `changeColor:`. Whenever either control is changed, this method adjusts the color of the screen accordingly. Listing 14.7 shows how.

Listing 14.7   Accessing a text field and a slider

```
- (IBAction)changeColor:(id)sender {

    int red; int green; int blue;

    if ([myText.text caseInsensitiveCompare:@"red"]
        == NSOrderedSame) {
            red = 1; green = 0; blue = 0;
    } else if ([myText.text caseInsensitiveCompare:@"blue"]
        == NSOrderedSame) {
            red = 0; green = 0; blue = 1;
    } else if ([myText.text caseInsensitiveCompare:@"green"]
        == NSOrderedSame) {
            red = 0; green = 1; blue = 0;
    } else {
            red = .5; green = .5; blue = .5;
    }
    float newShade = mySlider.value /
        (mySlider.maximumValue - mySlider.minimumValue);
    [self.view setBackgroundColor:
        [UIColor colorWithRed:red green:green blue:blue alpha:newShade]];
}
```

① Checks text input

② Calculates alpha percentage

The hardest part of working with a `UITextField` is setting it up, which you did in listing 14.6. Now that you've got input coming back, all you need to do is access the `text` property and do with it as you will **①**.

Meanwhile, by working with your three slider values you're able to easily generate a value from 0 to 1 **❷**. Putting that together with the color you generated from your text field input results in a background color that you can change in two ways. Figure 14.9 takes a final look at this new program.

Would this be better to do with a `UISegmentedControl` and a `UISlider`? Probably. But as is, it also offered a quick example of how a text field works. Furthermore, it showed how you can combine action management by letting multiple controls point to a single method, a technique that will be useful in more complex programs.

As usual, there's more information on both of these controls in the Apple class references, including lots of methods and properties that we didn't talk about.

14.6.4 Actions made easy

Throughout the latter half of this chapter we've seen controls that were tied to the fully fledged target-action mechanism. In the next chapter, that's going to change a bit when we see the same idea in a somewhat simplified form.

Figure 14.9 A text field and a slider conspire to set the color of the iPhone's background.

Sometimes buttons or other controls are built into other classes of objects (such as the button that can be built into the navigation bar). These controls will have special methods that allow them to automatically create a target-action pair. As a result, you don't have to go through the nuisance of calling the `addTarget:action:forControl-Events:` method separately.

We'll point this technique out when we encounter it as part of the navigation controller.

14.6.5 Actions in use

There are numerous control objects that we've opted not to cover here, mainly because they use the same general principles as those we've already talked about. Nonetheless, they'll remain an important factor throughout the rest of this book.

In particular, controls represent one of the main ways that users can offer input to your programs, and we'll discuss them when we talk about data in chapter 16. We'll also be offering more complex programs that use a variety of controls from chapter 16 on. Through those examples, the majority of the UI controls will receive some coverage in this book.

14.7   *Introducing notifications*

As we mentioned in chapter 10, there's one other way that a program can learn about events: through notifications. When directly manipulating events or actions, as we have throughout this chapter, individual objects receive events because the events occurred in their view, because the events occurred in a subview, or because the events occurred in a view that has delegated to them.

Notifications step outside this paradigm. Now, an object registers to receive notice when certain events occur. These are often events that lie beyond the standard view hierarchy, such as information when a network connection closes or when the iPhone's orientation changes. Notably, these notifications are also broadcast messages: many different objects can be notified when the event occurs.

All notifications occur through the NSNotificationCenter. You must create a copy of this shared object to use it:

```
[NSNotificationCenter defaultCenter]
```

Afterward, you may use the addObserver:selector:name:object: method to request a certain notification. The Observer: is the object that will receive the notification method (usually, self), the selector: is the method that will be called in the observer, the name: is the name of the notification (which will be in the class reference), and the object: can be used if you want to restrict which objects you receive notification from (but it is usually set to nil).

For example, to receive the UIDeviceOrientationDidChangeNotification notification that we're going to talk about in chapter 17, you might use the following code:

```
[[NSNotificationCenter defaultCenter] addObserver:self
  selector:@selector(deviceDidRotate:)
  name:@"UIDeviceOrientationDidChangeNotification" object:nil];
```

Overall, notification programming tends to have four steps. First, you learn that there's a notification by reading the appropriate class reference (UIDevice in this case). Second, you may need to explicitly turn on the notification (as is indeed the case for UIDeviceOrientationDidChangeNotification). Third, you write a method that will respond to the notification (in this case, deviceDidRotate:). Fourth, you connect up the notification to the method with the NSNotificationCenter.

There is considerably more power in the notification system. Not only can you set up multiple observers, but you can also post your own notifications. If you want more information on these advanced features, you should read the class references on NSNotificationCenter, NSNotification, and NSNotificationQueue.

14.8   *Summary*

The iPhone OS includes an extensive set of frameworks that takes care of lots of the details for you, making your iPhone programming as painless as possible. We've seen this to date in everything we've done, as sophisticated objects appear on our screens with almost no work.

The same applies to the iPhone's event system. There is a complex underlying methodology. It centers on a responder chain and granular reporting of touches and allows us to follow precise user gestures. You may occasionally have to manipulate events via these more complex means.

However, the iPhone also supports a higher-level action system that lets your program respond to specific actions applied to controls rather than more freeform gestures. We've explained how to use both, but it's the target-action mechanism that you're more likely to rely on when programming.

With actions and events now out of the way, we're ready to look at the final fundamental building block of the SDK. We've already encountered views, controls, and basic view controllers, but there's another category of object that's critical for most SDK programming: the advanced view controller that allows for navigation over multiple screens of content.

That's going to be the basis of our next chapter.

Creating advanced
view controllers

This chapter covers

- Tab-based interfaces
- Navigation-based interfaces
- The flipside controller

When we started our look at view controllers in chapter 13, we promised that we'd return to the more advanced view controllers that manage several pages of content at once. That's the purpose of this chapter: to introduce you to the final fundamental building block of the iPhone OS that allows you to build complex multipage applications.

In this chapter we'll take an in-depth look at two view controllers: the tab bar controller and the navigation controller. We'll also take a briefer look at the flipside controller that appears in one of Xcode's templates and talk about some modal controllers that we'll see in part 4 of this book.

As in our previous chapter on view controllers, we'll offer some more skeletal examples since our main purpose is to provide you with the reusable programming frameworks that will allow you to use these controllers in your own programs. Let's kick off our discussion with the tab bar view controller.

15.1   *The tab bar view controller*

Of the multipage view controllers, the tab bar is the easiest to use because it supports simple navigation between several views. As with all of the advanced view controllers, it has a complex underlying structure incorporating several objects that work in tandem.

15.1.1   *The anatomy of a tab bar controller*

To function, a tab bar view controller requires a hierarchy of at least six objects:

- One `UITabBarController`
- A minimum of two `UIViewControllers`
- One `UITabBar`
- A minimum of two `UITabBarItems`

This hierarchy of objects is depicted in figure 15.1.

The tab bar controller and its associated view controllers are the heart of this setup. Essentially the tab bar controller switches off between different pages, each of which uses a view controller to manage its events. In Xcode you'd have to create and hook up these view controllers by hand, while in Interface Builder (which is what we'll be using) it's automated. In either case, you'll need to fill in the controllers' views once your controllers are ready to go.

The tab bar itself is created automatically when you instantiate a tab bar controller. It displays a set of radio buttons that go at the bottom of the page. Each of those buttons is a tab bar item (which Interface Builder also creates automatically). Each tab bar item then links to an individual view controller. Usually you shouldn't have to mess with the tab bar at all; you can do all the modifications you require through either the tab bar controller or the view controllers.

The connection between the tab bar controller and its tab bar is a simple delegation, as we've seen in use in previous chapters. The tab bar has a `delegate` property that is hooked up to the controller, which must respond to the `UITabBar-Delegate` protocol.

The tab bar controller can also designate a `delegate`. The controller's delegate must follow the `UITabBarControllerDelegate` protocol. This protocol requires response to two types of high-level events: when the tab bar is rearranged and when a view controller is selected.

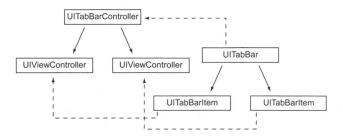

Figure 15.1   A collection of six objects (at minimum) is required to create a functioning tab bar controller.

15.1.2 *Creating a tab bar controller*

Each of the advanced view controllers has its own Xcode template that you can use to immediately instantiate the controller. Since this is our first advanced view controller, though, we'll look at how you'd create it by hand before we move over to simpler, template-driven object creation.

CREATING YOUR TAB BAR CONTROLLER BY HAND

To create a tab bar controller manually, begin with the Window-Based Application template. Use it to create a project imaginatively called "tabex."

Once you've created your project, you should pop straight over to Interface Builder by clicking on the MainWindow.xib file.

To create a tab bar controller:

1 Drag the Tab Bar Controller object from the Library window (where you'll find it under Controllers) to the nib display window.
2 Drop the Controller down next to your window object. When you do that, a tab bar controller Main display window should appear.
3 Dismiss your old Main display; you won't need it anymore. Instead you'll create new objects as subviews of your tab bar controller.

The results are shown in figure 15.2.

Believe it or not, that's it. All six objects of note have been created. The tab bar controller is accessible from the nib display window. The other five objects are accessible from the black bar at the bottom of the Main display window. Click a button once to get its `UIViewController` and a second time to get its `UITabBarItem`. Click in the middle of the strip (between the buttons) to access the `UITabBar`. By selecting these items, you can set their attributes, connections, size, and identity.

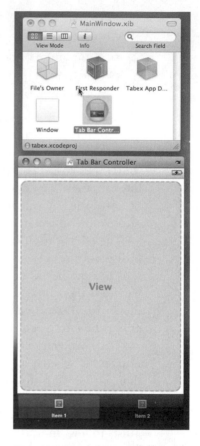

Figure 15.2 Just dragging a tab bar controller to the nib display window creates the whole tab bar interface.

We took this slight diversion into the "harder" side of tab bar controller design to show what all the objects look like in Interface Builder. If you've been following on a computer, we suggest clicking around for a while to see how everything works. Once you've seen all of the fundamental objects that are created as part of an advanced view controller, we've played the Window-Based Application template's last trick. In the future we're just going to jump straight to the appropriate template for each sort of view controller—starting with the tab bar controller template.

CREATING YOUR TAB BAR THROUGH A TEMPLATE

It's even easier to create a tab bar controller using the existing tab bar template. Just select Tab Bar Application when you create a new project. This template will set you up with a tab bar controller much like the one you just created by hand, except it does three additional things:

- The template defines the tab bar controller as an IBOutlet, giving the app delegate access to the object IBOutlet UITabBarController *tabBarController;
- The template creates the view controller for the first window as part of a special FirstViewController class. You'll probably want to have an individual view controller class for each tab to take care of events on a per-page basis, but that is easy enough to change by adding class files and adjusting the Identity tab for the view controllers. For now, leave things as they are so that we can examine how to work with the default template setup.
- The template associates a second .xib file with the second view. It does this in a way we've seen before, by defining a nib Name for the view controller inside Interface Builder.

For the rest of this section, we're going to assume that you're working with this pre-built tab bar controller template as your "tabex" project.

With a working tab bar controller in hand, we can now start programming multiple pages of screens.

Tab bars and toolbars

The UIKit supports two very similar interfaces, the UITabBar and the UIToolBar. They each include a strip of icons that goes along the bottom of the screen. Their main difference is in functionality.

The UITabBar is intended as a modal interface that changes the selections when they're tapped (usually with a permanent highlight). The purpose of the UIToolBar is to provide a menu of possible actions that don't change the appearance of the selection when tapped (except with a temporary highlight).

Despite their similar appearance, the two items share no inheritance other than a common ancestor in UIView. Consider it convergent evolution.

We'll present a fully functional example of a UIToolBar in chapter 18, section 18.4.

15.1.3 Building a tab bar interface

At this point you've got a tab bar controller that contains two tabs, each of which has relatively empty content. You've also got tabs on your tab bar without pictures and without meaningful names. To build your tab bar interface, you'll want to adjust all of these things.

ADDING MORE TABS

Inside Interface Builder you add tabs to the tab bar by going to the Attributes tab of the tab bar controller and clicking the plus sign (+) in its view controller area. A tab bar item and related view controller will be added to the right-hand side of your bar. Go ahead and create a third tab by clicking the +.

To allow for easy access to this new controller's view, you'll probably want to create a new .xib file and connect the view controller to that .xib file. Both of these procedures were described at the end of chapter 12.

CONNECTING VIEWS

Once you have the right number of tabs, you can then connect views to each of the tab bar's view controllers. This can be done in three major ways:

- You can input views through .xib files, as noted earlier.
- If a view controller has its own class file, you can add views through the `load-View` or `viewDidLoad` method for that class.
- If a view controller doesn't have its own class file, you can load views elsewhere, such as in the app delegate's `applicationDidFinishLaunching:`.

We've already offered several examples for the first two ways to load views (including plenty of Interface Builder examples in chapter 12 and `viewDidLoad:` examples in chapter 13), so we're not going to repeat those methods here. Instead, since the latter two view controllers don't have their own class files, you'll see how you can create their views using `applicationDidFinishLaunching:`. Honestly, it'd probably be simpler to create their views in Interface Builder, but this example will demonstrate how you can use the tab bar controller.

Although you don't have outlets for the controllers themselves, you can link to them straight from the tab bar controller object, which you *do* have access to, thanks to that `IBOutlet` that we're already seen. This relates to a concept that we discussed when talking about basic view controllers in chapter 13; since view controllers have to do MVC management, they should give you easy access to related objects. Within the tab bar controller is a `viewControllers` property, which is an `NSArray` list of the view controllers that a tab bar controller contains.

Listing 15.1 shows how to access this information and programmatically build a couple of views for the second and third controller within tabexAppDelegate.m. This is the skeleton of a simple program that would let you edit a text view in the first window, keep a count of what you've written in the second, and search in the third.

Listing 15.1   A tab bar controller setup

```
- (void)applicationDidFinishLaunching:(UIApplication *)application {

    UIViewController *secondController =
        [tabBarController.viewControllers objectAtIndex:1];
    UIViewController *thirdController =
        [tabBarController.viewControllers objectAtIndex:2];

    UITextView *secondView = [[UITextView alloc]
        initWithFrame:[[UIScreen mainScreen] bounds]];
    secondView.text = @"A word count would appear here.";
    secondView.editable = NO;
```

❶ Retrieves view controllers

```
secondController.view = secondView;

UITextView *thirdView = [[UITextView alloc]
    initWithFrame:[[UIScreen mainScreen] bounds]];
thirdView.text = @"A search function would go here.";
thirdView.editable = NO;

thirdController.view = thirdView;

[window addSubview:tabBarController.view];

[secondView release];
[thirdView release];
}
```

② Sets views

③ Displays tab bar controller

To access the view controllers, you just pull elements out of an array using the appropriate `NSArray` calls **①**. You then associate views with each view controller, as you've done in the past **②**. Finally you link the tab bar controller to the window, using a call that was already sitting in your file when you loaded it **③**.

You now have three modal pages (including that first controller's page, which we assume was taken care of in its class files, provided by default by the template). Each does what you want, and navigation between them is easy. But you can still do some work to make your tab bar look better.

MODIFYING THE BUTTONS

Although we have views associated with each button, the buttons just say First, Second, and Third, rather than providing any useful clue as to a button's purpose. You can change three things on each button to improve their usability: the icon, the title, and the badge. Figure 15.3 shows the goal, which is to fill out some or all of this information for each of our tab buttons.

The *icon* is the image that appears on the tab bar item. This image can only be set when you create a tab bar item. If you were creating the tab bar programmatically, you'd use the `initWithTitle:image:tag:` method when creating the tab bar item. More likely, you'll just go into Interface Builder and load a small PNG graphic that you want to use.

This process is similar to incorporating the image into your project in chapter 12. You should create a transparent PNG that's approximately 30 x 30. If your image is too big, the SDK will resize it, but it's better to start off at the right size. After you drag the image into your project, you'll be able to load it up in Interface Builder. We used a Wingdings font to create the simple images that appeared in figure 15.3.

Figure 15.3   You can customize tab bars to make navigation clear and simple.

The *title* is the word that appears on the tab bar. You'll probably set that in Interface Builder too, which just involves going to the tab in question and changing the text there.

If you want to later change the title during runtime, it is accessible in Xcode. The catch is that these titles aren't found in the tab bar controller. Instead, they follow the overarching idea of MVC: since a view controller is responsible for an individual view, it's the controller that actually sets the title of the page. This is done with the view controller's `title` property, which we've mentioned before and which we'll meet again:

```
secondController.title = @"Word Count";
```

The *badge* is the little red circle that appears above the title and over the icon on the tab bar. As always, you could change this in Interface Builder, but Xcode is where you'll generally want to do this work. That's because the information in a badge is meant to be dynamic, changing as the view changes and alerting a user to new content during runtime. It's badges, for example, that tell you when you have new mail or new voicemail.

Getting to the badge property is a two-step process. You'll start with your view controller. From there you access `tabBarItem`, which is a property of the controller that links you to its connected tab bar item, and then `badgeValue`, which is a property of the tab bar item. Fortunately, this can all be done as one nested line:

```
secondController.tabBarItem.badgeValue = @"16";
```

The 16, as it happens, is the initial character count of the main text view. If you were building a live program, you could change this count over the course of your program's runtime.

Table 15.1 summarizes the three main elements of the tab bar and how to customize them.

Table 15.1 From your view controllers, it's easy to customize the associated tab bar items.

Property	Summary	Interface builder	Xcode
badge	Tab bar info	Yes	`viewcontroller.tabBarItem.badgeValue`
icon	Tab bar picture	Yes	only at `init`
title	Tab bar words	Yes	`viewcontroller.title`

There's one more way to change both the icon and the title of a tab bar item simultaneously: by creating a tab bar item with the `initWithTabBarSystemItem:tag:` method. This creates a tab bar using one of the constants defined under `UITabBarSystemItem`, each of which relates to a standard iPhone function and correlates a name and a picture with that function.

You'll probably be doing this in Interface Builder, where you select a specific "Identifier" instead of entering a title and a picture. Since your third tab allows searches, you can initialize it as a `UITabBarSystemItemSearch` button, which gives it the title of "Search" and the picture of a magnifying glass, as shown in figure 15.3.

Once you've got the tab bar all set up, you're ready to start using the controller.

15.1.4 *Using your tab bar controller*

The main function of a tab bar is to allow navigation between multiple pages in your application. This is an implicit function of the object and you don't need to do anything more to enable it. The rest of the tab bar controller's functionality goes beyond our basic overview of the topic, but we'll mention it briefly. There are two main elements you want to consider: customization and delegation.

TAB BAR CUSTOMIZATION

One of the neat things about a tab bar is that users can customize them to contain exactly the tab bar items that interest them. You can allow this by setting the `customizableViewControllers` property to include a list of view controllers that can be changed by the user:

```
tabBarController.customizableViewControllers =
    tabBarController.viewControllers;
```

The `UITabBar` reference contains all the information you'll need on managing customization itself.

TAB BAR CONTROLLER DELEGATION

As we noted, you can set a delegate for your tab bar controller to hand off the scant amount of event management that it requires. The delegate object must follow the `UITabBarControllerDelegate` protocol, which is a fancy way of saying that it'll respond to two specific events: one when a view controller is selected and another when the tab bar controller is customized. There's a protocol reference that covers the specifics of this.

Two methods are associated with these protocols. `tabBarController:didEndCustomizingViewControllers:changed:` reports the end of tab bar customization and `tabBarController:didSelectViewController:` reports when the user switches between controllers. The latter is probably more generally useful. For example, you might use it in your word count example to recalculate the word count totals whenever a user jumps to the word count page.

But now that you've got a basic example of how to navigate with a tab bar, you're ready for the next advanced controller: the navigation controller.

15.2 *The navigation controller*

The navigation controller is probably the most-seen user-interface item on the iPhone. Whenever you have a hierarchical set of pages where you can move up or down through the hierarchy, that's the navigation controller at work. It appears in the Text, Calendar, Photos, and Notes iPhone utilities, just to name a few.

Working with the navigation controller is a bit harder than working with the tab bar controller, because you have to manage your hierarchy of views live as the user interacts with your program, but the SDK still keeps it simple.

As with our previous view controllers, we'll look at an overview of the class, and then examine how to create, build, and use a navigation controller. Let's get started with an overview of its hierarchy.

15.2.1   *The anatomy of a navigation controller*

As with the tab bar controller, the navigation controller involves a whole hierarchy of items. The `UINavigationController` sits atop a stack of `UIViewControllers` that can be pushed or popped as a user moves up and down through it.

Each of these controllers also has an associated `UINavigationItem`, which will sit in the `UINavigationBar` when it's active. Each `UINavigationItem` may also contain one or more `UIBarButtonItems`, which allow for additional action items to sit on the navigation bar.

To tie things back together, the `UINavigationBar` is also linked to the `UINavigationController` so that navigation items and view controllers stay in sync over the course of a program's runtime. Whenever a `UIViewController` loads into the `UINavigationController`, its `UINavigationItem` also loads into the `UINavigationBar`.

A minimalistic navigation controller has just four objects in it: the `UINavigationController`, the `UINavigationBar`, a stack containing a single `UIViewController`, and a `UINavigationItem` (which is placed into the `UINavigationBar`). Presumably more view controllers and associated navigation items will be added to the stack as the program runs.

This is all illustrated in figure 15.4.

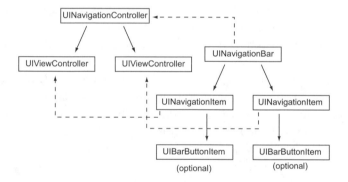

Figure 15.4   The navigation controller will contain at least four objects, and may be built into a complex web of interconnections.

You'll note how similar this diagram of navigation controller parts looks to figure 15.1, our diagram of tab bar controller parts. This is not an accident in our drawing, nor do we expect that it was an accident in Apple's design. The navigation controller works quite a bit like the tab bar controller, and thus we'll see familiar elements, such as the title of the view controller creating the title within the navigator itself.

The biggest difference is that where the tab bar controller presented a modal paradigm, entirely organized by the controller itself, the navigation controller instead creates a hierarchical paradigm. The navigation controller doesn't have any particular sense of the organization of the entire structure. Instead, a linked list is created, with each navigation item only knowing about the pages on either side of it.

A NOTE ON TABLE VIEWS

The standard iPhone paradigm is to do hierarchical navigation through table views, each of which contains lists of many different subpages that you can go to. As a result, despite the fact that any `UIViewController` could sit beneath a `UINavigation-Controller`, it'll usually be a `UITableViewController`.

This is exactly the setup we'll see in the navigation-based template.

15.2.2 *Creating a navigation controller*

To create a navigation controller, create a new project (which we're calling "navex") using the Navigation-Based Application template. You can page through the .xib file and the Xcode listing to see what you've been given. Let's start with the .xib files, whose content you can see in figure 15.5.

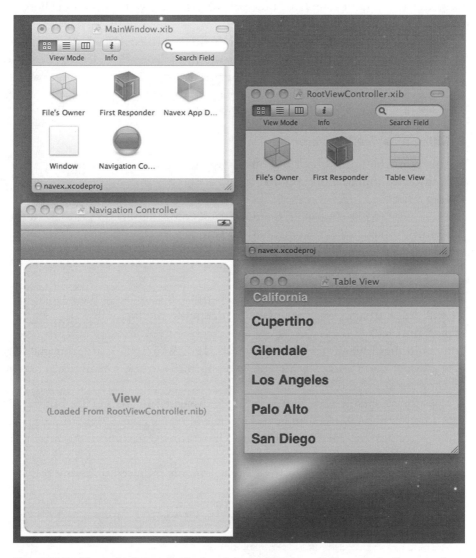

Figure 15.5 The navigation controller template contains two .xib files, one for the main view (left) and one for what appears inside the controller (right).

Mainwindow.xib contains a `UINavigationController` in the nib window with a `UINavigationBar` hidden under it. The main display window contains a `UINavigationItem` and a `RootViewController`. The latter is a subclass of `UITableViewController` created through Xcode, just as when you designed your own table controller in chapter 13. Note that this sets up the standard iPhone paradigm of navigation controllers being built atop table controllers. The table view controller's contents are instantiated through a second .xib file, RootViewController.xib, as shown in the table view controller's attributes window.

RootViewController.xib is a boring .xib file because it just contains a table view. Consider it a good example of how pairing .xib files with view controllers can keep your program well organized.

Finally, if you look at your Xcode files created by the template, you can see that the navigation controller gets linked to your window in the app delegate file. Among the other default files are the `RootViewController` class files you'd expect to see. Since you're working with a table view controller, you know the `RootViewController` class files will be important when you input the table view's data.

15.2.3 *Building a navigation controller*

At this point we'll do three things to complete the navigation controller: add a title, add navigation links, and (optionally) add action buttons.

ADDING A TITLE

Just like the tab bar controller, the navigation controller takes its title from the title of the individual page's view controller. All you have to do is define `title` in your table view controller file:

```
self.title = @"Color List";
```

This turns out to be a critical bit of data, because it's also what the navigation controller will use as a back button when you're deeper in the hierarchy.

ADDING THE LINKS

You could theoretically use whatever method you want to link to additional pages via a navigational controller. The default mechanism is to use a table list, and that's the method we'll use in this example.

Design your table view controller as we discussed in chapter 13, but this time you should give each table cell an accessory view of type `UITableViewCellAccessoryDisclosureIndicator`. That's the standard chevron used to indicate hierarchical navigation.

Listing 15.2 includes all of the major elements required to define this navigation table inside RootViewController.m.

Listing 15.2   A table for a navigator

```
- (id)initWithCoder:(NSCoder *)decoder {
    self = [super initWithCoder:decoder];
```

```
        if (self) {
            self.title = @"Color List";
            colorList = [NSArray arrayWithObjects:
                [NSDictionary dictionaryWithObjectsAndKeys:
                    @"Red",@"titleValue",
                    [UIColor redColor],@"colorValue",nil],
                [NSDictionary dictionaryWithObjectsAndKeys:
                    @"Green",@"titleValue",
                    [UIColor greenColor],@"colorValue",nil],
                [NSDictionary dictionaryWithObjectsAndKeys:
                    @"Blue",@"titleValue",
                    [UIColor blueColor],@"colorValue",nil],
                nil];

            [colorList retain];
        }
        return self;
}

- (NSInteger)numberOfSectionsInTableView:(UITableView *)tableView {

    return 1;
}

- (NSInteger)tableView:(UITableView *)tableView
    numberOfRowsInSection:(NSInteger)section {

    return [colorList count];
}

- (UITableViewCell *)tableView:(UITableView *)tableView
    cellForRowAtIndexPath:(NSIndexPath *)indexPath {

    static NSString *MyIdentifier = @"MyIdentifier";

    UITableViewCell *cell = [tableView
        dequeueReusableCellWithIdentifier:MyIdentifier];

    if (cell == nil) {
        cell = [[[UITableViewCell alloc] initWithFrame:CGRectZero
            reuseIdentifier:MyIdentifier] autorelease];
    }

    cell.text = [[colorList objectAtIndex:indexPath.row]
        objectForKey:@"titleValue"];
    cell.textColor = [[colorList objectAtIndex:indexPath.row]
        objectForKey:@"colorValue"];
    cell.accessoryType = UITableViewCellAccessoryDisclosureIndicator;

    return cell;
}
```

There's nothing new here, but we've included it to clarify the rest of our discussion of the navigation controller.

ADDING ACTIONS

If you want, you can now move right on to using your navigation controller. Alternatively, you can do some extra work with buttons. Besides the standard navigation controls, you could also add buttons to the navigation bar. This is done through the leftBarButtonItem and the rightBarButtonItem properties of the UINavigation-Item. A left button will replace your back button, while a right button just sits in the usually blank right-hand side of your navigation bar.

As we've noted, each view controller is linked to its own navigation item. A view controller can access its navigation item through the navigationItem property at any time.

When you set a button, you must set it to be a UIBarButtonItem object, which you'll have to create. There are four init methods you can use, as shown in table 15.2. You will probably use the initWithCoder: method, the same place that you should be initializing your array for use with your table view.

Table 15.2   You can create bar button items using a variety of methods to get precisely what you want.

Method	Summary
initWithBarButtonSystemItem:target:action:	Creates a standard button drawn from UIButtonSystemItem
initWithCustomView:	Creates a special button
initWithImage:style:target:action:	Creates a button with a picture
initWithTitle:style:target:action:	Creates a button with a word

Note that all the buttons except for the custom view button come with their own target and action links. These are the simpler target-action mechanisms that we alluded to in the previous chapter. They work exactly like the more complex target-action mechanisms but are built in.

Here's how you could create a button as part of the page represented by your UITableViewController:

```
self.navigationItem.rightBarButtonItem = [[UIBarButtonItem alloc]
    initWithBarButtonSystemItem:UIBarButtonSystemItemAdd
    target:self action:@selector(changeTitle)];
```

As you can guess from our title, this button press just enacts a very innocuous title change, but it would be easy to redraw your table list or even to integrate that button with the navigation itself, perhaps using it as a home button.

At this point you've got a navigation controller that does precisely nothing, other than showing a gray bar with a title, and perhaps a working button. Unlike with the other controllers that you've met so far, you're going to need to do some runtime work to get your navigation controller actually working.

15.2.4   *Using your navigation controller*

A navigation controller has one core job: to allow a user to move up and down through a hierarchy of pages.

NAVIGATING FORWARD

To allow a user to navigate to a page deeper in your hierarchy, you need to use the navigation controller's interface to push a new view controller on top of the navigation controller's stack, which will then cause that new view controller's view to become the visible view in your program. This is shown in listing 15.3, which continues to expand upon RootViewController.m.

Listing 15.3 Activating a navigation controller

```
- (void)tableView:(UITableView *)tableView                           ❶
    didSelectRowAtIndexPath:(NSIndexPath *)indexPath {

    UIViewController *colorController = [[UIViewController alloc] init];   ❷
    colorController.title =
        [[tableView cellForRowAtIndexPath:indexPath] text];               ❸

    colorController.view = [[UIView alloc] init];    ❹
    colorController.view.backgroundColor =                               ❺
        [[tableView cellForRowAtIndexPath:indexPath] textColor];

    [self.navigationController                                           ❻
        pushViewController:colorController animated:YES];

    [colorController release];
}
```

To navigate using tables, you must modify the table view controller's `tableView:did-SelectRowAtIndexPath:` method ❶, which you first met in chapter 13. Clearly, if you're activating your navigation controller by some other method, you'll use different means.

When your users select an item that should lead them to the next page, you'll have to create the page they'll be going to. Start off by creating a view controller ❷. Remember to set the `title` since it'll be the title that appears in your new view controller's navigation item. Matching the title to your table cell's text ❸ will probably be a common way to set this property.

Once you've created a view controller, you need to decide how to create its default view. Here you're creating a plain view ❹. Prefer to create your view in Interface Builder? No problem. Just use the `initWithNibName:` method when you create your view controller, as we discussed in chapter 12.

Each view should have different content based on what your user selected. Here, you'll be looking at the text color of the table cell's text and then setting the whole view to that color ❺. More often you'll probably look up an `NSDictionary` element from the same array that you used to fill in your table and use that information to generate a unique page. For example, it'd be easy to pull a nib name out of a dictionary.

Once you've set up your new page, you just send one message to the navigation controller to switch over to it ❻. Note that you can find a reference to your navigation controller by using the view controller's `navigationController` property, another of many object links available in the view controller. The actual push command is simple: it just adds a new page to the top of the navigation controller's stack, and thus sends your user over to it.

NAVIGATING BACKWARD

Once you've loaded a new page onto a navigation controller's stack, it appears with all the peripherals you'd expect, including a titled navigation bar with a titled back button (each based on the `title` property of the appropriate controller). This is all shown in figure 15.6.

You also don't have to worry about coding the backward navigation. Clicking the back bottom will automatically pop the top controller off the stack, without any work on your part.

OTHER TYPES OF NAVIGATION

Navigation doesn't have to be just forward and backward. You can also do some slightly more complex things, either during setup or at runtime.

At setup you can choose to create a navigational hierarchy and push a user into it before he or she takes any actions. You can see this in action in various iPhone programs. Mail always returns you to the last mailbox you were at, while Contacts always gives you a back button to return to the "groups" page.

You can do fancy things during runtime using three navigation controller methods: `popToRoot-`

Figure 15.6   With a few simple commands, a navigation controller's setup is largely automated.

`ViewControllerAnimated:` (which brings you back to the top of your stack), `pop-ToViewController:Animated:` (which returns you to a specific view controller), or `popViewControllerAnimated:` (which just pops the top controller off the stack).

They're quite powerful, though you have to take care when changing the standard navigation paradigm so that you don't confuse your users. But, as an example, you could place a `UIBarButtonItem` in your nav bar that returns you to home from deep in your hierarchy. Alternately, you might pop the top page automatically after a user takes some action on the page that concludes its usefulness.

NAVIGATORS AND DATABASES

So far we've built all of our table view controllers—including the one embedded in this navigation controller—using arrays. This is a perfectly acceptable technique for a small, static table. But if you want to have a bigger or a more dynamic table, you'll probably want to use a database as your data back end. We'll have a complete example of how to do so in chapter 16, when we cover the SQLite database package.

OTHER METHODS AND PROPERTIES

There's very little else to be done with the navigation controller, though you'll find a few other properties in the class reference. You can set those properties to modify the look of individual `UIBarButtonItems` and to set your nav bar to be hidden.

We've now covered the two most important advanced view controllers, but before we finish our discussion of the topic, let's take a brief look at the flipside controller, which exists only as a template, not as a class within the UIKit framework. The template instead creates a subclass of ViewController within your program.

15.3　*Using the flipside controller*

To create a flipside controller, choose the Utility Application template when you start a new project. It will create a small hierarchy of objects, as shown in figure 15.7.

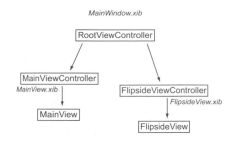

The flipside controller contains three view controllers and two views. Each of the view controllers is a subclass of UIViewController, while the views are each a subclass of UIView.

The main view controller is called the Root-ViewController. It's loaded through Main-

Figure 15.7　Several objects are created in a flipside controller.

Window.xib. Much of the template's work is done, as you'd expect, through its class files. The RootViewController.m file loads the MainViewController (using the initWith-NibName: method to load its unique nib file), and then creates a special toggleView method for when the info button at the bottom of the page is pushed. When this happens, the FlipsideViewController also loads. Listing 15.4 shows this standard method, which you shouldn't have to modify at all.

Listing 15.4　The flipside toggler

```
- (IBAction)toggleView {

    if (flipsideViewController == nil) {
        [self loadFlipsideViewController];
    }
    UIView *mainView = mainViewController.view;
    UIView *flipsideView = flipsideViewController.view;

    [UIView beginAnimations:nil context:NULL];
    [UIView setAnimationDuration:1];
    [UIView setAnimationTransition:([mainView superview] ?
        UIViewAnimationTransitionFlipFromRight :
        UIViewAnimationTransitionFlipFromLeft) forView:self.view
            cache:YES];

    if ([mainView superview] != nil) {
        [flipsideViewController viewWillAppear:YES];
        [mainViewController viewWillDisappear:YES];
        [mainView removeFromSuperview];
        [infoButton removeFromSuperview];
        [self.view addSubview:flipsideView];
        [self.view insertSubview:flipsideNavigationBar
            aboveSubview:flipsideView];
        [mainViewController viewDidDisappear:YES];
        [flipsideViewController viewDidAppear:YES];
    } else {
        [mainViewController viewWillAppear:YES];
```

```
            [flipsideViewController viewWillDisappear:YES];
            [flipsideView removeFromSuperview];
            [flipsideNavigationBar removeFromSuperview];
            [self.view addSubview:mainView];
            [self.view insertSubview:infoButton
               aboveSubview:mainViewController.view];
            [flipsideViewController viewDidDisappear:YES];
            [mainViewController viewDidAppear:YES];
        }
        [UIView commitAnimations];
    }
```

Because you shouldn't need to modify it, we're not going to cover all of this code, but if you read through it, you'll find that it includes some nice nuances, including the ability to work with `UIView`'s simple animation and some different ways to call `insertSubview:`. This template provides a great example of how to connect multiple Xcode class files and multiple nib files, and reading through it can serve as a great tutorial for more advanced work you're going to do yourself.

For example, take a look at the MainWindow.xib file. You'll note there that connections are made to two different files, as shown in figure 15.8. The app delegate file contains a link to the root view controller object, while the root view controller file contains links to the info button object and its action. This shows the sort of organization you'll want to consider for you own projects.

Figure 15.8   Objects in Interface Builder can connect to a variety of different files.

Given that, how do you actually use the flipside controller? All you need to do is lay out objects in the two .xib files and/or make changes to their accompanying class files. Then you can build controller actions, events, and other activities into the two controller files.

For example, you could use Interface Builder to make your main view red; then you could go into the FlipsideViewController.m file and change the default background color to `greenColor` (instead of its current `flipsideColor`) to create a simple red and green flash card, which can be used to express your interest in a conference topic. We'll also show how to use a flipside controller to create local preferences on the backside of your program in chapter 16. If you ever need a two-sided application, the flipside controller is a great place to get started.

So far we've explored those view controllers that you might use as the building blocks of your own views. Tab bars, navigators, and flipsides are ultimately tools that you'll use to construct other sorts of programs. However, there's a different type of view controller that exists to accomplish very specific tasks: the modal view controller.

15.4 *Modal view controllers*

Technically, a modal view refers to a temporary view that's placed on top of all of the elements of an existing view and then later dismissed. A modal view controller is a view controller that manages such a modal view.

Practically, the modal views already available in the iPhone OS are all "helper" programs, which allow you to start up a complex graphical interface that's been preprogrammed by Apple while only managing the responses. You get the advantage of lots of programming (and a standardized interface), and you don't have to do much yourself.

Whenever you want to display a modal view, you use the `UIViewController`'s `presentModalViewController:animated:` method to start it up:

```
[self presentModalViewController:myPicker animated:YES];
```

Later you'll dismiss it using another `UIViewController` method:

```
[self dismissModalViewControllerAnimated:YES];
```

You could design your own modal view controllers for when you want to have users make a choice before returning them to their regular program. More commonly, you'll use picker controllers that are intended to be run as modal view controllers. In chapter 16 you'll meet the Address Book UI people picker (as well as some related controllers that run inside a navigator), and then in chapter 18 you'll meet the image picker.

15.5 *Summary*

At this point we've finished with what we consider our basic introduction to the SDK. Since this is an introductory SDK book, it's been our main goal to show you all the fundamentals before we set you lose in the wilds of iPhone programming so that you have the building blocks that you need when you begin programming on your own.

To briefly review them all:

- The SDK is built on top of Objective-C and is supported by a large set of frameworks called the iPhone OS. (See chapter 10.)
- Programming can be done in either Xcode or Interface Builder, supporting two powerful ways to create objects. (See chapters 11 and 12.)
- Basic view controllers take the controller role of the MVC model and allow you to administer your views in a rational way. (See chapter 13.)
- Events provide low-level methods for seeing what a user is doing, while actions provide more sophisticated connections to buttons, sliders, text fields, and other tools. (See chapter 14.)
- Advanced view controllers provide you with a variety of ways to navigate among pages. (See chapter 15.)

Although we've completed our introduction to the iPhone SDK, we're not done yet. We're going to finish this book by looking at some of the neat tools that are available inside the SDK, including a look at internet-related tools, bringing us full circle to the web techniques that opened this book. Our first stop, however, is an in-depth look at the many ways to input data into an iPhone program.

Part 4

Programming
with the SDK Toolkit

Programming the iPhone can go well beyond the fundamentals that you learned in part 3 of this book. There are numerous frameworks that can give you access to complex, preprogrammed functionality. Part 4 of this book will highlight some of the most important possibilities.

These include accessing data, such as text inputs, preferences, files, databases, and the Address Book (chapter 16); using positioning technologies, including the accelerometers and the GPS (chapter 17); working with media, such as images, movies, and sounds (chapter 18); drawing graphics with Quartz, Core Animation, and OpenGL (chapter 19); and connecting to the internet using URLs, web views, XML, POSTs, and social web technologies (chapter 20).

16

Data: actions, preferences, files, SQLite, and addresses

This chapter covers

- Accepting user input through controls
- Allowing user choice through preferences
- Accessing and creating files
- Using the SQLite library
- Manipulating the Address Book

In part 3 of this book, we offered a tutorial on the most important features of the SDK: we outlined Objective-C and the iPhone OS; we explored the two main tools, Xcode and Interface Builder; we examined view controllers of all types; and we looked at the standard event and action models for the iPhone. In the process, we tried to provide the strong foundation that you need to do any type of iPhone programming. Armed with that knowledge, and with the extensive documentation available online (or as part of Xcode), you should be able to start programming right away.

But we also want to offer you some additional information on many of the SDK's best features—that's the purpose of the fourth and final part of this book. In these five chapters, we're going to touch upon five major categories of SDK tools and show you how to use them.

In the process, we're going to go over some ground covered by Apple in its own documentation for each of these tools. As usual, we're going to add value by approaching things in a tutorial manner and by offering specific examples of how each of the tools can be used in a real program.

We'll also be expanding on our sample programs a bit. Having completed the introduction to SDK, we can take advantage of your knowledge of Objective-C to incorporate at least one in-depth example in each chapter; our intent is to show how different objects can work together to create a more complex Objective-C project. We can't give you full iPhone App Store programs, because of the breadth of what we're covering here, but expect to see some code examples that are more than a page long, and which typically include some off-topic elements.

This chapter will kick off our look at the SDK toolkit with a discussion of data, which will describe many of the ways you can deal with information generally (and text-specifically). We've broken this into a few broad categories. First, we'll look at the ways users can input data into your program, focusing on actions and preferences. Second, we'll examine ways that you can store and retrieve internal data, including using files and the built-in SQL database. In our long example for this chapter, you'll build table views from SQLite data. Third, we'll discuss the Address Book—a comprehensive iPhone system that allows for the simple input *and* retrieval of contact information that can be shared among multiple programs.

16.1   *Accepting user actions*

The simplest way to accept new data from a user is through `UIControls`, a topic that we covered in some depth in the latter half of chapter 14 and that we're looking at again here for completeness' sake. Table 16.1 includes some notes on the controls that you can use to accept user data.

Table 16.1   Various controls allow you to accept user input, most using simple interfaces.

Control	Summary
`UIButton`	Offers simple functionality when the user clicks a button. See section 14.5 for an example.
`UIPageControl`	A pure navigation object that allows users to move between multiple pages using a trio of dots.
`UIPickerView`	Not a `UIControl` object, but allows the user to select from a number of items in a "slot machine" selection. It includes the subclass `UIDatePicker`.

Table 16.1 Various controls allow you to accept user input, most using simple interfaces. *(continued)*

Control	Summary
UISearchBar	Not a UIControl object, but offers similar functionality to a UITextField. It provides an interface that includes a single-line text input, a search button, a cancel button, and a bookmark button. It could theoretically be used in any program where a user would want to save search results, though it's obviously specialized for a web browser. See section 16.5.3 for an example.
UISegmentedControl	A horizontal bar containing several buttons. See section 17.4.2 for an example.
UISlider	A slider that allows users to input from a range of approximate values. See section 14.6.2 for an example.
UISwitch	An on-off button of the sort used in preferences. See section 16.2.1 for an example.
UITextField	A single-line text input, and probably the most common control for true user input. It requires some work to make the keyboard relinquish control. See section 14.6.1 for complete discussion and an example.
UITextView	Not a UIControl object, but does allow the user to enter longer bits of text. As with a text field, you must have it resignFirstResponder status to return control to the program when the user is done typing. As shown in the iPhone Notes utility, this is typically done with a separate Done button at the top of the interface, because the Return key is used to input returns. See section 16.3.4 for an example.
UIToolBar	Not a UIControl object. Instead, it's a bar meant to hold a collection of UIBarButtonItems, each of which can be clicked to initiate an action. The bar is easy to configure and change. See section 18.4 for an example.

Clearly, these controls serve a variety of purposes. Many exist for pure user-interface purposes, which we covered pretty extensively in chapter 14. What's of more interest to us here are the text input controls (UISearchBar, UITextField, and UITextView) that you're likely to use in conjunction with files and databases. We'll look particularly at UISearchBar and UITextView, the two text inputs that we hadn't previously given much attention to, over the course of this chapter.

Not included in this table are the integrated controller pickers that allow users to input data and make choices using complex prebuilt systems. These include the Address Book UI Picker (which is discussed in section 16.5.4) and the image picker (which is discussed in section 18.3).

Controls will be central to any real-life program, so you'll see them throughout the upcoming chapters. Because we'll be seeing lots of examples of their use, we can now move on to the next method of user data input: preferences.

16.2   *Maintaining user preferences*

Preferences are the way that an iPhone program maintains user choices, particularly from one session to another. They're a way to not only accept user input, but also to save it. You can use your own programmatic interface to maintain these preferences, or you can use the Settings interface provided in the iPhone SDK.

If your program includes preferences that might change frequently, or if it would be disruptive for a user to leave your program to set a preference, you can create a preferences page within your program. This type of program-centric preference page is seen in the Stocks and Maps programs, each of which has settings that can be changed on the backside of the main utility.

Alternatively, if your program has preferences that don't change that much, particularly if the defaults are usually OK, you should instead set them using the system's settings. This type of iPhone-centric setting can be seen in the iPod, Mail, Phone, Photos, and Safari applications, all of which have their settings available under the Settings icon on the iPhone screen.

Of the two, the latter is the Apple-preferred way of doing things, but we'll touch upon both, starting with creating your own preferences page. You should feel free to use either method, based upon the needs of your program, but you should most definitely *not* mix the two styles of preferences, because that's likely to be quite confusing for your users.

16.2.1   *Creating your own preferences*

Whenever you're writing iPhone programs, you should always do your best to match the look, feel, and methodology of Apple's existing iPhone programs. Looking through built-in iPhone programs can offer lessons about when and how to use personal preferences on your own. Here's what the personal preferences of those built-in programs can tell us:

- They're used infrequently.
- When they do appear, they are used in conjunction with a program that has only a single page of content (like Stocks) or one that has multiple identical pages of content (like Weather).
- They appear on backside of a flipside controller.
- The preferences appear in a special list view that includes cells inside cartouches.

You can easily accommodate these standards when building your own programs. We're going to do so over the next few examples, with the goal being to create the simple preferences table shown in figure 16.1.

Figure 16.1   This preferences page was built from scratch on the back of a flipside controller.

DRAWING THE PREFERENCES PAGE

If you're going to create a program that has built-in preferences, you should create it using the Utility Application template. As we've previously seen, this will give you access to a flipside controller, which will allow you to create your preferences on the backside of your application.

To create the special cartouched list used by preferences, you must create a table view controller with the special UITableViewGrouped style. This can be done by choosing the Grouped style for your table view in Interface Builder, or by using the init-WithStyle: method in Xcode. Listing 16.1 shows the latter method by creating the UITableViewController subclass (which we've called a PreferencesController) inside the flipside controller's viewDidLoad method.

Listing 16.1 Creating a grouped table in a flipside controller

```
- (void)viewDidLoad {

    PreferencesController *myTableView = [[PreferencesController alloc]
        initWithStyle:UITableViewStyleGrouped];
    [self.view addSubview:myTableView.view];

}
```

Once you've done this, you can then fill in your PreferencesController's table view using the methods we described in chapter 13. You'll probably make use of the cells' accessoryView property, because you'll want to add switches and other objects to the preference listing. Listing 16.2 shows the most important methods required to create a simple preferences page with two switches.

Listing 16.2 Follow the table view methods to fill out your preferences table

```
- (id)initWithStyle:(UITableViewStyle)style {
    if (self = [super initWithStyle:style]) {
        settingsList = [NSArray arrayWithObjects:          ❶ Creates
            [NSMutableDictionary dictionaryWithObjectsAndKeys:    an array
                @"Sounds",@"titleValue",
                @"switch",@"accessoryValue",              ❷ Embeds Booleans
                [NSNumber numberWithBool:YES],               with NSNumber
                    @"prefValue",
                @"setSounds:",@"targetValue",nil],
            [NSMutableDictionary dictionaryWithObjectsAndKeys:
                @"Music",@"titleValue",
                @"switch",@"accessoryValue",
                [NSNumber numberWithBool:YES],@"prefValue",
                @"setMusic:",@"targetValue",nil],nil];
        [settingsList retain];

        switchList = [NSMutableArray arrayWithCapacity:settingsList.count];
        for (int i = 0 ;
             i < [settingsList count] ;
             i++)    {
            if ([[[settingsList objectAtIndex:i]
                objectForKey:@"accessoryValue"] compare:@"switch"] ==
                    NSOrderedSame) {
```

```
                   UISwitch *mySwitch = [[[UISwitch alloc]
                      initWithFrame:CGRectZero] autorelease];
                   mySwitch.on = [[[settingsList objectAtIndex:i]
                      objectForKey:@"prefValue"] boolValue];
                   [mySwitch addTarget:self
                      action:NSSelectorFromString([[settingsList
                         objectAtIndex:i] objectForKey:@"targetValue"])
                            forControlEvents:UIControlEventValueChanged];
                   [switchList insertObject:mySwitch atIndex:i];
                } else {
                   [switchList insertObject:@"" atIndex:i];
                }
             }
          [switchList retain];

          CGPoint tableCenter = self.view.center;
          self.view.center = CGPointMake(tableCenter.x,tableCenter.y+22);

       }
       return self;
    }

    - (NSInteger)numberOfSectionsInTableView:
      (UITableView *)tableView {

       return 1;
    }

    - (NSString *)tableView:(UITableView *)tableView
      titleForHeaderInSection:(NSInteger)section {

       return @"Audio Preferences";
    }

    - (NSInteger)tableView:(UITableView *)tableView
      numberOfRowsInSection:(NSInteger)section {
       return settingsList.count;
    }

    - (UITableViewCell *)tableView:(UITableView *)tableView
      cellForRowAtIndexPath:(NSIndexPath *)indexPath {

       static NSString *MyIdentifier = @"MyIdentifier";

       UITableViewCell *cell = [tableView
          dequeueReusableCellWithIdentifier:MyIdentifier];
       if (cell == nil) {
          cell = [[[UITableViewCell alloc] initWithFrame:CGRectZero
             reuseIdentifier:MyIdentifier] autorelease];
       }

       cell.text = [[settingsList objectAtIndex:indexPath.row]
          objectForKey:@"titleValue"];
       if ([switchList objectAtIndex:indexPath.row]) {
          cell.accessoryView =
             [switchList objectAtIndex:indexPath.row];
       }
       return cell;
    }
```

3 Prepares switch array

4 Moves table down

5 Counts sections

6 Names section

7 Counts rows

8 Creates cells

9 Puts switch in accessory view

This example generally follows the table view methodology that you learned in chapter 13. You use an array to set up your table view ❶. Besides a title, these (mutable) dictionaries also include additional info on the switch that goes into the table view, including what it should be set to and what action it should call. This example shows one nuance we mentioned before: only NSObjects can be placed in an NSDictionary, so you have to encode a Boolean value in order to use it ❷.

Your initWithStyle: method must do two other things. First, it must create a mutable array to hold all your switches for later access. You do all of the creation here ❸, based upon settingsList (or on whatever other means you might have used to pull in preferences data), because if you wait until you get to the table view methods, you can't guarantee the order in which they'll be created. If you didn't fill the switch list here, you could get an out-of-bounds error, if, for example, the switch in row 1 was created before the switch in row 0. Note also that these switches are created with no particular location on the screen, because we'll be placing them later.

Second, it must move your table down a little bit to account for the navigation bar at the top of the flipside page ❹.

The methods that define the section count ❺, the section head ❻, and the row count ❼ are all pretty standard. It's the method that defines the contents of the rows ❽ that's of interest, primarily because it contains code that takes advantage of the accessoryView property that we touched upon in chapter 13. In this method you read back the appropriate switch from your array and input it ❾.

There's no real functionality in this preferences page—that will ultimately be dependent upon the needs of your own program. But this skeleton should give you everything you need to get started. Afterward, you'll need to build your methods (here, setMusic: and setSounds:) which should access the switchList array, and then do the appropriate thing for your program when the switches are toggled.

Switches are the most common element of a preferences page. The other common feature that you should consider programming is the *select list*. That's usually done by creating a subpage with a table view all its own. It should be set in UITableView-Grouped style, like this table was. You'll probably allow users to checkmark one or more elements in the list.

SAVING USER PREFERENCES

We're leaving one element out of this discussion: what to do with your users' preferences after they've set them. It's possible that you'll only want to save user preferences for the length of a single session, but it's our experience that that can often be confusing and even annoying to users. More commonly, you should save preferences from one session to another. We offer three different ways to do so:

- *Save the preferences in a file*—Section 16.3 talks about file access. You can either save the preferences in plain text, or else use a more regulated format like XML, which is covered in chapter 20.
- *Save the preferences in a database*—Section 16.4 covers this.
- *Save the preferences using* NSUserDefaults—This option is discussed next.

As noted, we're going to cover the more general methods later in this chapter. NSUserDefaults is a storage mechanism that's specific to user preferences, though, so we're going to cover it here.

Generally, NSUserDefaults is a persistent shared object that can be used to remember a user's preferences from one session to another. It's sort of like a preferences associative array. There are four major methods, listed in table 16.2.

Table 16.2   Notable methods for NSUserDefaults

Method	Summary
standardUserDefaults	Class method that creates a shared defaults object.
objectForKey:	Instance method that returns an object for the key; there are numerous variants that return specific types of objects such as strings, Booleans, etc.
setObjectForKey:	Instance method that sets a key to the object; there are numerous variants that set specific types of objects such as strings, Booleans, etc.
resetStandardUserDefaults	Class method that saves any changes made to the shared object.

It would be simple enough to modify the previous preferences example to use NSUserDefaults. First, you'd change the init method to create a shared defaults object, then read from it when creating the settingListing array, as shown in listing 16.3.

Listing 16.3   Preferences setup with NSUserDefaults

```
NSUserDefaults *myDefaults = [NSUserDefaults standardUserDefaults];
settingsList = [NSArray arrayWithObjects:
    [NSMutableDictionary dictionaryWithObjectsAndKeys:
        @"Sounds",@"titleValue",
        @"switch",@"accessoryValue",
        [NSNumber numberWithBool:[myDefaults
          boolForKey:@"soundsValue"]],@"prefValue",      ⟵┐  Extracts and
        @"setSounds:",@"targetValue",nil],                   sets sound value
    [NSMutableDictionary dictionaryWithObjectsAndKeys:
        @"Music",@"titleValue",
        @"switch",@"accessoryValue",
        [NSNumber numberWithBool:[myDefaults
          boolForKey:@"musicValue"]],@"prefValue",       ⟵┐  Extracts and
        @"setMusic:",@"targetValue",nil],nil];               sets music value
```

The lines in which the prefValues are set are the new material here. The information is extracted from the NSUSerDefaults first.

The methods called when each of these switches are moved can set and save changes to the default values. You'll want to do other things here too, but the abbreviated form of these methods is shown in listing 16.4.

Listing 16.4   Setting and saving NSUserDefaults

```
-(void)setMusic:(id)sender {

    NSUserDefaults *myDefaults = [NSUserDefaults standardUserDefaults];
```

```
    UISwitch *musicSwitch = [switchList objectAtIndex:1];

    [myDefaults setBool:musicSwitch.on forKey:@"musicValue"];
    [NSUserDefaults resetStandardUserDefaults];
}
-(void)setSounds:(id)sender {

    NSUserDefaults *myDefaults = [NSUserDefaults standardUserDefaults];
    UISwitch *soundsSwitch = [switchList objectAtIndex:0];

    [myDefaults setBool:soundsSwitch.on forKey:@"soundsValue"];
    [NSUserDefaults resetStandardUserDefaults];
}
```

This functionality is simple. You call up the NSUserDefaults, set any values you want to change, and then save them. If you call up your program again, you'll find that the two switches remain in the position that you set them last time you ran the program.

Once you decide how to save your personal preferences, you'll have a skeleton for creating your own preferences page, and if that's appropriate for your program, you're done. But that's just one of two ways to let users add preference data to your program. More commonly, you'll be exporting your settings to the main Settings program. So, how do you do that?

16.2.2 Using the system settings

When you created a personal preferences page in the last section, you used all the SDK programming skills that you've been learning to date, creating objects and manipulating them. Conversely, using the system settings is much easier: it just requires creating some files.

About bundles

Xcode allows you to tie multiple files together into a coherent whole called a *bundle*. In practice, a bundle is just a directory. Often a bundle is made opaque, so that users can't casually see its contents; in this case, it's called a *package*.

The main advantage of a bundle is that it can invisibly store multiple variants of a file, using the right one when the circumstances are appropriate. For example, an application bundle could include executable files for different chip architectures or in different formats.

When working with Xcode, you're likely to encounter three different types of bundles: framework bundles, application bundles, and settings bundles. All frameworks appear packaged as framework bundles, though that's largely invisible to you. An application bundle is what's created when you compile a program to run on your iPhone; we'll talk about how to access individual files in a bundle in the next section, when we talk about files generally. Finally, the settings bundle contains a variety of information about system settings, a topic that we'll be addressing now.

More information on how to access bundles can be found in the NSBundle and CFBundle classes.

To begin using the System Settings, you must create a settings bundle. This is done in Xcode through the File > New File option. To date, we've only created new files using the Cocoa Touch Classes option (starting in section 11.3). Now you should instead choose Settings in the sidebar, which will give you the option to create just one sort of settings file: Settings Bundle. When you do this, Settings.bundle will be added to your current project.

This bundle will, by default, contain two files: Root.plist and Root.strings. Root.strings contains localized content, which means words in the language of your choice. (Localization is a topic that we've generally omitted in this book.) Root.plist is what defines your program's system settings.

EDITING EXISTING SETTINGS

Root.plist is an XML file, but as usual you can view it in Xcode, where it'll appear as a list of keys and values. You'll first want to change the `Title` to your project's name. All of the rest of your settings appear under the `PreferenceSpecifiers` category, as shown in figure 16.2.

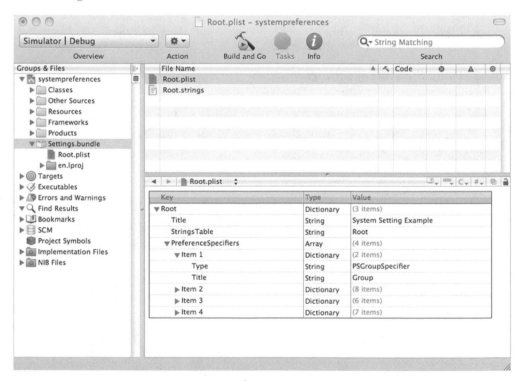

Figure 16.2   This look at system settings reveals some of Root.plist's `PreferenceSpecifiers`.

There are seven types of data that you can enter into the Settings plist file, each of which will create a specific tool on the Settings page. Of these, four appear by default in the plist file at the time of this writing, and are thus the easiest to modify. All seven options are all shown in table 16.3.

Table 16.3 Different preference types let you create different tools on the Settings page.

Preference	Summary	Default ✓
PSChildPaneSpecifier	Points to a subpage of preferences	
PSGroupSpecifier	Contains a group header for the current table section	✓
PSMultiValueSpecifier	Points to a subpage containing a select list	
PSSliderSpecifier	A UISlider	✓
PSTextFieldSpecifier	A UITextField	✓
PSTitleValueSpecifier	Shows the current, unchangeable value of the preference	
PSToggleSwitchSpecifier	A UISwitch	✓

The plist editor is simple to use and will allow you to easily do the vast majority of work required to create the settings for your program. You can cut and paste the existing four preferences (noted by checkmarks in table 16.3) to reorder them or create new instances of the four existing preference types. Then you fill in their data to create preferences that look exactly like you want.

For any setting, the Type string always describes which sort of preference you're setting. Other settings define what you can change. For example, to change the text that appears in a PSGroupSpecifier, you adjust the Title string inside the PSGroupSpecifier dictionary. Changing the PSSliderSpecifier, PSTextFieldSpecifier, and PSToggleSwitchSpecifier are equally easy. The only thing to note on those is the Key string, which sets the name of the preference's variable. You'll need that name when you want to look it up from inside your program (a topic we'll return to).

CREATING NEW SETTINGS

The remaining three preferences are a bit harder to implement because you don't have a preexisting template for them sitting in the default Root.plist file. But all you have to do is create a dictionary that has all the right values.

When you click individual rows in the plist editor, you'll see some iconic options to help in creating new preferences. At any time, you can create new PreferenceSpecifiers (which is to say, new preferences) by clicking the plus (+) symbol to the right of the current row. You can likewise add to dictionaries or arrays by opening them up, then clicking the indented row symbol to the right of the current row.

A PSTitleValueSpecifier is an unchangeable preference. It shows the preference name and a word on the Settings page. Its dictionary includes a Type (string) of PSTitleValueSpecifier, a Title (string) that defines the name of the preference, a Key (string) that defines the variable name, and a DefaultValue (string).

A PSMultiValueSpecifier is a select list that appears on a subpage. Its dictionary contains a Type (string) of PSMultiValueSpecifier, a Title (string), a Key (string), a DefaultValue (string), a Titles (array) that contains a number of String items, and a matched Values (array) that contains Number items.

▼ Item 5	Dictionary	(4 items)
Type	String	PSTitleValueSpecifier
Title	String	CPU Speed
Key	String	title_preference
DefaultValue	String	default
▼ Item 6	Dictionary	(6 items)
Type	String	PSMultiValueSpecifier
Title	String	Animation Type
Key	String	options_preference
DefaultValue	String	0
▼ Values	Array	(3 items)
Item 1	Number	0
Item 2	Number	1
Item 3	Number	2
▼ Titles	Array	(3 items)
Item 1	String	Line Drawings
Item 2	String	Filled Polygons
Item 3	String	Full CGI

Figure 16.3   This display shows how `PSTitleValueSpecifier` **and a** `PSMultiValueSpecifier` **look in Xcode.**

Figure 16.3 shows what these two items look like, laid out in Xcode.

The last sort of setting, `PSChildPaneSpecifier` does something totally different: it lets you create additional pages of preferences.

CREATING HIERARCHICAL SETTINGS

If necessary, you can have multiple pages of settings. To create a subpage, use the `PS ChildPaneSpecifier` type. It should contain a `Type` (string) of `PSChildPaneSpecifier`, a `Title` (string), and a `File` (string) that contains the new plist file without extension.

Once you've done this, you need to create your new plist file. There is currently no "Add plist" option, so we suggest copying your existing Root.plist file, renaming it, and going from there.

We've put together an example of all seven preference types in figure 16.4. It shows the types of preference files that you can create using Apple's built-in functionality.

Now you know everything that's required to give your users a long list of preferences that they can set. But how do you use them from within Xcode?

ACCESSING SETTINGS

Settings end up encoded as variables. As you saw when looking through the plist editor, each individual preference is an `NSString`, `NSArray`, `NSNumber`, or Boolean. These variables can be accessed using the shared `NSUserDefaults` object. We already discussed this class in the last section; it so happens that Apple's settings

Figure 16.4   In order, a `Group`, **a** `TextField`, **another** `Group`, **a** `Switch`, **a** `TitleValue`, **a** `MultiValue`, **a** `ChildPane`, **a third** `Group`, **and a** `Slider`.

bundle uses it, as we suggested you might. The functionality remains the same. You can create it as follows:

```
[NSUserDefaults standardUserDefaults];
```

Once you've done that, you can use `NSUserDefaults`' `objectForKey:` methods, like `arrayForKey:`, `integerForKey:`, and `stringForKey:`, as appropriate to access the information from the settings. For example, the following code applies a string from the settings to a label:

```
myLabel.text = [[NSUserDefaults standardUserDefaults]
    stringForKey:@"name_preference"];
```

Similarly, you can save new settings if you want by using the various `setObjectForKey:` methods—although we don't think this is a particularly good idea if users are otherwise modifying these values in Settings.

There is one considerable *gotcha* that you must watch for: if your user has not yet accessed the settings for your program, then all settings without default values will have a value of `nil`. This means that you'll either need to create your preferences by hand or build defaults into your program, as appropriate.

Most of the time, you'll only need to retrieve the setting values, as described here, but if more is required, you should look at the class reference for `NSUserDefaults`.

That concludes our look at the two ways to create preferences for your programs, and also at how users can input data into your program; but user input represents just one part of the data puzzle. Certainly, a lot of important data will come from users, but data can also come from various files and databases built into your program or into the iPhone itself. Retrieving data from those sources is the topic of the latter half of this chapter.

16.3 Opening files

When we talked about bundles earlier in this chapter, we took our first look at how the iPhone arranges its internal information for programs. That arrangement becomes vitally important when trying to access files that you added to a project.

Fortunately, you can look at how your program's files are arranged when you're testing applications on the iPhone Simulator. Each time you run a program, the program will be compiled to a directory under ~/Library/Application Support/iPhone Simulator/User/Applications. The specific directory will have a hexadecimal name, but you can search a bit to find the right one. Figure 16.5 shows an example of the directory for the sample program that we used to set up our system preferences example (though the subdirectories will be the same for any basic program).

As shown, there are four directories of files for this one simple program. The majority of the content appears in the *application bundle*, which in this example is called systempreferences.app. There you'll find everything that you added to your project, including text files, pictures, and databases. The other three directories that you can use are Documents, Library, and tmp.

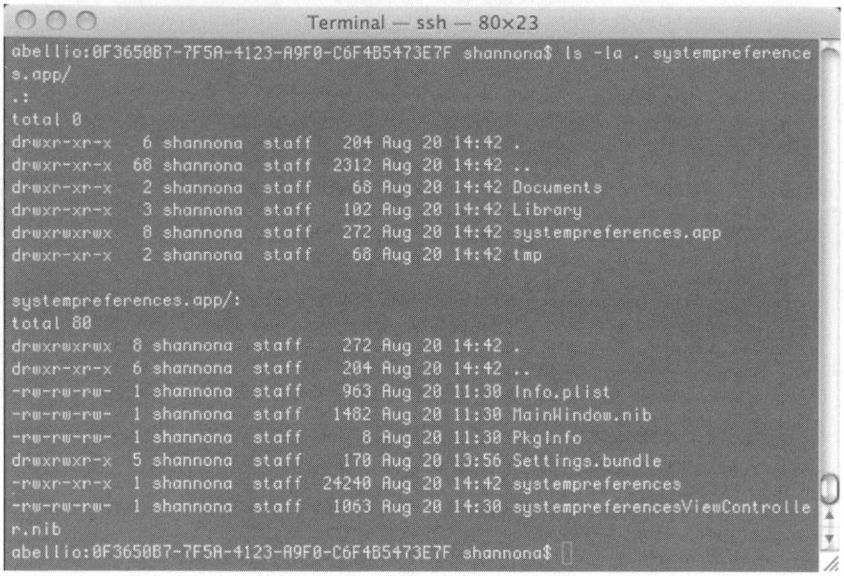

```
abellio:0F3650B7-7F5A-4123-A9F0-C6F4B5473E7F shannona$ ls -la . systempreference
s.app/
.:
total 0
drwxr-xr-x   6 shannona  staff   204 Aug 20 14:42 .
drwxr-xr-x  68 shannona  staff  2312 Aug 20 14:42 ..
drwxr-xr-x   2 shannona  staff    68 Aug 20 14:42 Documents
drwxr-xr-x   3 shannona  staff   102 Aug 20 14:42 Library
drwxrwxrwx   8 shannona  staff   272 Aug 20 14:42 systempreferences.app
drwxr-xr-x   2 shannona  staff    68 Aug 20 14:42 tmp

systempreferences.app/:
total 80
drwxrwxrwx   8 shannona  staff    272 Aug 20 14:42 .
drwxr-xr-x   6 shannona  staff    204 Aug 20 14:42 ..
-rw-rw-rw-   1 shannona  staff    963 Aug 20 11:30 Info.plist
-rw-rw-rw-   1 shannona  staff   1482 Aug 20 11:30 MainWindow.nib
-rw-rw-rw-   1 shannona  staff      8 Aug 20 11:30 PkgInfo
drwxrwxr-x   5 shannona  staff    170 Aug 20 13:56 Settings.bundle
-rwxr-xr-x   1 shannona  staff  24240 Aug 20 14:42 systempreferences
-rw-rw-rw-   1 shannona  staff   1063 Aug 20 14:30 systempreferencesViewControlle
r.nib
abellio:0F3650B7-7F5A-4123-A9F0-C6F4B5473E7F shannona$ []
```

Figure 16.5   Compiled programs contain several directories full of files.

These are all intended to be used for files that are created or modified when the program is run. Documents should contain user-created information (including new or modified text files, pictures, and databases), Library should contain more programmatic items (like preferences), and tmp should contain temporary information. Each of these directories starts out empty, other than the fact that the Library maintains a local copy of your system settings. We'll talk about how and why you'd fill them momentarily.

16.3.1   *Accessing your bundle*

In previous chapters, we've shown how easy it is to add files to your project. You drag the file into Xcode, and everything is correctly set up so that the file will become part of your program when it compiles. As we now know, that means the file is copied into your application bundle.

For many bundled files, you don't have to worry about anything beyond that. For example, when you worked with picture files, you entered the name of the file in Interface Builder, and the SDK automatically found it for you. But if you want to access a file that doesn't have this built-in link, you'll need to do a little bit more work.

Whenever you're working with the file system on the iPhone, access will be abstracted through objects. You'll send messages that tell the SDK what area of the file system you're looking for, and the SDK will then give you precise directory paths. The benefit of this abstraction is that Apple can reorganize the file system in future releases, and your program won't be affected at all.

The first files you'll want to access will probably be in your bundle: files that you included when you compiled your program. Accessing a bundle file is usually a two-step process, as shown in this database example (which we'll return to in the next section):

```
NSString *paths = [[NSBundle mainBundle] resourcePath];
NSString *bundlePath = [paths stringByAppendingPathComponent:dbFile];
```

In this example, `mainBundle` returns the directory path that corresponds to your application's bundle, and `resourcePath` expands that to be the directory path for the resources of your program (including, in this case, a database, but this could also be anything else you added to your program). Finally, you use `stringByAppendingPath-Component:` to add your specific file to the path. This `NSString` method makes sure that a path is constructed using slashes (/) as needed.

The result is a complete path that can be handed to other objects as needed. We'll see how that works with a database in the next section. You could likewise use it for `UImage`'s `imageWithContentsOfFile:` method or `NSFileHandle`'s `fileHandleFor-ReadingAtPath` method. We'll return to the latter shortly.

But there's one fundamental problem with accessing files inside the application bundle: you can't modify them. Apple generally suggests that you should treat the application bundle as read-only, and there's a real penalty if you don't: your program will stop working because it won't checksum correctly. This means that the application bundle is great for files that don't change, but if you want to modify something (or create something new), you need to use the other directories we mentioned, starting with the Documents folder.

16.3.2 *Accessing other directories*

When working with directories other than the bundle, you have to think about two things: how to access those files and how to move files among multiple directories.

RETRIEVING A FILE

Once a file is sitting in your Documents directory, you can retrieve it much like you retrieved files from the bundle directory:

```
NSArray *paths = NSSearchPathForDirectoriesInDomains(NSDocumentDirectory,
    NSUserDomainMask, YES);
NSString *documentsDirectory = [paths objectAtIndex:0];
NSString *docPath = [documentsDirectory
    stringByAppendingPathComponent:dbFile];
```

The magic here occurs in the `NSSearchPathForDirectoriesInDomains` function. The first argument is usually `NSDocumentDirectory` or `NSLibraryDirectory`, depending on which directory you want to get to. The other two arguments should always be the same for the iPhone. The result will be an array of strings, with each containing a path. The first path in the `NSArray` will usually be the right one, as shown here. You can then use the `stringByAppendingPathComponent:` method, like before, to build the complete path for your file. Voila! You've now used some slightly different methods to access a file in your Documents directory rather than the bundle directory.

COPYING A FILE

There's been a slight disconnect in our discussion of files and directories to date. When you compile your project, all of your files will be placed into your application bundle. But if you ever want to edit a file, it must be placed in a different directory, such as Documents. So how do you get a file from one place to the other? You use the NSFileManager:

```
NSFileManager *fileManager = [NSFileManager defaultManager];
success = [fileManager copyItemAtPath:bundlePath toPath:docPath
    error:&error];
```

The file manager is a class that allows you to easily manipulate files by creating them, moving them, deleting them, and otherwise modifying them. As is the case with many classes we've seen, you initialize it by accessing a shared object. You can do lots of things with the file manager, including copying (as we've done here) and checking for a file's existence (which we'll demonstrate shortly). You should look at the NSFile-Manager class reference for complete information.

As we'll see, the NSFileManager is one of numerous classes that you can use to work with files.

16.3.3 *Manipulating files*

It's possible that once you've built your file path, you'll be ready to immediately read the file's contents, using something like the UIImage methods (which we'll touch upon in chapter 19) or the functions related to SQLite (which we'll cover later in this chapter). But it's also possible that you'll want to manipulate the raw files, reading and parsing them in your code, as soon as you've created a file path. There are numerous ways to do this, as shown in table 16.4.

Table 16.4   A couple of ways to manipulate files using the SDK

Class	Method	Summary
NSHandle	fileHandleForReadingAtPath: fileHandleForWritingAtPath: fileHandleForUpdatingAtPath:	Allows you to open a file
NSHandle	readsDataofLength:	Returns an NSData containing the specified number of bytes from the file
NSHandle	readsDataToEndOfFile	Returns an NSData with the rest of the file's content
NSHandle	closeFile	Closes an NSHandle
NSFileManager	contentsAtPath:	Returns an NSData with the complete file's contents
NSData	initWithContentsOfFile:	Creates an NSData with the complete file's contents
NSData	writeToFile:atomically:	Writes the NSData to a file
NSString	stringWithContentsOfFile:encoding:error:	Returns an NSString with the complete file's contents

Table 16.4   A couple of ways to manipulate files using the SDK *(continued)*

Class	Method	Summary
NSString	initWithData:encoding:	Returns an NSString with the NSData's contents
NSString	writeToFile:atomically:encoding:error:	Writes the NSString to a file

As table 16.4 shows, you can access files in a huge variety of ways once you've created a file path. If you're a C programmer, opening a file handle, reading from that file handle, and finally closing that file handle is apt to be the most familiar approach. Or, you could use a shortcut and go straight to the NSFileManager and have it do the whole process. Even quicker is using methods from NSData or NSString to directly create an object of the appropriate type.

Any of these simpler methods is going to cost you the ability to step through a file byte by byte, which may be a limitation or a benefit, depending on your program. But with the simpler methods, you just need a single line of code:

```
NSString *myContents = [NSString stringWithContentsOfFile:myFile
    encoding:NSASCIIStringEncoding error:&error];
```

Table 16.4 also lists a few ways to write back to files, including simple ways to dump an NSData object or an NSString object to a file. There are also other ways. When you decide which set of methods you're most comfortable using, you should consult the appropriate class reference for additional details.

> **File content**
>
> In this section—and in our next example—we're largely assuming that files contain plain, unstructured text. But this doesn't have to be the case. XML is a great way to store local data in a more structured format. Chapter 20 covers how to read XML and includes an example of reading local XML data.

When you're working with files, you're likely to be doing one of two things. Either you have files that contain large blobs of user information or you have files that contain short snippets of data that you've saved for your program. To demonstrate how to use a few of the file objects and methods, we're going to tackle the first problem by building a simple notepad prototype.

16.3.4   Filesaver: a UITextView example

This program will allow you to maintain a text view full of information from one session to another. It's relatively basic, but you can imagine how you could expand it to mimic the iPhone's notepad program, with its multiple notes, toolbars, navigator, and image background.

Listing 16.5 shows this simple filesaver example. The objects, as usual, were created in Interface Builder: a UIToolBar (with associated UIBarButtonItem) and a UITextView.

Listing 16.5   A prototype notepad program that maintains a text field as a file

```
@implementation filesaverViewController

- (void)viewDidLoad {

    NSArray *paths =
        NSSearchPathForDirectoriesInDomains(NSDocumentDirectory,
            NSUserDomainMask, YES);
    NSString *documentsDirectory =
        [paths objectAtIndex:0];                          ← Creates  ❶
    filePath = [documentsDirectory                          file path
        stringByAppendingPathComponent:
            @"textviewcontent"];
    [filePath retain];

    NSFileManager *myFM =
        [NSFileManager defaultManager];                   ❷ Checks if
    if ([myFM isReadableFileAtPath:filePath]) {              file exists
        myText.text =
            [NSString stringWithContentsOfFile:filePath];   ❸ Fills text view
    }                                                         from file
    keyboardIsActive = NO;          ←   Sets instance
                                    ❹   variable
    [super viewDidLoad];
}

-(IBAction)finishEditing:(id)sender {      ←   Executes
                                           ❺   "Done" action
    if (keyboardIsActive == YES) {
        [myText resignFirstResponder];
    } else {
        exit(0);
    }
}

- (void)textViewDidBeginEditing:      ❻ Clears starting
    (UITextView *)textView {      ←      notepad

    if ([myText.text compare:@"Type Text Here."] == NSOrderedSame) {
        myText.text = [NSString string];
    }
    keyboardIsActive = YES;

}
                                                    ❼ Saves notepad
- (void)textViewDidEndEditing:(UITextView *)textView {  ←   contents

    [textView.text writeToFile:filePath atomically:YES
        encoding:NSASCIIStringEncoding error:NULL];
    keyboardIsActive = NO;

}
...
@end
```

This program shows how easy it is to access files. The hardest part is determining the path for the file, but that involves using the path creation methods we looked at a few sections back. When you've got your path, you save it as a variable so that you won't have to recreate the path later ❶. Next, you use NSFileManager to determine whether a file exists ❷. If it does, you can immediately fill your UITextField with its content ❸. Finally, you set a keyboardIsActive variable ❹, which you'll update throughout the program.

As we've previously noted, the objects that pull up keyboards are a bit tricky, because you have to explicitly get rid of the keyboard when editing is done. For UITextFields you could turn the Return key into a Done key to dismiss the keyboard, but for a UITextView you usually want your user to be able to enter returns, so you must typically create a bar at the top of the page with a Done button. Figure 16.6 shows this layout of items.

When the Done button is clicked, your finishEditing: method ❺ is called, which resigns the first responder, making the keyboard disappear (unless you're not editing, in which case it closes the program).

The last two methods are defined in the UITextFieldDelegate protocol. Whenever editing begins on the text field ❻, your program checks to see if the starting text is still there, and if so clears it. Whenever editing ends on the text field ❼, the content is saved to your file. Finally, the keyboardIsActive variable is toggled, to control what the Done button does in each state.

As you saw in table 16.4, there are numerous other ways to read files and save them. The methods in listing 16.5 are simple, but they allow you to make good use of your notepad's file.

Files are OK to use for saving one-off data, but if you're storing a lot of really large data, we suggest using a database when it's available. And on the iPhone, a database is *always* available.

Figure 16.6   A keyboard fills the screen of the Filesaver.

16.4   *Using SQLite*

The SDK's built-in database is SQLite, a public domain software package. You can find more information on it at http://www.sqlite.org, including documentation that's considerably more extensive than what we could include here. You need to know the SQL language to use SQLite, and we won't cover SQL syntax here at all. In addition, you must be familiar with the SQLite API. We'll show how to use it for some basic tasks here, but there's a much more extensive reference online.

SQLite has what we find to be two major limitations. First, there's no simple way to create your database. We think that there should be an integrated interface in Xcode, and we hope that there is sometime soon. Instead, you must create the database by hand for now. Second, there's no object-oriented interface to SQLite. Instead, you'll be using an API that falls back on C code, which we find less elegant and harder to use than the typical Objective-C class.

Given these limitations, we still think that using an SQL database is a better option than files for most situations, and we highly suggest that you learn enough about SQL to use it comfortably.

16.4.1 Setting up an SQLite database

Prior to using SQLite in your program, you must set up a database that contains all of your tables and the initial data that you want. We'll look at the general steps first, and then set up a database that can be used to drive a navigation menu.

CREATING AN SQLITE DATABASE

Creating an SQLite database typically is done from the command line, although it can also be done entirely programmatically. Programmatic creation of the database will not be covered here, but you can find documentation on the SQLite site for doing that. The steps for creating it from the command line are listed in table 16.5.

Table 16.5   Creating an SQLite database from the command line

Step	Description
1. Prepare your table.	Figure out the design of each table in your database. Create a file for the initial data of each table (if any) that has data cells separated with pipes (|) and data rows separated with returns.
2. Create your database.	Start SQLite with this command: `sqlite3 filename` Use a `CREATE TABLE` command to create each table.
3. Enter your initial info.	Use this command to fill each table: `.import table filename` Quit SQLite.
4. Add your database to the Xcode.	Inside Xcode, use the Add > Existing Files menu option to add your database to your project.

To show how all this works, we're going to put together a data file for a database-driven navigation controller. When we talked about tables in chapters 13 and 15, we created them from arrays and dictionaries. This is a fine technique when you're creating small, stable hierarchies, but what if you want to build something larger or something that can be modified by the user? In those cases, a database is a great backend for a navigation menu.

DESIGNING A NAVIGATION MENU

To support a database-driven menu, we've designed a simple database schema. Each row in the navigation hierarchy will be represented by one row in a database. Each of those rows will have five elements:

- `catid`—provides a unique (and arbitrary) ID for an individual row in the menu
- `parentid`—indicates what row in the database acts as the hierarchical parent of the current row, or lists 0 if it's a top-level row that would appear on the first page of the menu
- `title`—contains the printed text that will appear in the menu
- `entrytype`—specifies whether the row is a *category* (which opens up a submenu) or a *result* (which performs some action)
- `ordering`—lists the order in which the rows should appear on an individual page of the menu

Here's an example of what a data file might look like, with the five elements shown in the preceding order:

```
> cat nav.data
1|0|First|category|1
2|0|Third|category|3
3|0|Second|category|2
4|2|Submenu|category|1
5|0|Action #1|result|4
6|1|Action #1B|result|1
```

And here's how you'd create a table for that data and import it:

```
> sqlite3 nav.db
SQLite version 3.4.0
Enter ".help" for instructions
sqlite> CREATE TABLE menu (catid int(5),parentid int(5),title
   varchar(32),entrytype varchar(12), ordering int(5));
sqlite> .import nav.data menu
```

Afterward, you can add your now-complete database to Xcode using the normal procedures, a step that's not shown here. Once you've linked in your database the first time, you can go back and make changes to it, and the new version will always be used when you recompile your project.

You've now got a ready-to-run database, but you'll still need to prepare your Xcode to use SQLite. We'll look at that next.

16.4.2 Accessing SQLite

You'll have to link in some additional resources to use SQLite, as is typical for any major new functionality.

First, you need to add the framework, which you'll find under /usr/lib/libsqlite3.0.dylib, rather than in the standard Framework directory. Second, you must add the include file, which is sqlite3.h.

You've now got a database that's ready to use, and you've included the functionality that you need to use it. The next step is to access SQLite's functions.

16.4.3 Accessing your SQLite database

SQLite includes approximately 100 functions, about 20 object types, and a huge list of constants. We're going to cover the basics that you'll need to access the database you've created. Table 16.6 shows the most critical API commands. They generally revolve around two important concepts: the database handle (which is returned by `sqlite3_open` and is used by everything else) and the prepared statement (which is returned by `sqlite3_prepare` and which is used to run queries).

Table 16.6   The most important SQLite API commands

Function	Arguments	Summary
`sqlite3_open`	`filename, address of database`	Opens a database
`sqlite3_prepare`	`database, SQL as UTF-8, max length to read, address of statement, address of unread results`	Turns an SQL statement in UTF-8 format into a pointer to a prepared statement, which can be handed to other functions
`sqlite3_step`	`prepared statement`	Processes a row of results from a prepared statement, or else returns an error
`sqlite3_column_int`	`prepared statement, column #`	Returns an int from the active row; there are also several other simple functions that similarly return a specific column from the active row
`sqlite3_column_string`	`prepared statement, column #`	Returns a char *, which is to say a string, from the active row; there are also several other simple functions that similarly return a specific column from the active row
`sqlite3_finalize`	`prepared statement`	Deletes a prepared statement
`sqlite3_close`	`database`	Closes a database

These functions, in order, show the usual life cycle of an SQLite database:

1 Open the database.
2 Prepare statements, one at a time.
3 Step through a statement, reading columns.
4 Finalize the statement.
5 Close the database.

SQLite does include two convenience functions, `sqlite3_exec()` and `sqlite3_get_table()`, which simplify these steps; but the functions are built using the core functionality above, so that's what we've decided to highlight.

16.4.4 *Building a navigation menu from a database*

Now that you have a basic understanding of the SQLite functions, you can put together a prototype of a database-driven menu navigation system. What we'll do here is by no means complete, but it'll give you a great basis to build on. This example will also be one of the most complex in the book. It includes multiple classes of new objects designed to work either apart (in different programs) or together.

In this section we'll be covering the SKDatabase class (which abstracts database connections), the SKMenu class (which abstracts navigator menu creation), and the DatabaseViewController (which transforms a typical table view controller into a database-driven class). In the end, we'll hook everything together with the app delegate.

THE DATABASE CLASS

Because there aren't any preexisting object-oriented classes for the SQLite database functions, any program using a database should start off creating its own. Listing 16.6 contains the start of such a class, creating methods for the parts of the API that you'll need to create the database view controller.

Listing 16.6 SKDatabase, a new SQLite3 database class

```
#import "SKDatabase.h"              ◄———  ❶ Includes
                                          header file
@implementation SKDatabase                        ❷ Creates
                                                     database handle
- (id)initWithFile:(NSString *)dbFile {   ◄——┘

    self = [super init];

    NSString *paths = [[NSBundle mainBundle] resourcePath];
    NSString *path = [paths stringByAppendingPathComponent:dbFile];

    int result = sqlite3_open([path UTF8String], &dbh);
    NSAssert1(SQLITE_OK == result, NSLocalizedStringFromTable
      (@"Unable to open the sqlite database (%@).",
      @"Database", @""),
      [NSString stringWithUTF8String:sqlite3_errmsg(dbh)]);

    return self;
}                      ❸ Closes
                         database
- (void)close {   ◄——┘

    if (dbh) {
        sqlite3_close(dbh);
    }
}
                  ❹ Returns
                    database handle
- (sqlite3 *)dbh {   ◄——┘
    return dbh;
}
                                           ❺ Prepares SQL
- (sqlite3_stmt *)prepare:(NSString *)sql {   ◄——  statement

    const char *utfsql = [sql UTF8String];
    sqlite3_stmt *statement;

    if (sqlite3_prepare([self dbh],utfsql,-1,&statement,NULL)==SQLITE_OK) {
        return statement;
    } else {
        return 0;
    }
```

```
    }
- (id)lookupSingularSQL:(NSString *)sql forType:          ➏  Looks up
    (NSString *)rettype {                                       SQL results

    sqlite3_stmt *statement;
    id result;                                       ➐  Calls prepare
                                                         function      ➑  Steps through
    if (statement = [self prepare:sql]) {                                 rows
        if (sqlite3_step(statement) == SQLITE_ROW) {
            if ([rettype compare:@"text"] == NSOrderedSame) {
                result = [NSString stringWithUTF8String:
                    (char *)sqlite3_column_text          ➒  Gets results
                        (statement,0)];                      of text
            } else if ([rettype compare:@"integer"] == NSOrderedSame) {
                result = (id)sqlite3_column_int
                    (statement,0);                   ➓  Gets results of
            }                                            numbers
        }
    }
    sqlite3_finalize(statement);      ⟵
    return result;                        Destroys prepared
}                                     ⓫  statement

@end
```

The header file ➊, not shown, includes one variable declaration for dbh, the database handle. That's the one variable you want to always have available to your class, because it gives access to the database. Now you're ready to start work on the source code file.

The initWithFile: method ➋ uses some of the file commands that you learned in the previous section to find the database file, which is in the main bundle (but remember, you'll want to copy this to the Documents directory if you make changes to your database). It then opens the file using sqlite3_open, the first of several SQLite3 API commands. Note that the NSString for the path has to be converted with the UTF8String method. This must be done throughout the class, because the SQLite API does not use the Objective-C classes you're familiar with.

The next few methods are pretty simple. close signals the end of the database life cycle ➌, dbh is a getter for the class's one variable ➍, and prepare turns an SQL statement into a prepared statement ➎.

The lookupSingularSQL: method is where things get interesting, because it shows off the life cycle of a complete SQL function call ➏. Note that this function allows only a simple SQL call that returns one column from one row of information. That's all you need for the database view controller, but you'll doubtless need more complexity for a larger application.

The function starts off by turning the SQL statement into a prepared statement ➐. Then it steps to the first row ➑. Depending on the type of lookup, it either fetches a string ➒ or an int ➓. Finally, it cleans up the statement with a finalize ⓫.

In a more complex class, you'd doubtless want to write methods that execute SQL calls without any returns, that return multiple columns from a row, and that return multiple rows, but we're going to leave that for now (because we don't need any of those features for this example), and move on to the menu class. The SQLite API has more information on these features if you need them.

THE MENU CLASS

The next class, SKMenu, acts as an intermediary. At the front end, it accepts requests for information about the menu that will fill the table view. On the back end, it turns those requests into SQL queries. It's been designed in this way to create an opaque interface: a programmer will never have to know that a database is being used, simply that the SKMenu class returns results for a table view.

The simple code of SKMenu is shown in listing 16.7. It mainly illustrates how to use the SKDatabase class in listing 16.6.

Listing 16.7 SKMenu, an interface to the SKDatabase class

```
#import "SKMenu.h"                          ← ❶ Includes header file

@implementation SKMenu                           ❷ Sets up database for menu

- (id)initWithFile:(NSString *)dbFile {     ←

    self = [super init];
    myDB = [[SKDatabase alloc] initWithFile:dbFile];
    return self;

}                                                ❸ Counts rows in a page
- (int)countForMenuWithParent:(int)parentid {   ←

    int resultCount = 0;

    NSString *sql = [NSString stringWithFormat:
        @"SELECT COUNT(*) FROM menu WHERE parentid=%i",parentid];

    resultCount = (int)[myDB lookupSingularSQL:sql forType:@"integer"];
    return resultCount;

}
- (id)contentForMenuWithParent:(int)parentid        ❹ Gets text for row
    Row:(int)row content:(NSString *)contenttype {

    NSString *sql = [NSString stringWithFormat:@"SELECT %@ FROM menu WHERE
        parentid=%i AND ordering=%i",contenttype,parentid,row];
    return [myDB lookupSingularSQL:sql forType:@"text"];

}
- (int)integerForMenuWithParent:(int)parentid       ❺ Gets number for row
    Row:(int)row content:(NSString *)contenttype {

    NSString *sql = [NSString stringWithFormat:@"SELECT %@ FROM menu WHERE
        parentid=%i AND ordering=%i",contenttype,parentid,row];
    return (int)[myDB lookupSingularSQL:sql forType:@"integer"];

}
- (void)dealloc {           ←            ❻ Cleans up

    [myDB close];
    [myDB release];

    [super dealloc];

}

@end
```

Again, we haven't shown the include file ❶, but it includes one variable, myDB, which is a reference to the database object linked to the menu. The initWithFile: method ❷ initializes myDB by creating the database object.

The countForMenuWithParent: method is the first one to use the database ❸. It gets a sum of how many menu items there are at a particular level of the menu hierarchy. contentForMenuWithParent: ❹ and integerForMenuWithParent: ❺ are two other lookup functions. The first looks up database entries that will return strings, and the second looks up database entries that will return ints. This is required because, as you'll recall, SQLite has different database lookup functions for each of the variable types.

Finally, the dealloc method cleans up the database ❻, first closing it and then releasing the object. It's always important in Objective-C to keep track of which objects are responsible for which objects. Here, the menu is responsible for the database, so it does the cleanup.

THE DATABASE VIEW CONTROLLER

Now that you've got some menu methods that allow a program to figure out the contents of a hierarchy of menus, you can put together your table view controller, which will read that information and fill table views on the fly. Listing 16.8 shows how the menu functions are used.

Listing 16.8   DatabaseViewController, a database-driven table view controller

```
- (id)initWithParentid:(int)parentid        ❶ Sets up
  Menu:(SKMenu *)passedMenu {                   variables

  if (self = [super initWithStyle:UITableViewStylePlain]) {
    menuparentid=parentid;
    myMenu = passedMenu;
  }
  return self;
}

- (NSInteger)numberOfSectionsInTableView:     ❷ Counts
  (UITableView *)tableView {                      sections

  return 1;
}

- (NSInteger)tableView:(UITableView *)tableView    ❸ Counts
  numberOfRowsInSection:(NSInteger)section {          rows

  return [myMenu countForMenuWithParent:menuparentid];
}

- (UITableViewCell *)tableView:(UITableView *)tableView    ❹ Draws
  cellForRowAtIndexPath:(NSIndexPath *)indexPath {            cell

  static NSString *MyIdentifier = @"MyIdentifier";

  UITableViewCell *cell = [tableView
    dequeueReusableCellWithIdentifier:MyIdentifier];
  if (cell == nil) {
    cell = [[[UITableViewCell alloc] initWithFrame:CGRectZero
      reuseIdentifier:MyIdentifier] autorelease];
```

```
    }
    int thisRow = indexPath.row + 1;
    cell.text = [myMenu contentForMenuWithParent:menuparentid Row:thisRow
        content:@"title"];

    NSString *cellType = [myMenu contentForMenuWithParent:menuparentid
        Row:thisRow content:@"entrytype"];

    if ([cellType compare:@"category"] == NSOrderedSame) {
        cell.accessoryType = UITableViewCellAccessoryDisclosureIndicator;
    }
    return cell;
}
- (void)tableView:(UITableView *)tableView
    didSelectRowAtIndexPath:(NSIndexPath *)indexPath {

    int thisRow = indexPath.row + 1;
    NSString *cellType = [myMenu contentForMenuWithParent:menuparentid
        Row:thisRow content:@"entrytype"];

    if ([cellType compare:@"category"] == NSOrderedSame) {
        NSString *thisText = [myMenu contentForMenuWithParent:menuparentid
            Row:thisRow content:@"title"];
        int newParent = [myMenu integerForMenuWithParent:menuparentid
            Row:thisRow content:@"catid"];

        DatabaseViewController *newController =
            [[DatabaseViewController alloc]
                initWithParentid:newParent Menu:myMenu];
        newController.title = thisText;

        [self.navigationController pushViewController:newController
            animated:YES];

        [newController release];
    }
}
```

⑤ Pops up submenu

To properly understand how the database view controller works, you should remind yourself of the menu format that we introduced a few pages ago. Remember that each row of the menu has an individual ID (the catid), and a parentid that indicates what lies above it in the menu hierarchy. There's also a title, which lists what the menu row will say, a category, which indicates whether it leads to a new menu or is an end result, and an ordering variable. You'll use all that information in putting together your table view.

The database view controller will be called multiple times by your project, once per menu or submenu. Each time, the initWithParentid:Menu: method identifies what level of the hierarchy to draw from the menu that's enclosed ❶. For example, if the parentid is 0, the top-level menu is drawn; if the parentid is 2, the menu that lies under entry (catid) 2 is drawn. The sole purpose of the init is to save that information.

You then have to fill in the standard table view controller methods. The count of sections is always 1 ❷. The number of rows is calculated from the database, using the SKMenu's countForMenuWithParent: method ❸.

`tableView:cellForRowAtIndexPath:` is the first somewhat complex method **4**. After the standard setup of the cell, the method looks up the title to be placed in the menu row. It then determines whether the menu row is a category or not; this affects whether the chevron accessory is placed.

Finally, `tableView:didSelectRowAtIndexPath:` does the fancy work **5**. If the cell isn't a `category`, it doesn't do anything. (You will probably change this when creating another program, because you may want `results` to result in some action; this could be a great place to introduce a new protocol to respond when a `result` row is selected.)

If the cell *is* a category, magic happens. The database view controller creates a new database view controller, on the fly, using the same old menu, but the current `catid` becomes the new `parentid`, which means the new view controller will contain all of the rows that lie under the current row on the hierarchy. The new database view controller is then placed on the navigator controller's stack, using the navigation methods you learned in chapter 15.

Figure 16.7 shows how all this fits together, using the database that you created at the beginning of this section.

There's just one thing missing from this example—the app delegate.

THE APP DELEGATE

The app delegate needs to create the navigator, initialize the menu object, build the first level of the menu hierarchy, and clean things up afterward. Listing 16.9 shows the couple of steps required to do this.

Figure 16.7 This menu was created directly from a database.

Listing 16.9 The app delegate that glues together these classes

```
- (void)applicationDidFinishLaunching:          1  Sets
  (UIApplication *)application {                   things up

  myMenu = [[SKMenu alloc] initWithFile:@"nav.db"];
  DatabaseViewController *newController = [[DatabaseViewController alloc]
    initWithParentid:0 Menu:myMenu];
  newController.title = @"DB Menu";

  [self.navigationController pushViewController:newController
    animated:NO];
  [newController release];

  [window addSubview:[navigationController view]];
  [window makeKeyAndVisible];
}
                                                2  Cleans
- (void)dealloc {                                  up
```

```
    [myMenu release];
    [navigationController release];
    [window release];
    [super dealloc];
}
```

The `applicationDidFinishLaunching:` method sets things up ❶. After initializing the menu, it creates the first database view controller, and pushes it onto the navigation stack. The `dealloc` method later closes everything out ❷. Note that it releases the menu object, which in turn will close the database and release that, ending the menu's life cycle.

Not shown here is the Interface Builder file, which includes one object, a navigation controller. Its standard view controller should be deleted, because you'll be replacing it here.

Though it's relatively basic, you now have a hierarchical menu of tables built entirely from a database.

16.4.5 Expanding this example

This example not only showed how to use databases in a real application, but also how to put together a more complex project. Nonetheless, if you wanted to make regular use of the database and menu classes, you'd probably want to expand it more. We've already noted that `SKDatabase` could use more functionality, and that the database view controller will need to do something for the `result` pages that it arrives on.

Because this is all database driven, you can also hand off considerable power to the users. It would be easy to expand this example so that users could create their own rows in menus and reorder the existing ones.

With SQLite now covered to the depth we can give it, we're going to move on to the last major method of data retrieval, one of equal complexity: the Address Book.

16.5 Accessing the Address Book

Like SQLite, the Address Book is too complex to wholly document within the constraints of this chapter. It's made up of two different frameworks—the Address Book framework and the Address Book UI framework—and together they contain over a dozen references. Fortunately, Apple offers an extensive tutorial on the Address Book: "Address Book Programming Guide for iPhone OS."

In this section, we'll try to provide a basic reference that will supplement Apple's own tutorial, but we suggest you read their guide for more extensive information.

16.5.1 An overview of the frameworks

As noted, there are two frameworks for the Address Book. The Address Book framework contains what you'd expect: information on the data types that make up the Address Book and how to access them. The Address Book UI framework contains a bunch of handy interfaces that allow you to hand off the selection and editing of Address Book entries to modal view controllers that Apple has already written.

In order to use this functionality, you must include one or both frameworks, plus the appropriate include files: AddressBook/AddressBook.h and AddressBookUI/AddressBookUI.h.

Table 16.7 lists many of the most important classes in the frameworks.

Table 16.7 The Address Book classes

Class	Framework	Summary
ABAddressBook	Address Book	Interface for accessing and changing the Address Book; may not be required if you use the Address Book UI framework
ABNewPersonViewController	Address Book UI	Interface for entering new record manually
ABPeoplePickerNavigationController	Address Book UI	Interface for selecting users and properties
ABPersonViewController	Address Book UI	Interface for displaying and editing records
ABUnknownPersonViewController	Address Book UI	Interface for displaying "fake" contact and possibly adding it to Address Book
ABGroup	Address Book	Opaque type giving access to the records of groups
ABPerson	Address Book	Opaque type giving access to the records of individual people
ABRecord	Address Book	Record providing information on a person or group
ABMultiValue	Address Book	Type containing multiple values, each with its own label; its precise use is defined in ABPerson, where it's applied to addresses, dates, phone numbers, instant messages, URLs, and related names
ABMutableMultiValue	Address Book	An ABMultiValue whose values can be modified

Each of these classes contains numerous functions that can be used to build Address Book projects. We'll talk about a few important functions and point you to the class references for the rest.

16.5.2 *Accessing Address Book properties*

As we'll see shortly, the Address Book and Address Book UI frameworks ultimately provide different ways of accessing the iPhone's Contacts data information: you might be working with the Address Book programmatically, or a user might be making selections through fancy UIs. Ways to select individual contacts might vary, but once a contact has been selected, you'll generally use the same getter and setter functions to work with that record. These important functions are listed in table 16.8.

Table 16.8 Property setters and getters are among the most important functions in the Address Book.

Function	Arguments	Summary
ABRecordCopyValue	ABRecordRef, property	Looks up a specific property from a specific record
ABRecordSetValue	ABRecordRef, property, value, &error	Sets a property to a value in a record
ABMultiValueGetCount	ABMultiValue	Returns the size of a multi-value (which can contain one or more copies of a record, such as multiple phone numbers)
ABMultiValueCopyLabelAtIndex	ABMultiValueRef, index	Looks up the label of an entry in a multivalue
ABMultiValueCopyValueAtIndex	ABMultiValueRef, index	Looks up the content of an entry in a multivalue
ABCreateMutableCopy	ABMultiValueRef	Creates a copy of a multi-value
ABMultiValueReplaceLabelAtIndex	ABMutableMultiValueRef, label, index	Replaces a label at an index in a multivalue
ABMultiValueReplaceValueAtIndex	ABMutableMultiValueRef, value, index	Replaces a label at an index in a multivalue

Generally, when you're using the *getter* functions for contacts in the Address Book, you'll follow this procedure:

1 Select one or more contacts through either the Address Book or the Address Book UI framework.

2 To look at an individual property, like a name or phone numbers, use ABRecordCopyValue:

   – If it's a single value property, you'll immediately be able to work with it as a string or some other class.

   – If it's a multivalue property, you'll need to use the ABMultiValue functions to access individual elements of the multivalue.

We included the *setter* methods in table 16.8 to keep the methods all in one place, but you'll usually only be using the setters if you're working with the Address Book framework, not the Address Book UI framework. Here's how they work:

1 Make changes to individual properties or to multivalues (using the mutable multivalue).

2 Use ABRecordSetValue to save the value to your local copy of the Address Book.

3 Use ABAddressBookSave to save your local changes to the real Address Book database.

We won't be covering the setter side of things (which you can find out about in the "Address Book Programming Guide for iPhone OS"), but we're going to use many of the getter functions in the next section.

16.5.3 Querying the Address Book

Our first exploration of the Address Book will use the plain Address Book framework to access the Address Book and look up many of the values. This is shown in listing 16.10. It centers on a simple application with two objects built in Interface Builder: a UISearchBar and a UITextView (with an IBOutlet called myText).

We haven't used search bars before, but they're a simple way to enter search text. You set the search bar's delegate, and then respond to appropriate messages. In this case, our program responds to the searchBarSearchButtonClicked: delegate method, and then looks up the information that was entered.

Listing 16.10 An example of looking up information in the Address Book

```
- (void)searchBarSearchButtonClicked:(UISearchBar *)searchBar {

    [searchBar resignFirstResponder];

    ABAddressBookRef addressBook =          ① Copies
        ABAddressBookCreate();                 Address Book
    CFIndex abPCount =
        ABAddressBookGetPersonCount(addressBook);   ② Counts Address
    CFIndex abGCount =                                  Book entries
        ABAddressBookGetGroupCount(addressBook);
    CFArrayRef searchResults = ABAddressBookCopyPeopleWithName(addressBook,
        (CFStringRef)searchBar.text);       ◁—③ Searches Address Book

    myText.text = [NSString stringWithString:@"Possible Completions:"];

    for (int i=0; i < CFArrayGetCount(searchResults); i++) {
        ABRecordRef thisPerson =                    ④ Gets personal
            CFArrayGetValueAtIndex(searchResults, i);  record
        myText.text = [myText.text stringByAppendingFormat:@"\n\n%@",
            (NSString *)ABRecordCopyCompositeName
                (thisPerson)];               ⑤ Prints full name

        CFStringRef thisJob = ABRecordCopyValue(thisPerson,
            kABPersonJobTitleProperty);
        CFStringRef thisOrg = ABRecordCopyValue(thisPerson,   ⑥ Gets other
            kABPersonOrganizationProperty);                      properties

        if (thisJob != NULL && thisOrg != NULL) {
            myText.text = [myText.text stringByAppendingFormat:
                @"\n%@ of %@",thisJob,thisOrg];
        }
```

```
    ABMultiValueRef thisPhones = ABRecordCopyValue(thisPerson,
       kABPersonPhoneProperty);
                                          7  Gets phone multivalue
    if (thisPhones != NULL) {
       for (int j = 0; j <ABMultiValueGetCount(thisPhones); j++) {
         myText.text = [myText.text stringByAppendingFormat:
           @"\n%@: %@", (NSString *)
             ABMultiValueCopyLabelAtIndex(thisPhones, j),
           (NSString *)
             ABMultiValueCopyValueAtIndex          8  Prints individual
               (thisPhones, j)];                      phone number
       }
     }
   }
   myText.text = [myText.text stringByAppendingFormat:@"\n\nThere are %ld
      records and %ld groups in this address book.",abPCount,abGCount];

   CFRelease(searchResults);           9  Cleans up
   CFRelease(addressBook);                memory
}
```

You start out by running ABAddressBookCreate, which makes a local copy of the Address Book ❶. You'll need to do this whenever you're working manually with the Address Book. After that, you make use of a few general Address Book functions that let you do things like count your number of contacts and groups ❷. But it's the search function that's most important ❸. This is one of two ways that you might extract contacts from the Address Book by hand, the other being ABAddressBookCopyArray-OfAllPeople. Note the typing of searchBar.text as CFStringRef. This is a Core Foundation class equivalent to NSString *; there's more information on the details of Core Foundation in the "Using Core Foundation" sidebar.

The preceding steps are the major ones that differentiate working with the Address Book manually from working with it through a UI. With the Address Book framework, your program does the selection of contact records; with the UI framework, the user would do it through a graphical interface. Beyond that, things work similarly via either methodology.

Once you've got a list of contacts, you need to extract individuals from the array ❹. There are numerous functions that can then be used to look at their properties. ABRecordCopyCompositeName gives you a full name already put together ❺, and ABRecordCopyValue lets you pick out other properties ❻. The list of properties and returned values is in the ABPerson reference.

Multivalues are only a little more difficult to use than simple properties. You use ABRecordCopyValue as usual ❼, but then you have to work through the entire multivalue, which is effectively an associative array. The easiest thing to do is extract all the individual labels and values ❽. This program displays the slightly awkward label names (for your reference), but you probably wouldn't usually want to show off words like $!<Mobile>!$, and it's easy enough to strip them out.

The program ends by cleaning up some of the Core Foundation objects, using the standard Core Foundation memory management functions ❾. When you run it, this program displays some of the data from names that you search for, as shown in figure 16.8.

There's lots more that can be done with the Address Book, but this should outline the basics of how to access its several classes.

16.5.4   *Using the Address Book UI*

There are definitely times when you'll want to work with the low-level Address Book functions that we've seen so far. But you also don't want to reinvent the wheel. If you need to let a user select, edit, or insert a new contact, there's no need to program any of the UI. Instead, you can use the Address Book UI framework, which has all of that functionality preprogrammed.

The Address Book UI framework contains only the four classes that we summarized in table 16.7: `ABPeoplePicker-`

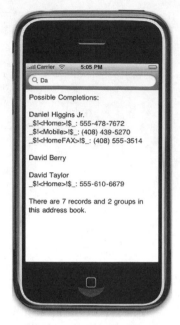

Figure 16.8   The Address Book framework gives you low-level access to contact information.

`NavigationController`, `ABNewPersonView-Controller`, `ABPersonViewController`, and `ABUnknownPersonViewController`. Each of these UI objects is—as the names suggest—a view controller. To be precise, they're highly specialized modal controllers that each assist you in a single Address Book–related task. Each controller also has a delegate protocol, which is how you link to a class that's already pretty fully realized. We're going to touch upon each of these classes, but we're only going to give a lot of attention to the people-picker (`ABPeople-PickerNavigationController`).

THE PEOPLE-PICKER VIEW CONTROLLER

To demonstrate the people-picker, we're going to put together a quick utility with substantially identical functionality to the previous Address Book example. But rather than searching for multiple users using the Address Book framework, the user will instead select a specific user using the Address Book UI framework.

This program is built with a couple of Interface Builder–created objects. A `UIToolBar` with a single button allows the user to activate the program via the `select-Contact:` method, and text will once more be displayed in a non-editable `UITextView` called `myText`. The program is shown in listing 16.11.

Listing 16.11   People-picker: a simple, graphical way to select contacts

```
-(IBAction)selectContact:(id)sender {

    ABPeoplePickerNavigationController *myPicker =
        [[ABPeoplePickerNavigationController alloc]
            init];
```

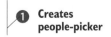 **Creates people-picker**

```
    myPicker.peoplePickerDelegate = self;          ◀─── ❷ Sets
    [self presentModalViewController:myPicker            delegate
       animated:YES];
                                          Displays  ❸
    [myPicker release];                 people-picker
}

- (BOOL)peoplePickerNavigationController:
    (ABPeoplePickerNavigationController *)peoplePicker
    shouldContinueAfterSelectingPerson:
        (ABRecordRef)thisPerson {         ◀─── Runs person-selection
                                          ❹ delegate routine
    CFIndex abPCount =
       ABAddressBookGetPersonCount
          (peoplePicker.addressBook);      ❺ Gets overall
    CFIndex abGCount =                        counts
       ABAddressBookGetGroupCount
          (peoplePicker.addressBook);

    myText.text = [NSString stringWithString:@"Selected Contact:"];
    myText.text = [myText.text stringByAppendingFormat:@"\n\n%@",
        (NSString *)ABRecordCopyCompositeName(thisPerson)];

    CFStringRef thisJob = ABRecordCopyValue(thisPerson,
       kABPersonJobTitleProperty);
    CFStringRef thisOrg = ABRecordCopyValue(thisPerson,
       kABPersonOrganizationProperty);

    if (thisJob != NULL && thisOrg != NULL) {
       myText.text = [myText.text stringByAppendingFormat:@"\n%@ of
          %@",thisJob,thisOrg];
    }

    ABMultiValueRef thisPhones = ABRecordCopyValue(thisPerson,
       kABPersonPhoneProperty);

    if (thisPhones != NULL) {
       for (int j = 0; j < ABMultiValueGetCount(thisPhones) ; j++) {
          myText.text = [myText.text stringByAppendingFormat:@"\n%@: %@",
             (NSString *)ABMultiValueCopyLabelAtIndex(thisPhones, j),
             (NSString *)ABMultiValueCopyValueAtIndex(thisPhones, j)];
       }
    }

    myText.text = [myText.text stringByAppendingFormat:@"\n\nThere are %ld
       records and %ld groups in this address book.",abPCount,abGCount];

    [self dismissModalViewControllerAnimated:YES];   ◀─── Dismisses
    return NO;   ◀───                                ❻ controller
}              ❼ Stops browsing

- (BOOL)peoplePickerNavigationController:
    (ABPeoplePickerNavigationController *)peoplePicker
    shouldContinueAfterSelectingPerson:(ABRecordRef)person
    property:(ABPropertyID)property
    identifier:(ABMultiValueIdentifier)identifier {   ◀─── Runs property-
                                                    ❽ selection routine
    return NO;
}
```

```
- (void)peoplePickerNavigationControllerDidCancel:
    (ABPeoplePickerNavigationController *)
        peoplePicker {

    [self dismissModalViewControllerAnimated:YES];
}
```

9 Runs cancellation routine

In order to instantiate a modal view controller, you follow three simple steps that are executed when the user clicks on the appropriate button in the toolbar. You create the controller **1**, set its delegate **2**, and use `UIViewController`'s `presentModalViewController:animated:` method to place it at the top of your user's screen **3**. You then don't have to worry about how the modal view controller looks or works; you just have to respond to the messages listed in the protocol reference.

The fully featured interface that's available to you as soon as you pop up the controller is shown in figure 16.9.

You'll do most of the work in the `peoplePicker-NavigationController:shouldContinueAfterSelectingPerson:` method **4**. This is called whenever a user selects an individual contact. Note that you can use a property of the `peoplePicker` variable to access the Address Book itself **5**, which allows you to use many of the `ABAddressBook` functions without needing to create the Address Book manually. Beyond that, the people-picker sends you an `ABRecordRef` for the contact that the user selected; from there, you work with it exactly like

Figure 16.9   A people-picker view controller

you worked with the `ABRecordRefs` that you looked up in listing 16.10.

In this example, users can only select individual contacts, so when the method is done, you dismiss the modal view controller **6**, and then return `NO` **7**, which tells the people-picker that you don't want to take the standard action for selecting the contact (which would be to call up a subpage with all of that contact's properties).

If you'd wanted to let a user select a specific property from within a contact, you'd fill in the `peoplePickerNavigationController:shouldContinueAfterSelecting-Person:property:identifier:` method **8**.

The third method defined by the `ABPeoplePickerNavigationController` protocol is `peoplePickerNavigationControllerDidCancel:` **9**, which here causes the program to (again) dismiss the people-picker.

You can do a little more with the people-picker. As we already noted, you could have opted to let a user select an individual property by returning `YES` for the first `should-Continue` method, and then filling in the second one. You could also choose the individual properties that display on a contact page. Information on these possibilities is available in the `ABPeoplePickerNavigationController` and `ABPeoplePickerNaviga-tionController-Delegate` class references.

Using Core Foundation

The Address Book framework is the first framework you've worked with that requires you to use Core Foundation, a non-Cocoa library. This means you'll have to program slightly differently, as we promised would be the case back in chapter 10. The biggest differences are how variables and memory allocation work.

Core Foundation variables use different classes, such as `CFStringRef` replacing `NSString *`. Remember that the Core Foundation variable types usually have equivalents in Cocoa that you can freely switch between by casting, as is done in listing 16.10 when moving between the Address Book records and the `UITextView` text. When you're using the Core Foundation variables natively, you'll have to use Core Foundation functions, such as `CFArrayCount`, to deal with them.

You'll also have to deal with memory management a little differently. Core Foundation memory management uses the same general approach as Cocoa Touch. There's a reference count for each object that is increased when it's created or retained and decreased when it's released. You just have to remember slightly different rules for when you have a reference. If you create an object with a function using the word(s) *create* or *copy*, you own a reference to it and must `CFRelease` it. If you create an object in another way, you do not have a reference, and you must `CFRetain` the object if you want to keep it around. Some classes of objects may have their own release and retain functions. The "Memory Management Programming Guide for Core Foundation" tutorial at developer.apple.com has more information.

Core Foundation will show up again in chapter 18, where it controls some audio services, and in chapter 19, where it's used for the Quartz 2D graphics package.

There are three other view controllers that you can use to allow users to interact with the Address Book.

THE OTHER VIEW CONTROLLERS

The other three view controllers work much like `ABPeoplePickerNavigationController`, with one notable difference: they must each be built on top of a navigation controller. Technically, they're probably not modal view controllers, because they go inside a navigation controller, but you can treat the navigation controller as a modal view controller once everything is loaded up, as you'll see in our example.

The `ABNewPersonViewController` allows a user to enter a new contact. You can prefill some of the info by recording it in an `ABRecordRef` and setting the `displayedPerson` property, but this is purely optional (and probably won't usually be done). Once you've created the controller, you'll need to respond to a method that tells you when the user has entered a new contact. You don't have to do anything with it except dismiss the modal controller, because the controller automatically saves the new contact to the Address Book. You can see what info the user entered, though, and do something with it if you want. Listing 16.12 shows how to deploy a new-person view on top of a navigation controller, and how to respond to its single method.

Listing 16.12   Functionality required to call up a new-person view controller

```
- (IBAction)newContact:(id)sender {

    ABNewPersonViewController *myAdder =
        [[ABNewPersonViewController alloc] init];
    myAdder.newPersonViewDelegate = self;

    UINavigationController *myNav = [[UINavigationController alloc]
        initWithRootViewController:myAdder];
    [self presentModalViewController:myNav animated:YES];

    [myAdder release];
    [myNav release];
}

- (void)newPersonViewController:
    (ABNewPersonViewController *)newPersonViewController
    didCompleteWithNewPerson:(ABRecordRef)person {

    [self dismissModalViewControllerAnimated:YES];
}
```

The other two view controllers work the same way, except for the specifics about what methods each protocol defines.

The `ABPersonViewController` displays the information for a specific user. You'll need to set the `displayedPerson` property to an `ABRecordRef` before you call it up. This `ABRecordRef` might have been retrieved from the Address Book search functions or from the people-picker, using the functions we've already discussed. The person view controller can optionally be editable. There's one method listed in the protocol, which activates when an individual property is selected.

Finally, the `ABUnknownPersonViewController` allows you to display the `ABRecordRef` defined by `displayedPerson` as if it were a real contact. Optionally, the user can create that information as a new contact, add it to an existing contact, or take property-based actions, like calling a number or showing a URL. It's a great way to give users the option to add contact info for your software company to their Address Book.

You should now understand the basics of how to use the Address Book in your own programs.

16.6   *Summary*

In this chapter, we covered a variety of ways that you can import primarily text-based data into your iPhone program.

User action is one of the most important methods, and one well covered by previous sections. Besides the `UITextFields`, `UITextViews`, and `UISearchBars`, there are any number of non-textual interface options.

Preferences mark the other major way that users can influence your program. You can either program them manually or use the System Setting bundle.

Ultimately, user input is going to be somewhat limited because of the slow typing speed of the iPhone. If you're dealing with piles of text, you more frequently want to pull that data from an existing resource on the iPhone.

Files are the traditional way to do it. We'll return to files as we deal with photos and sounds in the future. *Databases* are frequently an easier way to access data, particularly if the data is well organized. Finally, the *Address Book* gives you a way to share contact information among many different applications, and it even includes its own data entry routines.

There's only one data-input method that we've largely ignored: the *internet*. We consider that so important for the iPhone that we'll cover it in its own chapter at the end of the book.

The data-input and retrieval methods discussed in this chapter will form a foundation for much of the other work you do with the iPhone, because ultimately *everything's* data. You'll need to retrieve data when you work with images and sounds. Similarly, you may want to save data from your accelerometer, Core Location, or when you've created a graphic. Keep what you've learned here in your back pocket as you move on to the rest of the iPhone toolbox.

We're now ready to move on to what we expect are two of the most anticipated topics in this book: how to work with the iPhone's accelerometers and its GPS to determine locations.

17
Positioning: accelerometers and location

This chapter covers

- Measuring gravity
- Gauging movement
- Determining location

When we first introduced the iPhone, we highlighted a number of its unique features. Among them were two components that allow the iPhone to figure out precisely where it is in space: a trio of accelerometers, which give it the ability to sense motion in three dimensions, and a locational device (using either GPS or faux GPS), which lets it figure out where in the world it is.

Other than accessing some basic orientation information, we haven't done much with these features. That's because most of the functionality isn't available to the web interface, and because it lies beyond the basic concepts of the SDK that we've covered so far. But now we'll dive into these positioning technologies and examine how to use them in your programming.

We're going to start off with some new ways to look at orientation data, and then we'll expand into some original possibilities.

17.1 The accelerometer and orientation

The easiest use of the accelerometers is to determine the iPhone's current orientation. We already used the view controller's interfaceOrientation property, back in chapter 13. As we mentioned at the time, though, you can also access orientation information through the UIDevice object. It can provide more information and real-time access that isn't available using the view controller.

There are two ways to access the UIDevice information: through properties and through a notification.

17.1.1 The orientation property

The easy way to access the UIDevice's orientation information is to look at its orientation property. You must first access the UIDevice itself, which you can do by calling a special UIDevice class method, pretty much the same way you access the UIApplication object:

```
UIDevice *thisDevice = [UIDevice currentDevice];
```

Once you've done this, you can get to the orientation property. It will return a constant drawn from UIDeviceOrientation. This looks exactly like the results from a view controller's orientation property except there are three additional values, shown in table 17.1.

Table 17.1 UIDeviceOrientation **lists seven values for a device's orientation.**

Constant	Summary
UIDeviceOrientationPortrait	iPhone is vertical, right side up
UIDeviceOrientationPortraitUpsideDown	iPhone is vertical, upside down
UIDeviceOrientationLandscapeLeft	iPhone is horizontal, tilted left
UIDeviceOrientationLandscapeRight	iPhone is horizontal, tilted right
UIDeviceOrientationFaceUp	iPhone is lying on its back
UIDeviceOrientationFaceDown	iPhone is lying on its screen
UIDeviceOrientationUnknown	iPhone is in an unknown state

These three additional values are one reason you might want to access the UIDevice object rather than examine orientation using a view controller.

17.1.2 The orientation notification

The UIDevice class can also give you instant access to an orientation change when it occurs. This is done through a notification (a topic we introduced in chapter 14). Listing 17.1 shows how to access this information.

Listing 17.1 Notification can give you INSTANT access to orientation changes

```
[[UIDevice currentDevice]
    beginGeneratingDeviceOrientationNotifications];          ❶ Begins
                                                               notifications

[[NSNotificationCenter defaultCenter] addObserver:self
    selector:@selector(deviceDidRotate:)                     ❷ Observes
    name:@"UIDeviceOrientationDidChangeNotification"            notification
    object:nil];
```

This is a two-step process. First you alert the iPhone that you're ready to start listening for a notification about an orientation change ❶. This is one of a pair of `UIDevice` instance methods, the other being `endGeneratingDeviceOrientationNotifications`. You generally should only leave notifications on when you need them, because they take up CPU cycles and will increase your power consumption.

Second, you register to receive the `UIDeviceOrientationDidChangeNotification` messages ❷, our first live example of the notification methods we introduced in chapter 14. Then, whenever an orientation change notification occurs, your `deviceDidRotate:` method will be called. Note that you won't receive notification of what the new orientation is; you'll simply know that a change happened. For more details, you'll have to query the `orientation` property.

We've now seen the two ways in which an iPhone's orientation can be tracked with the `UIDevice` object, providing more information and more rapid notification than you received when using the view controller. But that only touches the surface of what you can do with the iPhone's accelerometers. It's probably the raw data about changes in three-dimensional space that you'll really want to access.

17.2 *The accelerometer and movement*

When you're using the iPhone's orientation notification, the frameworks are doing your work for you: they're taking low-level acceleration reports and turning them into more meaningful events. It's similar to the concept of iPhone actions, which turn low-level touch events into high-level control events.

WARNING   Accelerometer programs can't be tested on the iPhone Simulator. Instead you'll need to have a fully provisioned iPhone to test out your code. See appendix C for information on provisioning your iPhone.

Notifications aren't going to be sufficient for many of you, who would prefer to program entire interfaces that effectively use the iPhone's movement in three-dimensional space as a new user-input device. For that, you'll need to access two iPhone classes: `UIAccelerometer` and `UIAcceleration`.

17.2.1 *Accessing the UIAccelerometer*

The `UIAccelerometer` is a class that you can use to register to receive acceleration-related data. It is a shared object, like `UIApplication` and `UIDevice`. The process of using it is shown in listing 17.2.

Listing 17.2 Accessing the UIAccelerometer takes just a few steps

```
- (void)viewDidLoad {

    UIAccelerometer *myAccel =                              ❶ Creates shared
        [UIAccelerometer sharedAccelerometer];                accelerometer
    myAccel.updateInterval = .1;
    myAccel.delegate = self;                               ❷ Updates 10
                                                             times a second
    [super viewDidLoad];          Delegates
}                                 accelerometer
                              ❸  protocol
```

The first step is to access the accelerometer, which is done with another call to a shared-object method ❶. Having this step on its own line is probably unnecessary, because you could perform the other two steps as nested calls, but we find this a lot more readable.

Next, you select your update interval ❷, which specifies how often you'll receive information on acceleration. This is hardware-limited, with a current default of 100 updates per second. That's probably just right if you're creating a game using the accelerometer, but excessive for other purposes. We've opted for 10 updates per second, which is an updateInterval of .1. You should always set the lowest acceptable input to preserve power on the iPhone.

Finally, you must set a delegate for the accelerometer ❸, which is how you receive data on accelerometer changes. The delegate will need to respond to only one method, accelerometer:didAccelerate:, which sends a message containing a UIAcceleration object whenever acceleration occurs (to the limit of the updateInterval).

17.2.2 Parsing the UIAcceleration

The UIAcceleration information can be used to accurately and easily measure two things: the device's relationship to gravity and its movement through three-dimensional space. These are both done through a set of three properties, x, y, and z, which refer to a three-dimensional axis, as shown in figure 17.1.

The *x*-axis measures along the short side of the iPhone, the *y*-axis measures along the long side of the iPhone, and the *z*-axis measures through the iPhone. All values are measured in units of "g", which is to say g-force. A value of 1 g represents the force of gravity on Earth at sea level.

The thing to watch for when accessing your accelerometer is that it measures two types of force applied to the device: both the force of movement in any direction *and* the force of gravity, measured in units of g. That

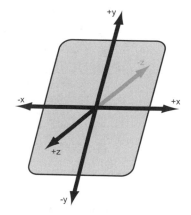

Figure 17.1 The accelerometers in the iPhone measure in three-dimensional space.

means that an iPhone at rest will always show an acceleration of 1 g toward the Earth's core. This may require filtering if you're doing more sophisticated iPhone work.

Filtering and the accelerometer

It might seem like the iPhone acceleration data is all mushed together, but it's easy to isolate exactly the data that you need using basic electronics techniques.

A low-pass filter passes low-frequency signals and attenuates high-frequency signals. That's what you'll use to reduce the effects of sudden changes in your data, such as would be caused by an abrupt motion.

A high-pass filter passes high-frequency signals and attenuates low-frequency signals. That's what you'll use to reduce the effects of ongoing forces, such as gravity.

We'll see examples of these two filtering methods in the upcoming sections.

17.2.3 Checking for gravity

When the accelerometers are at rest, they naturally detect gravity. This may be used to detect the precise orientation that an iPhone is currently held in, going far beyond the four or six states supported by the orientation variables.

READING ACCELERATION INFORMATION

Listing 17.3 shows how the accelerometers could be used to modify redBall, a UIImage picture of a red ball, created in Interface Builder and initially set in the middle of the screen.

Listing 17.3 A short program that causes an image to respect gravity

```
- (void)accelerometer:(UIAccelerometer *)accelerometer       ❶ Answers
    didAccelerate:(UIAcceleration *)acceleration {              delegate

    CGPoint curCenter = [redBall center];       ❷ Measures current center

    float newX = 3 * acceleration.x + curCenter.x;      ❸ Adds in
    float newY = -3 * acceleration.y + curCenter.y;         change

    if (newX < 25) newX = 25;
    if (newY < 25) newY = 25;                ❹ Keeps in
    if (newX > 295) newX = 295;                 bounds
    if (newY > 455) newY = 455;
                                          ❺ Moves
    redBall.center = CGPointMake(newX,newY);       center

}
```

Any accelerometer program begins with the accelerometer:didAccelerate: method ❶, which is accessed by setting the current program as a delegate of the Accelerometer shared action. You then need to mark the current position of the redBall ❷.

In order to access the accelerometer, all you do is look at the x and y coordinates of the UIAcceleration object ❸ and prepare to modify the redBall's position based on those. The acceleration is multiplied by three here to keep the ball's movement from being snail-like. There's also a z property for the third axis and a timestamp property indicating when the UIAcceleration object was created, none of which you'll need in this example.

You'll note that we're not yet talking about filtering; we'll address that topic after we finish this example. Movement will have a pretty limited effect on our example anyway, because an abrupt movement won't change the ball's slow roll much.

After acquiring your gravitic information, you should also make sure that the 50x50 red ball will stay within the bounds of the iPhone screen ❹. If you wanted to be fancy, you could introduce vectors and bounce the ball when it hits the edge, but that's beyond the scope of this example. After that check, you can move your ball ❺.

With a minimal amount of work, you've created a program that's acted on by gravity. This program could easily be modified to act as a leveler tool for pictures (by having it only move along one of the three axes) or could be turned into a game where a player tries to move a ball from one side of the screen to the other, avoiding pits on the way.

Now, what would it take to do this example *totally* right by filtering out all movement? The answer, it turns out, is not much more work at all.

FILTERING OUT MOVEMENT

To create a low-pass filter that will let through gravitic force but not movement, you need to average out the acceleration information that you're receiving, so that at any time the vast majority of your input is coming from the steady force of gravity. This is shown in listing 17.4, which modifies the example in listing 17.3.

Listing 17.4   A low-pass filter isolates gravity in the accelerometers

```
gravX = (acceleration.x * kFilteringFactor)
    + (gravX * (1 - kFilteringFactor));          ❶ Average
gravY = (acceleration.y * kFilteringFactor)          movement
    + (gravY * (1 - kFilteringFactor));

float newX = 3 * gravX + curCenter.x;            ❷ Apply
float newY = -3 * gravY + curCenter.y;               averaged data
```

This example depends upon three predefined variables: kFilteringFactor is a constant set to .1, which means that only 10 percent of the active movement will be used at any time; gravX and gravY each maintains a cumulative average for its axis of movement as the program runs.

You filter things by averaging 10 percent of the active movement with 90 percent of the average ❶. This smooths out any bumps, which means that sudden acceleration will be largely ignored. This example does this for the *x*- and *y*-axes because that's all that was used in the example. If you cared about the *z*-axis, you'd need to filter that too.

Afterward, you use the averaged acceleration instead of the raw acceleration when you're changing the position of your ball ❷. The gravity information can be extracted from what looked like an imposing mass of data with a couple of lines of code.

As we'll see, looking at only the movement is just as easy.

17.2.4 *Checking for movement*

In the previous example, you isolated the gravitic portion of the accelerometer's data by creating a simple low-pass filter. With that data in hand, it's trivial to create a high-pass filter. All you need to do is subtract out the low-pass filtered data from the acceleration value; the result is the pure movement data.

This process is shown in listing 17.5

> **Listing 17.5   Subtracting out the gravity data leaves you only the movement data**

```
gravX = (acceleration.x * kFilteringFactor)
      + (gravX * (1 - kFilteringFactor));
gravY = (acceleration.y * kFilteringFactor)
      + (gravY * (1 - kFilteringFactor));

float moveX = acceleration.x - gravX;
float moveY = acceleration.y - gravY;
```

This filter doesn't entirely stop gravitic movement, because it takes several iterations for the program to cut out gravity completely. In the meantime, your program will be influenced by gravity for a few fractions of a second at startup. If that's a problem, you could tell your program to ignore acceleration input for a second after it loads and after an orientation change. We'll show the first solution in our next example.

With that exception, as soon as you start using these new moveX and moveY variables, you'll be looking at the filtered movement information rather than the filtered gravity information. But when you start looking at movement information, you'll see that it's a bit trickier to use than gravity information. There are two reasons for this.

First, movement information is a lot more ephemeral. You'll see it appear for a second, and then it will be gone again. If you're displaying some type of continuous movement, as with the red ball example, you'd need to make your program much more sensitive to detect the movements. You'd have to multiply your moveX and moveY values by about 25 to see movement forces applied to that ball in any recognizable manner.

Second, it's a lot noisier. As you'll see when we look at some real movement data, motion occurs in a multitude of directions at the same time, forcing you to parse out the exact information that you want.

Ultimately, to interpret movement, you have to be more sophisticated, recognizing what are effectively gestures in three-dimensional space.

17.2.5 *Recognizing simple accelerometer movement*

If you want to write programs using acceleration gestures, we suggest that you download the Accelerometer Graph program available from Apple's developer site. This is a nice, simple example of accelerometer use, but, more importantly, it also provides you with a clear display of what the accelerometers report as you make different gestures. Make sure you enable the high-pass filter to get the clearest results.

Figure 17.2 shows what the Accelerometer Graph looks like in use (but without movement occurring). As you move around an iPhone, you'll quickly come to see how the accelerometers respond.

Here are some details you'll notice about how the accelerometers report information when you look at the Accelerometer Graph:

- Most gestures will cause all three accelerometers to report force; the largest force should usually be in the axis of main movement.
- Even though there's usually a compensating stop force, the start force will typically be larger, and will show the direction of main movement.
- Casual movement usually results in forces of .1 g to .5 g.
- Slightly forceful movement usually tops out at 1 g.
- A shake or other more forceful action will usually result in a 2 g force.
- The accelerometers can show things other than simple movement. For example, when you're walking with an iPhone, you can see the rhythm of your pace in the accelerometers.

Figure 17.2   The Accelerometer Graph shows movement in all three directions.

All of this suggests a simple methodology for detecting basic accelerometer movement: you monitor the accelerometer over the course of movement, saving the largest acceleration in each direction. When the movement has ended, you can report the largest acceleration as the direction of movement.

Listing 17.6 puts these lessons together in a program that could easily be used to report the direction of the iPhone's movement (which you could then use to take some action).

Listing 17.6   A movement reporter that could be applied as a program controller

```
- (void)accelerometer:(UIAccelerometer *)accelerometer
  didAccelerate:(UIAcceleration *)acceleration {

  accelX = ((acceleration.x * kFilteringFactor)
      + (accelX * (1 - kFilteringFactor)));
  accelY = ((acceleration.y * kFilteringFactor)
      + (accelY * (1 - kFilteringFactor)));
  accelZ = ((acceleration.z * kFilteringFactor)
      + (accelZ * (1 - kFilteringFactor)));

  float moveX = acceleration.x - accelX;
  float moveY = acceleration.y - accelY;
  float moveZ = acceleration.z - accelZ;
```

① Gathers filtered info

② Measures movement

```
    if (!starttime) {
        starttime = acceleration.timestamp;           ➌  Marks
    }                                                      start time

    if (acceleration.timestamp > starttime + 1 &&
        (fabs(moveX) >= .3 ||
         fabs(moveY) >= .3 ||          ➍  Notes movement
         fabs(moveZ) >= .3)) {             continuing

        if (fabs(moveX) > fabs(moveVector)) {
            moveVector = moveX;
            moveDir = (moveVector > 0 ? @"Right" : @"Left");
        }
        if (fabs(moveY) > fabs(moveVector)) {
            moveVector = moveY;                                    ➎  Saves
              moveDir = (moveVector > 0 ? @"Up" : @"Down");           largest
        }                                                            movements
        if (fabs(moveZ) > fabs(moveVector)) {
            moveVector = moveZ;
            moveDir = (moveVector > 0 ? @"Forward" : @"Back");
        }
        lasttime = acceleration.timestamp;
    } else if (moveVector && acceleration.timestamp      ➏  Notes movement
        > lasttime + .1) {                                   ending
        myReport.text =
            [moveDir stringByAppendingFormat:    ➐  Reports
                @": %f.",moveVector];                movement
        moveDir = [NSString string];     ➑  Cleans up
        moveVector = 0;
    }
}
```

As usual, you start by creating a low-pass filter ➊ and then taking the inverse of it ➋ in order to get relatively clean movement data. Because the data can be a little dirty at the start, you don't accept any acceleration data sent in the first second ➌. You could cut this down to a mere fraction of a second.

You start looking for movement whenever one of the accelerometers goes above .3 g ➍. When that occurs, you save the direction of highest movement ➎, and keep measuring it until movement drops below .3 g. Afterward, you make sure that at least a tenth of a second has passed ➏, so that you know you're not in a lull during a movement.

Finally, you do whatever you want to do with your movement data. Listing 17.6 reports the information in a label ➐, but you'd doubtless do something much more intricate in a live program. Cleanup is required to get the next iteration of movement reporting going ➑.

This sample program works well, unless the movement is very subtle. In those cases, it occasionally reports the opposite direction because of the force when the device stops its motion. If this type of subtlety is a problem for your application, more work would be required. To resolve this, you'd need to make a better comparison of the start and stop forces for movements; if they're similar in magnitude, you'll usually want to use the first force measured, not necessarily the biggest one. But for the majority of cases, the code in listing 17.6 is sufficient. You've now got an iPhone application that can accurately report (and take action based upon) direction of movement.

Together gravity and force measurement represent the most obvious things that you can do with the accelerometers, but they're by no means the *only* things. We suspect that using the accelerometers to measure three-dimensional gestures will be one of their best (and most frequent) uses as the iPhone platform matures.

17.3 The accelerometer and gestures

Three-dimensional gestures are one of the coolest results of having accelerometers inside your iPhone. They can allow your users to manipulate your iPhone programs without ever having to touch (or even look at) the screen.

To recognize a gesture, you must do two things. First, you must accurately track the movements that make up a gesture. Second, you must make sure that in doing so you won't recognize a random movement that wasn't intended to be a gesture at all.

Technically, recognizing a gesture requires only the coding foundation that we've discussed already. But we're going to show one example that puts that foundation into real-world use by creating a method that recognizes a shake gesture.

We're defining a shake as a rapid shaking back and forth of the iPhone, like you might shake dice in your hand before you throw them. As usual, Apple's Accelerometer Graph is a great tool to use to figure out what's going on. It shows a shake as primarily having these characteristics, presuming a program that's running in portrait mode:

- Movement is primarily along the *x*-axis, with some movement along the *y*-axis, and even less along the *z*-axis.
- There are at least three peaks of movement, with alternating positive and negative forces.
- All peaks are at least +/-1 g, with at least one peak being +/-2 g for a relatively strong shake.

We can use the preceding characteristics to define the average requirements for a shake. If we wanted to tighten them up, we'd probably require four or more peaks of movement, but for now, this will do. Alternatively, we might want to decrease the g-force requirements so that users don't have to shake their iPhone quite as much (and, in fact, we will). We've detailed the code that will watch for a shake in listing 17.7.

Listing 17.7 Shake, shake your iPhone

```
- (BOOL)didShake:(UIAcceleration *)acceleration {

    accelX = ((acceleration.x * kFilteringFactor)
        + (accelX * (1 - kFilteringFactor)));
    float moveX = acceleration.x - accelX;

    accelY = ((acceleration.x * kFilteringFactor)
        + (accelY * (1 - kFilteringFactor)));
    float moveY = acceleration.x - accelY;
```

 Filters data

```
if (lasttime && acceleration.timestamp > lasttime + .25) {          ②  Waits after
                                                                         last shake
    BOOL result;

    if (shakecount >= 3 && biggestshake >= 1.25) {          ③  Measures
        result = YES;                                            for shake
    } else {
        result = NO;
    }

    lasttime = 0;
    shakecount = 0;
    biggestshake = 0;
                              ④  Looks for
    return result;                movement      ⑤  Checks X
} else {                                           movement
    if (fabs(moveX) >= fabs(moveY)) {                        ⑥  Measures
        if ((fabs(moveX) > .75) && (moveX * lastX <= 0)) {        this move

            lasttime = acceleration.timestamp;
            shakecount++;
            lastX = moveX;
            if (fabs(moveX) > biggestshake) biggestshake = fabs(moveX);
        }
    } else {             ⑦  Checks Y movement
        if ((fabs(moveY) > .75) && (moveY * lastY <= 0)) {       ⑧  Measures
                                                                     this move
            lasttime = acceleration.timestamp;
            shakecount++;
            lastY = moveY;
            if (fabs(moveY) > biggestshake) biggestshake = fabs(moveY);
        }
    }
    return NO;
}
}
```

In this code, you're generally following the logic we used when viewing the accelerometer graph, though with increased sensitivity, as promised. The didShake: method registers a shake if it sees three or more movements of at least .75 g, at least one of which is 1.25 g, with movements going in opposite directions.

You start by removing gravity from the accelerometer data ①, as you did in previous examples. This time you don't worry about the quirk at the beginning of data collection; it won't register as a shake, because it'll be a small fraction of a g.

The main work of the function is found in its latter half, which is called whenever movement continues to occur ④. First, you check whether the strongest movement is along the *x*-axis ⑤. If so, you register the movement if it's at least .75 g and if it's in the opposite direction as the last *x*-axis move ⑥. You do the latter check by seeing if the product of the last two moves on that axis is negative; if so, one must have been positive and the other negative, which means they were opposite each other.

If the strongest move was instead on the *y*-axis ⑦, you check for a sufficiently strong *y*-axis move that's in the opposite direction as the last *y*-axis move ⑧. We could have written a more restrictive shake checker that only looked for *x*-axis movement, or a less restrictive checker that also looked for *z*-axis movement, but we opted for this middle ground.

As long as movement continues without a break of more than a quarter of a second, the shakecount continues to increment, but when movement stops ❷, the program is ready to determine whether a shake occurred. You check this by seeing if the shake count equals or exceeds 3 and if the largest movement exceeded 1.25 g ❸. Afterward, all of the variables are reset to check for the next shake.

By building this shake checker as a separate method, it could easily be integrated into a list of checks made in the accelerometer:didAccelerate: method. Listing 17.8 shows a simple use that changes the color of the iPhone's screen every time a shake occurs. The nextColor method could be changed to do whatever you want.

Listing 17.8   Integrating didShake: is simple

```
- (void)accelerometer:(UIAccelerometer *)accelerometer
  didAccelerate:(UIAcceleration *)acceleration {

  if ([self didShake:(UIAcceleration *)acceleration]) {
    self.view.backgroundColor = [self nextColor];
  }

}
```

We expect that the shake is going to be the most common three-dimensional gesture programmed into the iPhone. With this code, you've already got it ready to go, though you may choose to change its sensitivity or to make it work in either one or three dimensions.

Other gestures, such as a tilt, a fan, a circle, or a swat may be of use, depending on the specifics of your program, but we leave that up to your own ingenuity. For now, we've covered all of the main points of the accelerometers: orientation, gravity, movement, and gestures.

We're now ready to dive into the other major positioning-related tool, and one that we find a lot easier to program because the results are less noisy: Core Location.

17.4   All about Core Location

We've got only one unique iPhone feature left to look at: its ability to detect a user's location. Up until now, we haven't been able to show off the GPS, because it's not currently available when doing web development on the iPhone. While this may change in the future, right now the only way to access the GPS's Core Location data is through the SDK.

WARNING   You can only minimally test Core Location using the iPhone Simulator. Longitude and latitude will work, but they'll always report Apple's Cupertino headquarters. Altitude won't be displayed. For most realistic testing —particularly including distance or altitude—you must use a provisioned iPhone.

The original iPhone based its location information on the distance to cell phone towers and other set locations. Its accuracy could vary from a few blocks' radius to a few miles, even in an urban area. The iPhone 3G has a built-in GPS, but it still has limitations. The iPhone's antenna power is limited, which affects accuracy, and accuracy is further limited by concerns about power usage. As a result, even if you have an iPhone 3G, the device will make preferential use of cell tower data and will provide information on GPS locations using the minimal number of satellite contacts possible (though that minimum partially depends upon an accuracy requirement that you set).

With all that said, the iPhone 3G provides better location information than the original release, because cell tower triangulation is only one factor in its location information. But it may not be entirely accurate; in particular, altitude seems to be the least reliable information.

We offer this preamble both to describe a bit about how the iPhone's location information is created, and to introduce a bit of skepticism about the results. What you get should be good enough for 99 percent of your programs, but you don't want to do anything mission-critical unless you're quite careful.

The good news is that you don't have to worry about which type of iPhone device a user owns. The Core Location API will work identically whether they have a built-in GPS or not.

17.4.1 *The location classes*

The iPhone's location awareness is built into two SDK classes and one protocol. CLLocationManager gives you the ability to access location information in a variety of ways. It includes a delegate protocol, CLLocationManagerDelegate, which defines methods that can tell you when new location information arrives. Finally, the location information itself appears as CLLocation objects, each of which defines a specific location at a specific time.

Table 17.2 describes the most important methods and properties associated with each of these classes. For more details, you should, as usual, consult the Apple class references.

Table 17.2   The most important methods and properties for accessing location information

Class	Method/Property	Type	Summary
CLLocationManager	delegate	Property	Defines the object that will respond to CLLocationManagerDelegate
CLLocationManager	desiredAccuracy	Property	Sets the desired accuracy of location as a CLLocationAccuracy object
CLLocationManager	distanceFilter	Property	Specifies how much lateral movement must occur to cause a location update event

Table 17.2 The most important methods and properties for accessing location information *(continued)*

Class	Method/Property	Type	Summary
`CLLocationManager`	`location`	Property	Specifies the most recent location
`CLLocationManager`	`startUpdatingLocation`	Method	Starts generating update events
`CLLocationManager`	`stopUpdatingLocation`	Method	Stops generating update events
`CLLocationManagerDelegate`	`locationManager: didUpdateToLocation: fromLocation:`	Method	Delegate method that reports whenever an update event occurs
`CLLocationManagerDelegate`	`locationManager: didFailWithError:`	Method	Delegate method that reports whenever an update event fails to occur
`CLLocation`	`altitude`	Property	Specifies the height of location in meters
`CLLocation`	`coordinate`	Property	Returns the location's coordinates as a `CLLocationCoordinate2D` variable
`CLLocation`	`timestamp`	Property	Specifies an `NSDate` of when the location was measured

There are a number of additional properties and methods that you should examine (particularly for the `CLLocation` class), but we're staying with the basics here.

Generally, you'll see that the location information is generated much like the accelerometer information. You access a shared object (`CLLocationManager`) and set some standard properties for how you want it to work, including how often to update (`distanceFilter`). Like the accelerometer, you also have to explicitly turn location updating on (`startUpdatingLocation`). Afterward, you keep an eye on certain methods (as defined by `CLLocationManagerDelegate`). These methods generate an object (`CLLocation`) when location changes; you read the object to get the specifics.

With those generalities out of the way, let's see how `CLLocation` works in a real example.

17.4.2 An example using location and distance

Listing 17.9 shows an example of using Core Location to record a starting location, monitor the current location, and calculate the distance between them. It's the first of two longer examples in this chapter.

Listing 17.9   An application of Core Location for distances

```
- (void)viewDidLoad {

  [super viewDidLoad];

  myLM = [[CLLocationManager alloc] init];          ❶ Starts location
  myLM.delegate = self;                                manager
  myLM.desiredAccuracy =
    kCLLocationAccuracyNearestTenMeters;            ❷ Sets standard
  myLM.distanceFilter = 100;                           properties

  [myLM startUpdatingLocation];     ❸ Starts location updates
}

- (void)locationManager:(CLLocationManager *)manager
  didUpdateToLocation:(CLLocation *)newLocation       ❹ Waits for
  fromLocation:(CLLocation *)oldLocation {               updates

  if (startLoc == nil) {               ❺ Sets starting location
    startLoc = newLocation;
    [self updateLocationFor:startLabel toLocation:newLocation];
    [startLoc retain];
  }

  [self updateLocationFor:endLabel
    toLocation:newLocation];                          ❻ Calls label
  [self updateDistanceLabel:newLocation];                updates
}

- (IBAction)setEnd:(id)sender {
                                       ❼ Forces location
  [myLM stopUpdatingLocation];            update
  [myLM startUpdatingLocation];
}
                                       ❽ Forces label
- (IBAction)controlChange:(id)sender {    update

  if (myLM.location) {
    [self updateDistanceLabel:myLM.location];
  }
}                                                     ❾ Updates
                                                         distance
- (void)updateDistanceLabel:(CLLocation *)newLocation {  label

  if (startLoc != nil) {
    CLLocationDistance traveled
      = [startLoc getDistanceFrom:newLocation] / 1000;

    if (segmentControl.selectedSegmentIndex == 1) {
      traveled *= .62;
    }
    distanceLabel.text = [NSString stringWithFormat:@"%5.1f",traveled];
  }
}

- (void)updateLocationFor:(UILabel *)thisLabel        ❿ Updates
  toLocation:(CLLocation *)newLocation {                 location label

  CLLocationCoordinate2D curCoords = newLocation.coordinate;
  thisLabel.text = [NSString stringWithFormat:
    @"Lat: %2.4f; Long: %2.4f",curCoords.latitude,curCoords.longitude];
}
```

As usual, the foundation of this program is built in Interface Builder. Figure 17.3 displays the general setup that's used.

There are three labels: startLabel (at the top) and endLabel (at the bottom) each display information about a location; distanceLabel shows the distance between the two. There are two controls: a button control instantly updates the current location, and a segmented control chooses between miles and kilometers. They're each linked to an IBAction, which executes a method that we'll meet in the code.

This program generally follows the broad outline of steps that we've already discussed, but we'll go through each step in turn.

You start off initializing a CLLocationManager object ❶ and then set some standard properties ❷—here a delegate, the desiredAccuracy, and the distanceFilter. The desired accuracy of tens of meters and the update interval of every 100 meters might be more than this particular application requires, but you can tune these in your projects as seems appropriate. Remember that demanding more

Figure 17.3   This simple utility shows off locations and distance.

accuracy and updating more frequently will decrease the battery life of your user's iPhone. Finally, you have to start the CLLocationManager running ❸.

The locationManager:didUpdateToLocation:fromLocation: method is the workhorse of this program ❹. It should get called shortly after the LocationManager starts updating and every time your user walks 100 meters or so. First, it saves the current location as the starting location the first time it's called, updating the startLabel at the same time ❺. Then, every time it gets run, it updates the endLabel and the distanceLabel ❻. Note that you don't have to use the LocationManager's location property here (or at almost any other time in the program), because this method always provides you with the current location of the device, and seems to do so well before the location property is updated, based on our own tests. Caveat programmer.

The next few methods have to do with I/O. setEnd: gets run whenever the button control is pushed, to update the current location ❼. Unfortunately, there's no particularly clean way to ask for an update, so you must stop and start the location updates, as shown here. Letting the user force a location update is particularly important if you're using a high distanceFilter or if you're trying to measure altitude changes. In our altitude example, in the next section, we'll see an alternative way to do this, where the location manager usually isn't running at all. The controlChange: method gets run whenever the segmented control is updated ❽. It just updates the distanceLabel. Note that this is the one time that you depend on the location property, because there isn't a location event when you change the button.

The last few methods are utilities. The `updateDistanceLabel:` method makes use of an interesting `CLLocation` method that we haven't discussed, `getDistanceFrom:` **❾**. This measures the true distance between two locations, using complex calculations that correctly account for the curvature of the Earth. Your method also converts meters to kilometers and alternatively converts them to miles, depending on the status of the segmented control. Finally, `updateLocationFor:toLocation:` updates either the `start-Label` or the `endLabel` by extracting the latitude and longitude coordinates from the `CLLocation` object that it's passed **❿**.

The result is a program that can show a simple distance traveled, as the crow flies. If we were going to improve it, we'd probably save the starting location to a file, and perhaps even make it possible to record multiple trips. But for the purposes of showing how Core Location works, this is sufficient.

There's one thing that our example didn't show: how to measure altitude. It's just another `CLLocation` property, but we've written another short program to highlight this part of Core Location.

17.4.3 *An example using altitude*

Altitude is as easy to work with as longitude and latitude. It's just another property that can be read from a `CLLocation` object. The biggest problem is that it won't be available to all users. In particular, neither the iPhone Simulator nor the original iPhone report altitude at the time of this writing.

Apple suggests using following code to determine whether altitude is unavailable:

```
if (signbit(newLocation.verticalAccuracy)) {
```

If its return is non-zero, you need to bail out of checking for altitude information.

Even if a user has an iPhone 3G, there are two other gotchas that you must watch out for. First, altitude information can be perhaps 10 times more inaccurate than the rest of the location information. Adjust your `desiredAccuracy` accordingly. Second, remember that the Core Location information only updates when you move a certain distance, as determined by the `distanceFilter`, in a non-vertical direction. This probably means that you'll need to allow the user to update the distance by hand rather than depending on automatic updates.

Listing 17.10 repeats the techniques that we used previously, applying them to altitude. It also shows another nice integration of user input with a slightly more complex program. As usual, its core objects were all built in Interface Builder: three `UILabels`, one `UITextField`, two `UIImageViews`, and a `UIActivityIndicatorView`. The last is the most interesting, because we haven't seen it before; we'll talk about it in our quick discussion of the code. You should be able to pick out all of the objects other than the activity indicator in figure 17.4, which follows the code.

Listing 17.10   You can climb any mountain with your iPhone keeping track for you

```
@implementation altitudeViewController

- (void)viewDidLoad {
```

```
        destinationHeight.returnKeyType = UIReturnKeyDone;

        myLM = [[CLLocationManager alloc] init];
        myLM.delegate = self;
        myLM.desiredAccuracy = kCLLocationAccuracyBest;

        savedDestinationHeight = 0;

        [super viewDidLoad];
    }
- (BOOL)textFieldShouldReturn:(UITextField *)textField {
        [textField resignFirstResponder];
        return YES;
    }
- (IBAction)changeDestination:(id)sender {

        savedDestinationHeight = [destinationHeight.text intValue];
        [self resetGPS:sender];
    }
- (IBAction)resetGPS:(id)sender {

        if (savedDestinationHeight) {
            [myLM startUpdatingLocation];
            [myActivity startAnimating];
        }
    }

- (void)locationManager:(CLLocationManager *)manager
        didUpdateToLocation:(CLLocation *)newLocation
        fromLocation:(CLLocation *)oldLocation {

        if (savedDestinationHeight) {
            if (signbit(newLocation.verticalAccuracy)) {
                heightLabel.text = [NSString stringWithString:@"?? m."];
            } else {
                int currentHeight = 395 -
                    ceil((float)newLocation.altitude/savedDestinationHeight *
                        (401-65));

                heightLabel.text = [NSString stringWithFormat:@"%6.2f m.",
                    newLocation.altitude];
                heightButton.center = CGPointMake(176,currentHeight);
                heightLabel.center = CGPointMake(220,currentHeight);
            }
            [myLM stopUpdatingLocation];
            [myActivity stopAnimating];
        }
    }
...
@end
```

Annotations pointing to the code:

1 Prepares UITextfield

2 Prepares CLLocationManager

3 Dismisses keyboard from text field

4 Responds to text field

5 Starts location manager

6 Requests location updates

7 Animates activity icon

8 Receives location update

9 Shows altitude failure

10 Reports altitude information

11 Stops location update

12 Stops activity animation

Much of this code combines two SDK elements that you've already met: the flourishes necessary to make a UITextField work and the protocols that you must follow to use a location manager. You can see both of these elements in the viewDidLoad: method, which sets up the text field's return key **1** and then starts up the location manager **2**. Note that you don't start the location manager updating; you can't depend on it to update when you're only measuring vertical change, so it's best to have the user do it by hand. Next, you finish up the text field's functionality with the `textField-ShouldReturn:` method, which you've met before ❸.

This project contains two Interface Builder objects that can generate actions. When the text field is adjusted ❹, the project saves that destination height for future calculation, and then updates the current height using the `resetGPS:` method ❺. The latter method is also used when the Check Height button is pressed. Figure 17.4 shows these input devices, for clarity.

Note that `resetGPS:` does two things. First, it starts the location update ❻, which you only turn on for brief one-time uses. Besides being more appropriate for monitoring altitude, this also helps save energy. Second, it starts your activity indicator going ❼. This object was created in Interface Builder, where you should have marked it with the `hidesWhenStopped` property. Interface Builder also automatically hides the view so that it doesn't appear when the program is loaded. As a result, there's nothing on the screen until you start the animation here, at which time a little activity indicator appears and continues animating until it's stopped (which we'll see in a minute).

Figure 17.4 An altitude program measures how high you've climbed on a mountain of your choice.

The heavy lifting is done when the location manager reports back its information ❽. Here, you check whether you're getting altitude information at all ❾. If you are ❿, you move the red dot image and update its height label. To finish things off, you turn the location update back off ⓫, and then stop the animation of the activity indicator ⓬, which makes it disappear.

Voila! You have a working altitude monitor (if you have an iPhone 3G) and a nice combination of a few different SDK elements.

17.4.4 *Core Location and the Internet*

In this section, we've seen a few real-world examples of how to use location information in meaningful ways, but you'll find that you can make much better use of the information when you have an internet connection. When you do, you can feed longitudes and latitudes to various sites. For example, you can pull up maps with a site like Google Maps. You can also improve on the iPhone's relatively poor (or nonexistent) altitude information by instead requesting the geographic altitude of a location using a site like GeoNames. This won't be accurate if your user is in an airplane or a tall office building, but for the majority of situations, it'll probably be better than what the iPhone can currently deliver.

Rather than address these possibilities here, we've decided to save them for chapter 20, where we'll use them as examples of how to interact with the internet.

17.5 *Summary*

In this chapter, we've covered two of the most unique features available to you as SDK programmers.

The iPhone's accelerometers can give you access to a variety of information about where an iPhone exists in space. By measuring gravity, you can easily discover an iPhone's precise orientation. By measuring movement, you can see how an iPhone is being guided through space. Finally, you can build more complex movements into three-dimensional gestures, such as the shake.

We've usually talked about the iPhone's touch screen when discussing iPhone input, but the accelerometers provide another method for allowing users to make simple adjustments to a program. We can imagine game controls and painting programs both built entirely around the accelerometers.

The iPhone's GPS can give you information on an iPhone's longitude, latitude, and altitude. The horizontal information is the most reliable, though it'll prove more useful when you connect to the internet. The altitude information isn't available to everyone, and even then it has a higher chance of being incorrect, so use it with caution.

Though the accelerometer and GPS data is cool, we expect there will be many programs that don't use them at all. We can't say the same for the topics of our next chapter. There we'll be talking about media, highlighting pictures, videos, and sounds, which almost every iPhone project will incorporate.

18

Media:
images and sounds

This chapter covers

- Accessing and manipulating images
- Using the iPhone camera
- Playing sounds and video

So far, our focus on iPhone programming has mainly been on text. Sure, we've displayed the occasional UIImage, such as the mountain drawing in the last chapter, but we've only considered the simplest means for doing so.

The iPhone offers an experience that's potentially much richer and more engaging. A camera, a microphone, a complete photos library, and a speaker are just some of the utilities built into the iPhone. In this chapter, we're going to look at these features as part of a general exploration of media. We'll provide deep coverage on images, some information on the iPhone's media player, and a basic look at playing sounds on the iPhone.

More complex questions are beyond the scope of this chapter. We're saving the topic of image editing for the next chapter, when we look at the iPhone's graphic libraries. For more complex sound work, we'll offer pointers to Apple's extensive tutorials on the topic.

18.1   *An introduction to images*

We've already touched upon using images a few times, beginning in chapter 12 where we included an image in one of our earliest SDK examples. We've always created a UIImageView in Interface Builder, attached it to a filename, and not worried about the details.

We're now ready to consider the details. We'll look at some of the options you have available when you dive into Xcode, rather than depending upon Interface Builder's higher-level abstractions.

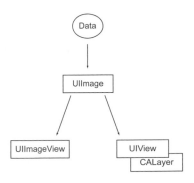

When you look more closely, you'll discover that using images is a two-step process. First, you load data into a UIImage, and then you make use of that UIImage via some other means. There are two major ways to make use of UIImages, as shown in figure 18.1.

We're going to explore the simpler methods of displaying images, using UIImageView, in this section, and in section 18.2 we'll examine the more complex means available for drawing images onto the back layer of a UIView.

Figure 18.1   Images can be shown in UIImageViews or in UIViews.

18.1.1 *Loading a UIImage*

The UIImage class offers seven different ways to create an instance of an image. The four factory methods are probably the easiest to use, and they're the ones we've listed in table 18.1. There are also some equivalent init methods that you can use if you prefer.

Table 18.1   Factory methods for creating a UIImage

Factory method	Summary
imageNamed:	Creates a UIImage based on a file in the main bundle
imageWithCGImage:	Creates a UIImage from a Quartz 2D object; this is the same as initWithCGImage:
imageWithContentsOfFile:	Creates a UIImage from a complete file path that you specify, as discussed in chapter 16; this is the same as initWithContentsOfFile:
imageWithData:	Creates a UIImage from NSData; this is the same as initWithData:

The image data can be of several file types, including BMP, CUR, GIF, JPEG, PNG, and TIFF. In this book, we've used mostly JPEGs (because they're small) and PNGs (because they look good and are accelerated on the iPhone hardware). You can also create a UIImage from a Quartz 2D object; this is the iPhone's fundamental graphics package, which we're going to talk about more in the next chapter. There's one suggested restriction when you're creating UIImages: the images shouldn't be larger than 1024x1024.

Once you import an image into your program, you can display it. If you're going to stay entirely within the simple methods of the UIKit, you'll want to use the UIImageView class to display the image.

18.1.2 *Drawing a UIImageView*

We've already used the UIImageView in our programs when displaying pictures. We're now ready to talk about the details of how it works.

There are two ways to initialize a UIImageView. First, you can use the initWithImage: method, which allows you to pass a UIImage, as follows:

```
UIImage *myImage1 = [UIImage imageNamed:@"sproul1.jpg"];
UIImageView *myImageView =
   [[UIImageView alloc] initWithImage:myImage1];
[self.view addSubview:myImageView];
```

Alternatively, you can use a plain initWithFrame: method and modify the object's properties by hand. Table 18.2 shows a few of the properties and methods that you're most likely to use when doing more extensive work with a UIImageView.

Table 18.2   A few properties and methods of note for UIImageView

Method or property	Type	Summary
animationDuration	Property	Specifies how often an animation cycles
animationImages	Property	Identifies an NSArray of images to load into the UIImageView
animationRepeatCount	Property	Specifies how many times to run an animation cycle
image	Property	Identifies a single image to load into a UIImageView
startAnimating	Method	Starts the animation
stopAnimating	Method	Stops the animation

To load a normal image, you could use the image property, but there's usually little reason to use it rather than the initWithImage: method—unless you're dynamically changing your image. If you want to create a set of images to animate, it's useful to take advantage of the other UIImageView methods and properties.

You can load an array of images into a UIImageView, declare how fast and how often they should animate, and start and stop them as you see fit. A simple example of this is shown in listing 18.1.

Listing 18.1 UIImageView allows for animated images

```
- (void)viewDidLoad {

   UIImage *myImage1 =
      [UIImage imageNamed:@"sproul1.jpg"];    │  Loads images
   UIImage *myImage2 =
```

```
        [UIImage imageNamed:@"sproul2.jpg"];
    UIImage *myImage3 =
        [UIImage imageNamed:@"sproul3.jpg"];      Loads images
    UIImage *myImage4 =
        [UIImage imageNamed:@"sproul4.jpg"];

    UIImageView *myImageView =
        [[UIImageView alloc]                   Creates
            initWithFrame:[[UIScreen           UIView
                mainScreen] bounds]];

    myImageView.animationImages =
        [NSArray arrayWithObjects:myImage1,
        myImage2,myImage3,myImage4,nil];       Starts
    myImageView.animationDuration = 4;         animation
    [myImageView startAnimating];

    [self.view addSubview:myImageView];
    [myImageView release];
    [super viewDidLoad];
}
```

Taking advantage of UIImageView's animation capability is one of the main reasons that you might want to load images by hand rather than do it through Interface Builder.

18.1.3 Modifying an image in the UIKit

Now you've seen how to create images and load them into image views programmatically. Surely, the next thing to do is to start modifying them.

Unfortunately, you have only limited capability to do so while working with UIImageView. You can make *some* changes, based on simple manipulations of the view. For example, if you resize your UIImageView, it'll automatically resize the picture it contains. Likewise, you can decide where to draw your UIImageView by setting its frame to something other than the whole screen. You can even layer multiple images by using multiple UIImageViews.

This all starts to get unwieldy pretty quickly, though, and you can't do anything fancier, like transforming your images or modifying how they stack through blending or alpha transparency options. To do that sort of work (and to start stacking graphics, not just views) you need to learn about Core Graphics.

UIImage offers some simple ways to access Core Graphics functionality that doesn't require going out to the Core Graphics framework (or learning about contexts or the other complexities that underlie its use). We're going to talk about those briefly here, but, for the most part, Core Graphics will wait for the next chapter, which concentrates on the entire Quartz 2D graphics engine.

18.2 Drawing simple images with Core Graphics

Although it doesn't give access to the entire Core Graphics library of transformations and other complexities, the UIImage class does include five simple methods that take advantage of the way Core Graphics works. They're described in table 18.3.

Table 18.3   Instance methods for drawing a `UIImage`

Method	Summary
`drawAsPatternInRect:`	Draws the image inside the rectangle, unscaled, but tiled as necessary
`drawAtPoint:`	Draws the complete unscaled image with the `CGPoint` as the top-left corner
`drawAtPoint:blendMode:alpha:`	A more complex form of `drawAtPoint:`
`drawInRect:`	Draws the complete image inside the `CGRect`, scaled appropriately
`drawInRect:blendMode:alpha:`	A more complex form of `drawInRect:`

The trick is that these methods *cannot* be used as part of `viewDidLoad:` or whatever other method you usually use to load up your objects. That's because they depend upon a graphical *context* to work. We're going to talk about contexts more in the next chapter, but a graphical context is a destination that you're drawing to, like a window, a PDF file, or a printer.

On the iPhone, `UIView`s automatically create a graphical context as part of their `CALayer`, which is a Core Animation layer associated with each `UIView`. You can access this layer by writing a `drawRect:` method for the `UIView` (or rather, for a new subclass that you've created). You'd usually have to capture a special context variable to do this type of work, but the `UIView` methods take care of this for you, to keep things simple.

Listing 18.2 shows how to collage together a few pictures using this method.

Listing 18.2   A `UIView`'s `drawRect:` allows you to use lower-level draw commands

```
- (void)drawRect:(CGRect)rect {

    UIImage *myImage1 = [UIImage imageNamed:@"sproul1.jpg"];
    UIImage *myImage2 = [UIImage imageNamed:@"sproul2.jpg"];
    UIImage *myImage3 = [UIImage imageNamed:@"sproul3.jpg"];

    [myImage1 drawAtPoint:CGPointMake(0,0) blendMode:kCGBlendModeNormal
        alpha:.5];
    [myImage2 drawInRect:CGRectMake(10, 10, 140, 210)];
    [myImage3 drawInRect:CGRectMake(170, 240, 140, 210)];
}
```

Note that the `drawAtPoint:` method gives you access to more complex possibilities, such as blending your pictures (using Photoshop-like options such as color dodge and hard light) and making them partially transparent. Here you're using a normal blend, but only 50 percent transparency (hence the use of the `drawAtPoint:` method). The rest of the code is standard enough. The simplicity of using these singular draw commands rather than going to the effort of creating multiple `UIImageView` objects speaks for itself (and it's presumably more efficient too).

There's still a lot that we can't do until we delve fully into the iPhone's Core Graphics framework, but for now we've got some control, which should be sufficient for most of your common media needs. If you need more control, skip right ahead to the next chapter.

We've talked lots about images, and we've presumed so far that you're loading them from your project's bundle. But what if you want to let a user select photographs? That's the topic of our next section.

18.3 Accessing photos

You can use the SDK to access pictures from an iPhone's photo library or its camera roll. You can also allow a user to take new photos. This is all done with the UIImage-PickerController, another modal controller that manages a fairly complex graphical interface without much effort on your part. Figure 18.2 shows what it looks like.

Figure 18.2 The image picker is another preprogrammed controller for your use.

18.3.1 Using the image picker

The UIImagePickerController is loaded up by creating the object, setting a few variables, and presenting it as a modal view controller. By default, the image picker controller will allow users to access (and optionally edit) the pictures in their photo library:

```
UIImagePickerController *myImagePicker =
    [[UIImagePickerController alloc] init];
myImagePicker.delegate = self;
myImagePicker.allowsImageEditing = NO;
[self presentModalViewController:myImagePicker animated:YES];
```

Once you've created your image picker controller, you need to have its delegate respond to two methods: imagePickerController:didFinishPickingImage:editingInfo: and imagePickerControllerDidCancel:. For the first method, you should dismiss the modal view controller and respond appropriately to the user's picture selection, and for the second, you only need to dismiss the controller.

Overall, the image picker controller is easy to use because you're mainly reacting to a picture that was selected. We've got a complete example of its use in the next section.

18.3.2 Taking photos

As we noted earlier, the UIImagePickerController has three possible sources, represented by these constants:

- UIImagePickerControllerSourceTypePhotoLibrary, a picture from the photo library
- UIImagePickerControllerSourceTypeSavedPhotosAlbum, a picture from the camera roll
- UIImagePickerControllerSourceTypeCamera, new picture taken by the camera

You should always make sure that the source is available before you launch an image picker controller, although this is most important for the camera. You can confirm that the source exists with the `isSourceTypeAvailable:` class method:

```
if ([UIImagePickerController
    isSourceTypeAvailable:UIImagePickerControllerSourceTypeCamera]) {
```

Once you've verified the existence of a source, you can tell the image picker to use it with the `sourceType` property. For example, to use the camera, do the following:

```
myImagePicker.sourceType = UIImagePickerControllerSourceTypeCamera;
```

Note that pictures taken in a program only go to that program. If you want them to go into the photo album, your program will have to save them there (as we'll discuss momentarily).

In our experience, the camera is a bit of a resource hog. We had it grind to a halt a few times during testing. More than anything else, this means that you need to think about saving your program's state when using the camera, because it could cause you to run out of memory.

We'll have an example of using the camera in our example in section 18.4.

18.3.3 *Saving to the photo album*

You may wish to save a new photograph to the photo album, or you may wish to place a graphic created by your program there. In either case, you use the `UIImageWriteTo-SavedPhotosAlbum` function. It has four variables: the first lists the image, and the other three reference an optional asynchronous notification function to call when the save has been completed. Usually you'll call the function like this:

```
UIImageWriteToSavedPhotosAlbum(yourImage,nil,nil,nil);
```

If you instead want to take advantage of the asynchronous notification, take a look at the `UIKit` function reference, which is where this function is hidden away, or look at our example in the next chapter.

You can use this function (and a bit of trickery) to save the `CALayer` of a `UIView` to your photo album, which, for example, will allow you to save those draw commands that you wrote straight to the `CALayer` earlier. This once more depends upon graphical contexts, which we'll explain in the next chapter, but here's how to do it:

```
UIGraphicsBeginImageContext(myView.bounds.size);
[myView.layer renderInContext:UIGraphicsGetCurrentContext()];
UIImage *collageImage = UIGraphicsGetImageFromCurrentImageContext();
UIGraphicsEndImageContext();
UIImageWriteToSavedPhotosAlbum(collageImage,nil,nil,nil);
```

In order for this to work correctly, you must link in the Quartz Core framework.

With all of the fundamentals of images now covered, we're ready to put them together in our "big" example for this chapter, which is a program that collages together multiple pictures, first selecting them with a `UIImagePickerController`, then allowing them to be moved about with a `UIImageView`, and finally drawing them to a `CALayer` that can be saved.

18.4   Collage: an image example

The collage program depends on three objects. The collageViewController, as usual, does most of the work. It writes out to a collageView object, which exists mainly as a CALayer to be written upon. Finally, you'll have a tempImageView object that allows the user to position an image after it's been selected but before it's permanently placed.

18.4.1   The collage view controller

The collage view controller is built on a few Interface Builder objects: the view controller itself; a toolbar called myTools, which will be filled over the course of the program; and the collageView UIView class, which exists as its own class file and is referred to in the program as self.view. You'll also need to add the Quartz Core framework to your project as you'll use that save-picture trick that we just discussed.

Listing 18.3 shows the complete view controller, which is the most extensive file in this program.

> **Listing 18.3   A view controller manages most of the collage's tasks**

```
@implementation collageViewController

- (void)viewDidLoad {          ◁────❶ Sets up objects

   UIBarButtonItem *picButton = [[UIBarButtonItem alloc]
      initWithBarButtonSystemItem:UIBarButtonSystemItemAction target:self
      action:@selector(choosePic:)];
   UIBarButtonItem *camButton - [[UIBarButtonItem alloc]
      initWithBarButtonSystemItem:UIBarButtonSystemItemCamera target:self
      action:@selector(takePic:)];
   UIBarButtonItem *saveButton = [[UIBarButtonItem alloc]
      initWithBarButtonSystemItem:UIBarButtonSystemItemSave target:self
      action:@selector(savePic:)];

   picButton.style = UIBarButtonItemStyleBordered;
   camButton.style = UIBarButtonItemStyleBordered;

   if ([UIImagePickerController
      isSourceTypeAvailable:UIImagePickerControllerSourceTypeCamera]) {
         origToolbar = [[NSArray alloc] initWithObjects:
            picButton,camButton,saveButton,nil];
   } else if ([UIImagePickerController
   isSourceTypeAvailable:UIImagePickerControllerSourceTypePhotoLibrary]) {
         origToolbar = [[NSArray alloc] initWithObjects:
            picButton,saveButton,nil];
   } else {
      exit(0);
   }
   [myTools setItems:origToolbar animated:NO];

   [picButton release];
   [camButton release];
   [super viewDidLoad];
}
                                      ❷ Activates
- (IBAction)choosePic:(id)sender {   ◁──┘  image picker
```

```
    UIImagePickerController *myImagePicker =
        [[UIImagePickerController alloc] init];
    myImagePicker.delegate = self;
    myImagePicker.allowsImageEditing = NO;

    [self presentModalViewController:myImagePicker animated:YES];
}

-(IBAction)takePic:(id)sender {         ←——❸ Activates camera

    UIImagePickerController *myImagePicker =
        [[UIImagePickerController alloc] init];
    myImagePicker.sourceType = UIImagePickerControllerSourceTypeCamera;
    myImagePicker.delegate = self;
    myImagePicker.allowsImageEditing = NO;

    [self presentModalViewController:myImagePicker animated:YES];
}

- (void)imagePickerController:(UIImagePickerController *)picker
    didFinishPickingImage:(UIImage *)image
    editingInfo:(NSDictionary *)editingInfo {    ←——┐
                                            ❹ Responds to
    [self dismissModalViewControllerAnimated:YES];    image selection
    [picker release];

    float percentage = [self scaleImage:image] / 2;
    startingSize = CGSizeMake(image.size.width*percentage,
        image.size.height*percentage);

    myImageView = [[tempImageView alloc]
        initWithFrame:CGRectMake(80,115,
            startingSize.width,startingSize.height)];
    myImageView.image = image;
    myImageView.userInteractionEnabled = YES;

    [self.view addSubview:myImageView];
    [myTools setItems:[NSArray arrayWithObject:[[UIBarButtonItem alloc]
        initWithBarButtonSystemItem:UIBarButtonSystemItemDone target:self
        action:@selector(finishPic:)]] animated:YES];

    mySlider = [[UISlider alloc] initWithFrame:CGRectMake(90,415,210,44)];
    mySlider.value = .5;
    [mySlider addTarget:self action:@selector(rescalePic:)
        forControlEvents:UIControlEventValueChanged];
    [self.view addSubview:mySlider];
}

- (void)imagePickerControllerDidCancel:      ❺ Responds to picker
    (UIImagePickerController *)picker {   ←——┘  cancellation

    [self dismissModalViewControllerAnimated:YES];
    [picker release];
}
                                    ❻ Resizes
-(void)rescalePic:(id)sender {     ←——┘  picture

    myImageView.frame = CGRectMake(myImageView.frame.origin.x,
        myImageView.frame.origin.y,
        startingSize.width * mySlider.value * 2,
        startingSize.height * mySlider.value * 2);
```

```
}
-(void)finishPic:(id)sender {        ◀──  ⑦ Adds picture
                                             to CALayer
    [self.view addPic:myImageView.image at:myImageView.frame];

    [myImageView removeFromSuperview];
    [myImageView release];
    [mySlider removeFromSuperview];
    [mySlider release];

    [myTools setItems:origToolbar animated:NO];
}
-(void)savePic:(id)sender {        ⑧ Saves collage
    UIGraphicsBeginImageContext(self.view.bounds.size);  ◀──
    myTools.hidden = YES;
    [self.view.layer renderInContext:UIGraphicsGetCurrentContext()];
    UIImage *collageImage = UIGraphicsGetImageFromCurrentImageContext();
    myTools.hidden = NO;
    UIGraphicsEndImageContext();

    UIImageWriteToSavedPhotosAlbum(collageImage,nil,nil,nil);
}

-(float)scaleImage:(UIImage *)image {        ◀──  Scales
                                               ⑨ image
    float toSize = 1.0;

    if (image.size.width * toSize > 320) {
        toSize = 320 / image.size.width;
    }
    if (image.size.height * toSize > 460) {
        toSize = 460 / image.size.height;
    }
    return toSize;
}
// ...
@end
```

Although long, this code is simple to follow in bite-sized chunks. It starts off with viewDidLoad:, which sets up the UIToolBar ①. We've long lauded Interface Builder, but we've also said that it might not be sufficient when you're creating more dynamic projects. That's the case here. You can't efficiently fill the UIToolBar in Interface Builder because you're going to be changing it based on the program's state. You're placing buttons on the toolbar that call three methods: choosePic:, takePic: (when a camera's available), and savePic:.

choosePic: ② and takePic: ③ are similar methods. Each calls up the image picker controller, but the first one accesses the photo library and the second one lets the user take a new picture. The wonder of these modal controllers is that you don't have to do a thing between the time when you create the picker to the point where your user either selects a picture or cancels.

When the user selects a picture, imagePickerControl:didFinishPicking-Image:editingInfo will be called ④, returning control to your program. Here you do four things:

- Dismiss the modal view controller.
- Look at the picture you've been handed and resize it to fill a quarter or less of the screen.
- Instantiate the image as a `tempImageView` object, which is a subclass of `UIImageView`.
- Change the toolbar so that there's a Done button available, along with a slider.

At this point, the user will be able to do three things:

- Use `UITouches` to move the image view (which is covered in the `tempImageView` class, because that's where the touches go, as we saw in chapter 14).
- Use the slider to change the size of the picture.
- Click Done to accept the image size and location.

The end results of what can be produced are shown in figure 18.3.

Note that if the user instead canceled the image picker, your `imagePickerControllerDidCancel:` method would correctly shut down the modal controller ❺.

The `UISlider` was hooked up to the `rescalePic:` method ❻. It redraws the frame of the `UIImageView`, which automatically resizes the picture inside. Meanwhile, the Done button activates the `finishPic:` method ❼. This sends a special `addPic:at:` message to the `collageView`, which is where the `CALayer` drawing is done, and which we'll return to momentarily. `finishPic:` also dismisses the `UISlider` and the `tempImageView` and resets the toolbar to its original setup.

That original toolbar had one more button that we haven't covered yet: Save. It activates the `savePic:` method ❽, which saves a `CALayer` to the photo library. Note that this method temporarily hides toolbar in the process. Because the toolbar is a subview of the `UIView`, it'd get included in the picture if you didn't do this.

The last method, `scaleImage:` ❾, is the utility that sets each image to fill about a quarter of the screen.

Figure 18.3   The collager in use displays many photos simultaneously.

This code has two dangling parts: the methods in the `tempImageView`, which allow a user to move the `UIImageView`, and the methods in the `collageView`, which later draw the image into a `CALayer`.

18.4.2 *The collage temporary image view*

The `tempImageView` class only has one purpose: to intercept `UITouches` that indicate that the user wants to move the new image to a different part of the collage. This simple code is shown in listing 18.4.

Listing 18.4   A temporary image can be moved about by touches

```
- (void) touchesMoved:(NSSet *)touches withEvent:(UIEvent *)event {

    UITouch *thisTouch = [touches anyObject];
    CGPoint thisPoint =
        [thisTouch locationInView:self];
    float newX = thisPoint.x+self.frame.origin.x;
    float newY = thisPoint.y+self.frame.origin.y;

  if (newX < 0) {
     newX = 0;
  } else if (newX > 320) {
     newX = 320;
  }
  if (newY < 0) {
     newY = 0;
  } else if (newY > 416) {
     newY = 416;
  }
  self.center = CGPointMake(newX,newY);
}
```

❶ Determines position in view

❷ Calculates overall position

This is similar to the touch code that you wrote in chapter 14, and it isn't worthy of a lot of additional commentary. Recall that locationInView: ❶ gives a CGPoint internal to the view's coordinate system and needs to be converted ❷ into the global coordinate system of the application.

In testing, we discovered that when run on an iPhone (but not in the iPhone Simulator) the result could sometimes be out of bounds, so you need to double-check your coordinates before you move the temporary image view.

18.4.3   *The collage view*

Last up, we have the collageView itself, which is the background UIView that needs to respond to the addPic:at: message and draw onto the CALayer with drawRect:. The code to do this is shown in listing 18.5.

Listing 18.5   A background view manages the low-level drawing once an image is set

```
-(void)addPic:(UIImage *)newPic at:(CGRect)newLoc {        ◄——❶ Saves into array

  if (! myPics) {
     myPics = [[NSMutableArray alloc] initWithCapacity:0];
     [myPics retain];
  }

  [myPics addObject:[NSDictionary dictionaryWithObjectsAndKeys:
     newPic,@"picture",
     [NSNumber numberWithFloat:newLoc.origin.x],@"xpoint",
     [NSNumber numberWithFloat:newLoc.origin.y],@"ypoint",
     [NSNumber numberWithFloat:newLoc.size.width],@"width",
     [NSNumber numberWithFloat:newLoc.size.height],@"height",
        nil]];

  [self setNeedsDisplay];
```

```
    }
- (void)drawRect:(CGRect)rect {          ❷ Draws onto
                                            CALayer
    if (myPics) {
        for (int i = 0 ; i < myPics.count ; i++) {

            UIImage *thisPic = [[myPics objectAtIndex:i]
                objectForKey:@"picture"];
            float xpoint = [[[myPics objectAtIndex:i]
                objectForKey:@"xpoint"] floatValue];
            float ypoint = [[[myPics objectAtIndex:i]
                objectForKey:@"ypoint"] floatValue];
            float height = [[[myPics objectAtIndex:i]
                objectForKey:@"height"] floatValue];
            float width = [[[myPics objectAtIndex:i]
                objectForKey:@"width"] floatValue];

            [thisPic drawInRect:CGRectMake(xpoint,ypoint,width,height)];
        }
    }
}
```

This code is broken into two parts. The addPic:at: method ❶ saves its information into an instance variable, adding a myPics dictionary to the NSMutableArray. Note that you have to convert values into NSNumbers so that you can place them in the dictionary. This method then calls setNeedsDisplay on the view. You should *never* call drawRect: directly. Instead, when you want it to be executed, call the setNeedsDisplay method, and everything else will be done for you.

drawRect: gets called shortly afterward ❷. It reads through the whole NSMutableArray, breaks it apart, and draws each image onto the CALayer using the techniques we learned earlier.

We haven't shown the few header files and the unchanged app delegate, but this is everything important needed to write a complete collage program.

18.4.4 *Expanding on this example*

This was one of our longer examples, but it could still bear some expansion to turn it into a fully featured application.

First, it's a little unfriendly with memory. It'd probably be better to maintain references to filenames, rather than keeping the UIImages around. In addition, the NSArray that the CALayer is drawn from should be saved out to a file, so that it wouldn't get lost if memory is low. But the program as it exists should work fine.

The program could be made more usable. An option to crop the pictures would be nice, but it would probably require access to Core Graphics functions. An option to move pictures around after they've been locked in would be relatively simple: you could test for touches in the collageView, and read backward through the NSArray to find which object the user was touching. Reinstantiating it as a UIImageView would then be simple.

In any case, you've seen how all of these pictorial fundamentals work together, so we're now ready to move on to the next major types of media: audio and video.

18.5 Using the Media Player framework

Audio and video both allow for considerably more complexity than images. Fortunately, there's a high-level framework that gives you access to both of them: the Media Player. If you don't need the audio or video to be tightly integrated into the rest of your application, it's a pretty good choice—it often works well for video, but it's a somewhat more questionable choice for audio.

The Media Player framework includes two classes: `MPMoviePlayerController` and `MPVolumeView`. They manage a full-page audio or video player that doesn't give you a lot of control over the specifics of how it works, but it does give you easy access to audio or video files.

To use either of these media player classes, you should add the Media Player framework and the MediaPlayer/MediaPlayer.h header file to your project.

18.5.1 The media player class

In order to use the media player itself, you need to init an `MPMoviePlayerController` object with the URL string for the file that you'll be calling up. This may be any .mp3, .mp4, .mov, or .3gp file or anything else supported by the iPhone. You can either start it playing immediately (which will cause the iPhone's wheels to spin for a bit, while it gets ready) or you can wait until you've received a notification that the file has loaded.

There are three notifications that you may wish to pay attention to when using the media player, as described in table 18.4.

Table 18.4 Notifications that tell you what the media player is doing

Notification	Summary
`MPMoviePlayerContentPreloadDidFinishNotification`	File has loaded
`MPMoviePlayerPlaybackDidFinishNotification`	Playback was completed
`MPMoviePlayerScalingModeDidChangeNotification`	Scaling mode for player changed

INVOKING THE MEDIA PLAYER

Listing 18.6 displays a simple invocation of the player. This program's construction begins in Interface Builder with a `UITextField` (for the input of a URL), a `UILabel` (for reporting status and errors), and a `UIActivityIndicatorView` (to show activity during a load). It depends on the notifications to keep track of how the media player is doing.

Listing 18.6 A simple invocation of the media player

```
- (void)viewDidLoad {        ◁——❶ Prepares text field

   myText.returnKeyType = UIReturnKeyDone;
}
- (BOOL)textFieldShouldReturn:           ❷ Dismisses
   (UITextField *)textField {     ◁——       keyboard
   [textField resignFirstResponder];
```

```
        return YES;
    }
    -(IBAction)chooseFile:(id)sender {

        myMP = [[MPMoviePlayerController alloc]
            initWithContentURL:
                [NSURL URLWithString:myText.text]];
        myMP.movieControlMode = MPMovieControlModeDefault;

        [[NSNotificationCenter defaultCenter] addObserver:self
            selector:@selector(movieDidLoad:)
            name:@"MPMoviePlayerContentPreloadDidFinishNotification"
            object:nil];
        [[NSNotificationCenter defaultCenter] addObserver:self
            selector:@selector(movieDidFinish:)
            name:@"MPMoviePlayerPlaybackDidFinishNotification"
            object:nil];

        errorLabel.text = @"File is loading ...";
        [myActivity startAnimating];
    }
    -(void)movieDidLoad:(id)sender {

        errorLabel.text = [NSString    string];
        [myActivity stopAnimating];
        [myMP play];
    }
    -(void)movieDidFinish:(id)sender {
        myText.text = [NSString string];
    }
```

❸ Loads media file

❹ Starts media player

❺ Sets player control mode

❻ Requests media notifications

❼ Placates user

❽ Starts movie when loaded

❾ Updates screen when done

Your project starts off simply enough by setting up your UITextField. This involves setting the Return key ❶ and writing its main delegate method, textFieldShould-Return: ❷, as you've done several times before.

It's the text field that really kicks things off. When text data is entered, the choose-File: method is called ❸, which is what loads up the player. You assume you're passed a URL (which we've used because of its simplicity, though we'll talk about local files momentarily), turn it into an NSURL, and then create the player ❹. There are a small number of properties that you can then set to specify how the player works ❺, all of which are listed in the class reference. You link into the player's notifications using the procedure you've seen before ❻. It's always good to let the user know that you're processing things, so the last few lines update some status info and start an activity indicator going ❼.

Once the file has been loaded, your movieDidLoad: method is notified ❽. It cleans up your updating info and then starts the player playing. Like the modal view controllers we've seen in previous chapters, the media player takes over at this point, and you don't have to worry about anything until its additional notifications come back. In this case, when it's finished ❾, you do some final cleanup.

LOADING FROM A FILE

If you prefer to load from a file rather than from the internet, you can include media files as part of your bundle. You create a path to those local files using the methods that we discussed in chapter 16, and you create your NSURL with the fileURLWithPath: factory method:

```
NSString *paths = [[NSBundle mainBundle] resourcePath];
NSString *mediaFile = [paths
    stringByAppendingPathComponent:@"yourfile.mp4"];
NSURL *mediaURL = [NSURL fileURLWithPath:mediaFile isDirectory:NO];
```

At the time of this writing, loading files is considerably more reliable than loading from the internet. Music files occasionally break for no reason when loaded from the internet, and streaming video doesn't appear to work at all. We expect these problems to be corrected soon, possibly by the time this book is published.

There's only one other bit of functionality that the media player supports: you can allow users to set the volume outside of the player itself.

18.5.2 *The volume view*

You can allow user adjustment of the volume by calling up the MPVolumeView item, which is done like this:

```
myVolume = [[MPVolumeView alloc]
    initWithFrame:CGRectMake(200, 100, 100, 200)];
[myVolume sizeToFit];
[self.view addSubview:myVolume];
```

You don't need to do any backend work; when the user changes volume control, the system volume will change immediately. If you prefer an alternative method, there are three general functions that can be used to call up a volume alert. They're listed in table 18.5.

Table 18.5 Setting volume with alerts

Function	Summary
MPVolumeSettingsAlertShow	Shows a volume alert
MPVolumeSettingsAlertHide	Hides a volume alert
MPVolumeSettingsAlertIsVisible	Returns a Boolean to show the status of the volume alert

Note that these are functions, not methods. They're not tied to any class, but rather are generally available once you've loaded the Media Player framework.

WARNING   At the time of this writing, the volume controls do not work in the iPhone Simulator.

18.5.3 *Better integrating the media player*

The biggest problem with the media player is that it calls up a stand-alone screen. As a result, it's difficult to use it to integrate music or video directly into your program.

Figure 18.4 The media player
doesn't integrate music well.

For music, the problem is relatively unsolvable at this time. As shown in figure 18.4, when music plays, the screen is taken over by a large QuickTime logo. We hope that future versions of the SDK will give you the option to define a background for when sounds are playing (or better, allow you to remain within your normal views, thus truly integrating the media player's audio capability).

For playing videos, the biggest problem is the controls, because you don't want users to manipulate the video if you're using it as a cut scene. You can resolve this by setting the `MPMoviePlayerController`'s `movieControlMode` property to `MPMovieControlModeVolumeOnly` (which only allows use of the volume control) or `MPMovieControlModeHidden` (which doesn't allow users to access any controls).

Given this ability to hide the movie controls, the media player should be all you ever require to display video, but it continues to come up short for audio use, forcing you to seek alternatives. Unfortunately, there are no high-level frameworks for playing audio, so you'll have to do quite a bit of work to get things going. A lot of these specifics lie beyond the scope of this book because of their complexity, but we're going to get you started with the simplest methods for dealing with audio outside of the media player.

18.6 *Playing sounds manually*

It's not entirely correct to say that there's no high-level framework for iPhone audio. There *is*, and it's called Celestial. Unfortunately, Celestial is one of many "private frameworks" on the iPhone, which means that it's being used internally at Apple but hasn't been made available to external developers. We've opted in this book not to talk about the "jailbreak" methods that you can use to access private frameworks, as they're likely to change at any time, and using them could make your program exceedingly vulnerable to OS upgrades. Instead, we need to fall back to the frameworks officially provided by Apple.

And there are a lot of them! The iPhone's Core Audio system contains over a half-dozen frameworks that give you access to audio files at a low level. These include Audio Queue Services, Audio File Stream Services, Audio File Services, OpenAL, Audio Session Services, and more. For an in-depth look at all this, refer to the "Audio & Video" section of the Apple reference library, beginning with "Getting Started with Audio & Video" and "Core Audio Overview."

These frameworks are all old enough that they haven't been pulled out of Core Foundation, so you're going to have to fall back on the lessons that you've learned from your occasional forays into those older programming styles.

We're only going to graze the surface of audio. We'll provide some examples of how to play simple sounds and vibrate the iPhone, but for the more complex Audio Queue Services, we're going to outline the process and point you to Apple's extensive tutorials on the subject.

18.6.1 *Playing simple sounds*

System Sound Services is a C interface that lets you play simple sounds and vibrate your iPhone. It's part of the Audio Toolbox framework and is declared in AudioToolbox/AudioServices.h.

This interface can only be used to play short audio files of 30 seconds or less in .aif, .caf, or .wav formats. To use System Sound Services, you create a system sound ID from a file, optionally create a callback for when the sound is done playing, and launch it. Table 18.6 shows the major functions.

Table 18.6 **Major functions of the System Sound Services**

Function	Arguments	Summary
`AudioServicesCreateSystemSoundID`	`URL, &SSID`	Creates a sound from a URL
`AudioServicesDisposeSystemSoundID`	`SSID`	Removes a sound when done
`AudioServicesAddSystemSoundCompletion`	`SSID, run loop, run loop mode, routine, data`	Registers a call-back for sound play completion
`AudioServicesRemoveSystemSoundCompletion`	`SSID`	Removes a call-back when done
`AudioServicesPlaySystemSound`	`SSID`	Plays a sound

There are some additional functions that deal with system sound properties; they can be found in the System Sound Services reference. Listing 18.7 shows how to use the most important functions.

Listing 18.7 The Audio Toolbox supports the playing of short audio

```
-(IBAction)playSimple:(id)sender {

  NSString *paths =
    [[NSBundle mainBundle] resourcePath];
  NSString *audioFile = [paths
    stringByAppendingPathComponent:
      @"sound_bubbles.wav"];
  NSURL *audioURL =
    [NSURL fileURLWithPath:audioFile isDirectory:NO];
```

❶ Prepares URL

```
SystemSoundID mySSID;
OSStatus error =
    AudioServicesCreateSystemSoundID          ② Creates sound
        ((CFURLRef)audioURL,&mySSID);       ◁
AudioServicesAddSystemSoundCompletion(mySSID,NULL,NULL,
    simpleSoundDone,NULL);     ◁
                              ③ Adds callback
if (error) {
    statusLabel.text = [NSString stringWithFormat:@"Error: %d",error];
} else {
    AudioServicesPlaySystemSound(mySSID);    ◁
}                                      ④ Plays sound
}

static void simpleSoundDone           ⑤ Cleans up sound
    (SystemSoundID mySSID, void *args) {   ◁

    AudioServicesDisposeSystemSoundID (mySSID);
}
```

As with the media player, you start out the System Sound Services interface by building a path to your file (using the lessons learned in chapter 16) and then turning that into a URL ①. Once you've done that, you can create your system sound ②, which requires bridging your NSURL * to a CFURLRef and handing off a pointer to a system sound ID.

Adding a callback function is optional ③, but if you want something to happen when your sound is done playing, this is how you do it.

Once you have your system sound ID, playing it is simple ④, but you should check that the system sound ID was created correctly, as we do here.

As usual, you should clean up your memory when you're done. That's what the callback function in this example does ⑤.

18.6.2 *Vibrating the iPhone*

There's another cool little feature that's implicit in the System Sound Services interface: you can use it to vibrate the user's iPhone. This is done by handing off a predefined system sound ID, as shown in listing 18.8.

Listing 18.8   Vibrating the iPhone requires one line of code

```
-(IBAction)playVibrate:(id)sender {

    AudioServicesPlaySystemSound(kSystemSoundID_Vibrate);
}
```

For an audio system that's at times difficult to use, this is simplicity itself. Unfortunately, we've now covered all of the easy work with audio.

Before we close out, we're going to outline what it takes to play an audio file that doesn't meet the requirements of the System Sound Services interface—either because it's too long or it's of a different file type.

18.6.3 *Playing complex sounds*

If you need to do something fancier than playing a short 30-second sound, you'll need to fall back on Audio Queue Services. This will allow you to play longer sounds, play sounds of other types than the narrow set of audio types supported by System Sound Services, and even record sounds. Apple provides two excellent sets of tutorial code on Audio Queue Services, which you can copy for your own use, so we haven't duplicated them here. The "Apple Queue Services Programming Guide" provides a thorough example of how to write a player in a procedural C-oriented environment. The "SpeakHere" sample code shows how to perform a similar task using primarily object-oriented Objective-C code. The player appears in the AudioPlayer.m file.

In order to clarify Apple's sample code, table 18.7 outlines the standard steps that you'll need to follow in order to play complex sounds. It depends upon a few core ideas:

- An *audio file ID* is similar to the system sound ID that you encountered in the previous section; it points to the audio contents of a file.
- An *audio queue* contains a number of buffers—usually at least three. These buffers are filled one at a time with audio content, usually from a file, and then are dispatched to a player, which is typically a speaker.
- *Audio queue buffers* are the individual units of sound being passed through the queue. Each one has a user-set size. You'll enqueue the buffers as you fill them with data; they'll later be played by the queue.
- An *audio queue callback* is a special function that gets called to deal with an audio queue. The callback occurs when the data is being pulled into the audio queue. It needs to fill up buffers, and then enqueue them.
- A *custom audio structure* is a user-created structure that contains all of the data the callback needs to know about the state of the audio file and the state of the queue. It's handed off to the callback as part of the function call.

Figure 18.5 depicts most of these concepts graphically.

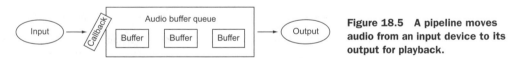

Figure 18.5 A pipeline moves audio from an input device to its output for playback.

With those definitions in hand, you should be able to parse Apple's code. Table 18.7 outlines the necessary steps.

Table 18.7 Steps for playing from an audio queue

Step	Description
1. Link in frameworks.	Link in the Audio Toolbox framework and the AudioToolbox/AudioQueue.h file.
2. Prepare an audio file.	Create a URL reference to your audio file. Call `AudioFileCreateWithURL` to return an `AudioID`.
3. Create an audio queue.	Use `AudioQueueNewOutput` to prepare your queue. Add three or more buffers to your queue. Each time, run `AudioQueue-AllocateBuffer` to set its size and then your callback function.

Table 18.7   Steps for playing from an audio queue *(continued)*

Step	Description
4. Prepare custom audio structure.	Define the structure that you'll use to pass all the queue and audio data to your callback function.
5. Write callback function.	Write your callback function to read data from your audio file into a buffer, and then enqueue the buffer. Playback will happen automatically once the data is in the queue.
6. Start playback.	Begin playback with `AudioQueueStart`.
7. End playback.	Once the audio file has been emptied, the callback function should end the playback. This is done with `AudioFileClose` and `AudioQueueStop`.

A lot of the rigmarole of the audio queues comes from its setup, which consists of steps 2–5 in table 18.7. Once you have those in place, you should have a good, repeatable methodology for playing audio with audio queues. As we said, Apple offers some complete examples of this code, which we could only duplicate here.

Once you've got an audio queue in place, you can use a variety of functions to operate it, as outlined in table 18.8.

Table 18.8   The main functions used to control an audio queue

Function	Summary
`AudioQueuePrime`	Prepares a queue for immediate use (optional)
`AudioQueueStart`	Begins playback (or recording)
`AudioQueuePause`	Pauses playback (or recording); restart it with `AudioQueueStart`
`AudioQueueFlush`	Ensures all buffers are emptied after the last one is enqueued (optional)
`AudioQueueStop`	Calls `AudioQueueReset` and then ends playback (or recording)
`AudioQueueReset`	Removes buffers from queue and resets to default state

Though we've concentrated on playback here, recording is much the same. A microphone will place data into an audio queue. All of the insertion of data into buffers is automatically dealt with. You then need to write a callback function for the backend that takes data out of a buffer, writes it to your file (or other output device), and adds the buffer back to the end of the queue, where it will eventually be filled with data again.

18.6.4 *Other audio frameworks*

Our focus on the System Sound Services framework and our quick summary of the Audio Queue Services framework represent just the tip of the iceberg for iPhone sounds. If you're serious about sounds, you'll also want to look at OpenAL, an API that allows for sound positioning.

Much more info on codecs, plug-ins, audio graphs, and the like can be found in the "Core Audio Overview" that we've already mentioned.

18.7 Summary

Dealing with media on the iPhone is a huge topic that probably could fill a book on its own. Fortunately, there are relatively easy (if limited) ways to utilize each major sort of media. For images, you can use `UIImageView`, for videos you can use `MPMoviePlayer`, and for sounds you can use System Sound Services.

If you want to go further, you'll have to dig into deeper levels of the iPhone OS. There isn't much else you can do with video, but for audio you can use more complex packages like Audio Queue Services or OpenAL. For images, you can do more complex work with Quartz 2D, which is the topic of our next chapter.

19
Graphics: Quartz, Core Animation, and OpenGL

This chapter covers

- Using Quartz 2D for drawing
- Understanding context, paths, and state
- Using Core Animation
- Learning about OpenGL ES

As we saw in the last chapter, creating and displaying images often isn't enough. In games and other more complex programs, you'll also want to manipulate those images in various ways at runtime. The iPhone OS offers two major ways to do this.

The first is through Quartz 2D, a two-dimensional drawing library that allows for complex line drawings, much as Canvas did on the web. It's also the heart of the Core Graphics frameworks. We already touched upon Quartz in the previous chapter, when we drew images straight to the CALayer of a UIView; it'll be the focus of the majority of this chapter. Quartz also supports Core Animation functions, which we'll address somewhat more briefly.

The second major way to manipulate images is through the OpenGL ES API. This cross-platform API, originally developed by Silicon Graphics, could be the topic of its own book, so we'll just show you how to get started with it.

But most of this chapter is going to be about Quartz, a topic that we're going to dive into immediately.

19.1 *An introduction to Quartz 2D*

Quartz 2D is a two-dimensional drawing library that's tightly integrated into the iPhone OS. It works well with all the relevant iPhone frameworks, including Core Animation, OpenGL ES, and the `UIKit`.

Fundamentally, Quartz's drawings depend upon three core ideas: context, paths, and state, each of which will be the topic of a future section.

- *Context* is a description of where the graphics are being written to, as defined by a `CGContextRef`. You'll usually be writing to a `UIView` or to a bitmap.
  - *Layers* are a little less important for this overview, but they're where Quartz drawing occurs. They can be stacked one on top of another, creating a complex result. When working with the iPhone, you'll often only have a single layer associated with each of your `UIKit` objects.
- *Paths* are what you'll typically be drawing within Quartz. These are collections of lines and arcs that are drawn in advance, and then are "painted" to the screen by either stroking or filling the path in question (or, possibly, by clipping it).
- *State* saves the values of transformations, clipping paths, fill and stroke settings, alpha values, other blending modes, text characteristics, and more. The current state can be stored with `CGContextSaveGState` and restored with `CGContextRestoreGState`, allowing for easy switching among complex drawing setups.

Quartz is built on the older Core Foundation framework that we've met a few times over the course of this part of the book. This means that you'll need to use older styles of variables to integrate with Cocoa Touch using toll-free bridging, and to respect Core Foundation's memory-management techniques. Take a look at the "Using Core Foundation" sidebar in chapter 16 if you need a refresher on these topics.

If you need more information on any Quartz topic, your should reference the "Quartz 2D Programming Guide" at Apple's developer website. It's a fine introduction to Quartz, though not as focused on the iPhone as you'd probably like, a deficiency that we'll correct in this chapter.

Using Quartz requires little special setup. It can be easily integrated into any template and any project that you want. Just be sure to include the Core Graphics framework and the CoreGraphics/CoreGraphics.h include file before you get started.

With that said, we're ready to dive into our first major Quartz topic: the context.

19.2 *The Quartz context*

A *graphical context* is a description of *where* Quartz will be writing to. This could include a printer, a PDF file, a window, or a bitmap image. On the iPhone, you're only likely to make use of two of these possibilities.

Most frequently, you'll work with the graphical context that is automatically associated with the `CALayer` (Core Animation layer) of each `UIView`. That means that you can use Quartz to draw to most `UIKit` objects. To do so, you override the `drawRect:` method and, inside the object in question, you use `UIGraphicsGetCurrentContext` to retrieve the current context.

You might alternatively create a bitmap context in order to create or modify an image that you'll use elsewhere in your program. You do this by using the `UIGraphicsBeginImageContext` and `UIGraphicsEndImageContext` functions.

Warning: inverse coordinate system ahead

By now, you should be familiar with the standard iPhone coordinate system. It has the origin at the top left of the screen, with the main axes running to the right and down. Quartz's default coordinate system is inverted, with the origin at the bottom left of the screen and the main axes running right and up.

This won't *usually* be a problem. The Cocoa Touch methods that you'll be using to create and write to graphical contexts will usually transform Quartz's default coordinates so that they look like iPhone coordinates to you.

Once in a while, though, you'll run into a situation where you'll draw to a UI-derived context and find your content flipped upside down (and in the wrong position). This is a result of accessing Quartz in a way that hasn't been transformed.

As of this writing, we're aware of two situations where you'll have to correct Quartz's coordinate system by yourself, even when using one of the UI-derived contexts: if you import images using the native Quartz functions (as opposed to the `UIImage` methods that we saw in the last chapter), and if you write text. We'll talk about each of these when we get to them.

Personally, we consider these coordinate inversions bugs, and it's our expectation that they'll eventually be corrected, perhaps even by the time this book is published.

If you create a context without using Cocoa Touch, expect *everything* to be inverted. This is something that we don't expect to change in the future.

There are a variety of Core Graphics functions that can be used to access other sorts of contexts—types that you won't usually use on an iPhone. The functions required to capture a PDF context are one such example. These have two deficits that you should be aware of: they depend more heavily on the Core Foundation frameworks and they use Quartz's inverted coordinate system.

One thing to note about graphical contexts is that they're created in a stack: when you create a new context, it's pushed on top of a stack, and when you're done with it, it's popped off. This means that if you create a new bitmap context, it'll be placed on top of any existing context, such as the one associated with your `UIView`, and will stay there until you're done with the bitmap.

Table 19.1 lists these context-related functions, including both the standard UI context functions and the older Core Graphics function that you're most likely to use—for PDFs.

Table 19.1   Methods for graphical context creation

Function	Arguments	Summary
UIGraphicsGetCurrentContext	(none)	Returns current context, which is usually the context of the current UIKit object, but could also be a context that you created by hand
UIGraphicsBeginImageContext	CGSize	Creates a bitmap context
UIGraphicsEndImageContext	(none)	Pops a bitmap context off the stack
UIGraphicsGetImageFromCurrentImageContext	(none)	Returns a bitmap as a UIImage *; used with a bitmap context only
CGPDFContextCreate	CGDataConsumerRef, CGRect, CGDictionaryRef	Creates a PDF context

We won't be covering PDFs in this book, but we're going to look at how to use each of the UIKit context styles, starting with the UIView.

19.2.1 Drawing to a UIView

In chapter 18, we offered an introductory example of how to write to a UIView graphical context using the drawRect: method. That example was somewhat simplified because the UIKit draw image commands mostly hide the idea of graphical contexts from you. They automatically write to the current context, which inside drawRect: is the context related to the UIView. For most other functions, you'll need to do a bit more work: retrieving the graphical context and passing that context along to any drawing commands that you use.

Listing 19.1 shows how to draw a simple abstract face using this technique.

Listing 19.1   A few arcs drawn inside an existing context

```
- (void)drawRect:(CGRect)rect {

    CGContextRef ctx = UIGraphicsGetCurrentContext();

    CGContextBeginPath(ctx);
```

```
    CGContextAddArc(ctx,110,50,30,0,2*M_PI,1);
    CGContextAddArc(ctx,210,50,30,0,2*M_PI,1);
    CGContextAddArc(ctx,160,110,15,0,2*M_PI,1);
    CGContextAddArc(ctx,160,210,25,0,2*M_PI,1);
    CGContextFillPath(ctx);
}
```

This example is fairly simple. You create a `UIView` subclass, and then you go to its `drawRect:` method. Once there, you capture the current context and use it to do whatever Quartz 2D drawing you desire.

The function calls won't be familiar to you, but they're calls to draw a bunch of circles; we'll discuss them in the next section. As shown in figure 19.1, the art ends up looking oddly abstract, which shows how Quartz draws continuous paths. You see lines connecting one circle to the next, as if the pencil never comes off the page, a topic we'll talk about more in the next section.

Leaving aside those specifics for a moment, this shows one of the two ways that you can use all of the Quartz functions described in this chapter: by painting a `UIView`. And remember that a `UIView` can be almost any `UIKit` object, due to inheritance.

Drawing to a `UIView` allows for on-screen picture creation, but you can also draw pictures without displaying them immediately. That's done with a bitmap.

Figure 19.1   The iPhone does abstract art.

19.2.2   *Drawing to a bitmap*

The main reason to create a bitmap rather than draw directly to a view is to use your graphic several times in your program—perhaps all at the same time. For example, Apple offers a sample program that draws the periodic table by creating a standard bitmap that's used for all the elements, and then repeating it. You might similarly create billiard balls using bitmaps if you were programming a billiards game. In chapter 17, we could have used Quartz to create the red dots that we used in our gravity and altitude programs as bitmaps, so that we didn't have to separately create them outside of the program.

The process of creating a bitmap and turning it into a `UIImage` is relatively simple. You create a graphical context, draw in that context, save the context to an image, and close the context. Listing 19.2 shows how to create a red dot image like the one you used in earlier programs.

Listing 19.2   A new context created to hold an image

```
- (void)viewDidLoad {
    [super viewDidLoad];

    UIGraphicsBeginImageContext(CGSizeMake(20,20));
```

① Creates bitmap context

```
CGContextRef ctx = UIGraphicsGetCurrentContext();

CGContextBeginPath(ctx);
CGContextAddArc(ctx,10,10,10,0,2*M_PI,1);
CGContextSetRGBFillColor(ctx, 1, 0, 0, 1);
CGContextFillPath(ctx);

UIImage *redBall =
    UIGraphicsGetImageFromCurrentImageContext();
UIGraphicsEndImageContext();

UIImageView *redBallView = [[UIImageView alloc] initWithImage:redBall];
redBallView.center = CGPointMake(160,330);
[self.view addSubview:redBallView];
}
```

2 Retrieves new context's pointer

3 Saves bitmap context to an image

4 Closes bitmap context

Again, this example is simple. You could do this work anywhere you wanted, but we've elected to use the `viewDidLoad` setup method. To start the process, you create an image context, which is to say a bitmap **1**, and you immediately retrieve that context's variable for use **2**. Following that, you do whatever drawing work you want. When you're done, you turn the bitmap into a `UIImage` **3** and close out your context **4**. You can then manipulate the image as you see fit; here it was turned into a `UIImageView`.

You now know two ways to use contexts in the Quartz environment. With that in hand, you're ready to dive straight into what Quartz can do, starting with paths, which will be the foundation of most Quartz work.

19.3 Drawing paths

The *path* is *what* Quartz will be drawing. If you're familiar with Canvas, this will look familiar, because both libraries use the same drawing paradigm. A path is a set of lines, arcs, and curves that are all placed continuously within a graphical context. You only "paint" a path when it's complete, at which point you can choose to either fill it or stroke it.

Many of the functions required to define and draw paths are listed in table 19.2.

Table 19.2 A variety of simple drawing functions that allow for vector-based graphics

Function	Arguments	Summary
CGContextBeginPath	context	Creates a new path.
CGContextAddArc	context, x, y, radius, startangle, endangle, clockwise	Creates an arc, with the angles defined in radians. A line will be drawn to the start point if there are previous entries in the path, and from the end point if there are additional entries. The more complex functions CGContextAdd-ArcToPoint, CGContextAddCurveTo-Point, and CGContextAddQuadCurveTo-Point allow for the creation of tangential arcs, Bezier curves, and quadratic Bezier curves.
CGContextAddEllipseInRect	context, CGRect	Creates an ellipse that fits inside the rectangle.

Table 19.2   A variety of simple drawing functions that allow for vector-based graphics *(continued)*

Function	Arguments	Summary
CGContextAddLineToPoint	context, x, y	Creates a line from the current point to the designated end point. The more complex CGContextAddLines function allows the addition of an array of lines.
CGContextAddRect	context, CGRect	Creates a rectangle. The more complex CGContextAddRects function adds a series of rectangles.
CGContextMoveToPoint	context, x, y	Moves to the point without drawing.

CGContextMoveToPoint is the one function that deserves some additional discussion. As you'll recall, we said that a path was a continuous series of lines and arcs that you draw without picking the pen up off the paper. But there *is* a way to pick the pen up, and that's with the CGContextMoveToPoint function, which is vital when you want to draw unconnected objects as part of a single path.

For example, to avoid drawing a line between the first two circles in listing 19.1, you'd use the following code:

```
CGContextAddArc(ctx,110,50,30,0,2*M_PI,1);
CGContextMoveToPoint(ctx, 240, 50);
CGContextAddArc(ctx,210,50,30,0,2*M_PI,1);
```

After drawing the first circle, you move your virtual pencil to the point where you'll begin drawing the arc of the second circle, which is 240, 50.

The rest of the functions are largely self-explanatory. We already saw the arc commands in some of our earlier examples, and the others work in similar ways. For more information on the more complex functions, take a look at the CGContext class reference. If you're unfamiliar with Bezier and quadratic curves, take a look at our explanation of the nearly identical Canvas functions in chapter 6 (section 6.2.2; particularly figure 6.4, which depicts what both sorts of curves look like).

We're going to move on from these simple drawing commands to the question of what you do once you have a path. There are several options, beginning with the simple possibility of closing it and drawing it.

19.3.1 Finishing a path

As we've already noted, the path functions define the points and lines that make up a drawing. When you've got that in hand, you have to do something with it. There are three main choices: stroke the path, fill the path, or turn it into a clipping path. These functions are all listed in table 19.3.

You'll usually either stroke (outline) a path or fill it when you're done. We used a fill in each of our previous examples, but a stroke could have been substituted; the difference is that our circles wouldn't have been filled in.

Table 19.3 Functions for finishing a path

Function	Arguments	Summary
CGContextClosePath	context	Draws a line from the end point of your path to the start point, and then closes it. This is an optional final command that's usually used when you're stroking a path.
CGContextFillPath	context	Closes your path automatically, and paints it by filling it in. CGContextEOFillPath is an alternative that does the filling in a slightly different way.
CGContextStrokePath	context	Paints your path by stroking it.
CGContextClip	context	Turns the current path into a clipping path.

A clipping path is a bit more complex, in that you don't draw something on the screen. Instead, you define an area, which corresponds to the area inside the path that you'd have filled in, and you only show later drawings that appear inside that clipping path. We'll talk about clipping paths more, and show an example, when we get to graphical states. For now, note that you create them from paths.

19.3.2 *Creating reusable paths*

So far, you've created paths by drawing them directly to a context, be it a UIView or a bitmap. But it's also possible to create reusable paths that you can quickly and easily apply later. This has many of the same advantages as creating a bitmap: you get reusability and multiplicity. Reusable paths will probably be particularly useful in animations and programs where you use the same graphic on multiple pages.

To create reusable paths, you use the CGPath commands rather than the CGContext commands. There are equivalents to many of the simple CGContext functions, as shown in table 19.4.

When you're working with reusable paths, you first use the CGPathCreateMutable function to create a CGPathRef, and then you use CGPath commands to add lines or

Table 19.4 CGPath commands and their CGContext equivalents

CGPath Function	CGContext Function
CGPathCreateMutable	CGContextBeginPath
CGPathAddArc	CGContextAddArc
CGPathAddEllipseInRect	CGContextAddEllipseInRect
CGPathAddLineToPoint	CGContextAddLineToPoint
CGPathAddRect	CGContextAddRect
CGPathMoveToPoint	CGContextMoveToPoint
CGPathCloseSubpath	CGContextClosePath

arcs to that `CGPathRef`. Your reusable path can include multiple, discrete subpaths that don't have to connect to each other. You can end one subpath and start another with the `CGPathCloseSubpath` function.

Note that there are no painting functions associated with the reusable paths. That's because they're storage devices. In order to use one, you add it onto a normal path with the `CGContextAddPath` function, which draws your stored path to your graphical context, where it'll abide by the normal rules.

Listing 19.3 shows how to use a mutable path to replace the `CGContext` commands that we previously used in listing 19.1 to draw an abstract face. A more realistic example would probably hold on to the path for use elsewhere; we released it here to remind you of how Core Foundation memory management works.

Listing 19.3   A drawing with `CGPath`

```
- (void)drawRect:(CGRect)rect {

    CGMutablePathRef myPath = CGPathCreateMutable();
    CGPathAddArc(myPath,NULL,110,50,30,0,2*M_PI,1);
    CGPathMoveToPoint(myPath,NULL, 240, 50);
    CGPathAddArc(myPath,NULL,210,50,30,0,2*M_PI,1);
    CGPathAddArc(myPath,NULL,160,110,15,0,2*M_PI,1);
    CGPathAddArc(myPath,NULL,160,210,25,0,2*M_PI,1);

    CGContextRef ctx = UIGraphicsGetCurrentContext();
    CGContextBeginPath(ctx);
    CGContextAddPath(ctx,myPath);
    CGContextStrokePath(ctx);

    CFRelease(myPath);
}
```

Of note here is the `NULL` that's constantly being sent as a second argument to the `CGPath` commands. This argument is intended to be a `CGAffineTransform` variable. It allows you to apply a transformation to the element being drawn, which is something we'll discuss shortly.

Now that we've looked at two different ways to create complex paths, we're going to take a step back and look at how to draw much simpler objects in a simpler way.

19.3.3 *Drawing rectangles*

Drawing paths takes some work, but if you want to draw a rectangle, Quartz makes it easy. All you have to do is use one of a few functions listed in table 19.5. These functions take care of the path creation, drawing, and painting for you in a single step.

Table 19.5   Specific functions allow you to draw rectangles

Function	Arguments	Summary
`CGContextClearRect`	context, CGRect	Erases a rectangle.
`CGContextFillRect`	context, CGRect	Draws a filled rectangle. The more complex variant `CG-ContextFillRects` allows you to fill a whole array of rectangles.

Table 19.5 Specific functions allow you to draw rectangles *(continued)*

Function	Arguments	Summary
`CGContextStrokeRect`	`context, CGRect`	Draws a stroked rectangle.
`CGContextStrokeRectWithWidth`	`context, CGRect, width`	Draws a stroked rectangle, with the stroke being the designated width.

The `CGContextClearRect` function can be particularly useful for erasing a window when you're ready to draw something new to it. Now that we've told you how to draw objects in the simplest way possible, we're ready to move on and start talking about how to draw objects in more complex ways—by modifying state.

19.4 Setting the graphic state

The graphic *state* is *how* Quartz will be drawing. It includes a variety of information such as what colors are being used for fills or strokes, which clipping paths constrain the current drawing path, what transformations are being applied to the drawing, and a number of other less important variables.

State is maintained in a stack. You can save a state at any time; it doesn't change how things are being drawn, but it does push that current state onto the top of a stack for later retrieval. Later, you can restore a state, which pops the top state off the stack, putting things back to how they were before the last save. We've mentioned these functions before, but we've also listed them here in table 19.6.

Table 19.6 State-related functions that help define how you draw

Function	Arguments	Summary
`CGContextSaveGState`	`context`	Pushes state onto a stack
`CGContextRestoreGState`	`context`	Pops state off of a stack

As we've already noted, there are a *lot* of things that you can store in graphic state. We're going to cover many of them here, starting with colors.

19.4.1 Setting colors

In Quartz, you select colors by setting the fill color, the stroke color, or both in the current graphical state. Once you've done this, any fill or stroke commands following the color commands will appear in the appropriate colors. Note that color is irrelevant while you are drawing the individual elements of a path—the color commands apply only to the painting of the complete path at the end.

You can select colors from a variety of *color spaces*, which are different ways to choose colors. They include RGB (red-green-blue), RGBA (red-green-blue-alpha), CMYK (cyan-magenta-yellow-black), and CGColor (the underlying Core Graphics color model). On the iPhone, you'll usually want to either use the RGBA color space or use a command that lets you select a color using standard `UIKit` methods. Table 19.7 lists the four most relevant of these functions.

Table 19.7   The most important of numerous coloring functions

Function	Arguments	Summary
CGContextSetRGBFillColor	context, red, green, blue, alpha	Sets the fill to the RGBA value
CGContextSetRGBStrokeColor	context, red, green, blue, alpha	Sets the stroke to the RGBA value
CGContextSetFillColorWithColor	context, CGColor	Sets the fill to the CGColor
CGContextSetStrokeColorWithColor	context, CGColor	Sets the stroke to the CGColor

The two RGB functions allow you to set a color using values from 0 to 1 for each of red, green, blue, and alpha transparency (opacity). We saw an example of this in listing 19.2:

```
CGContextSetRGBFillColor(ctx, 1, 0, 0, 1);
```

The last two functions in table 19.7 allow you to set the color using any CGColor, and you'll understand how useful that is when you realize that you can read a CGColor property from any UIColor you create:

```
CGContextSetFillColorWithColor(ctx, [[UIColor redColor] CGColor]);
```

Given that you're already familiar and comfortable with the UIColors, we expect that this latter function will be a popular one.

Having now covered the main ways to apply colors to your graphic state, we're ready to move on to the next topic: how to change how you draw through graphical state transformations.

19.4.2   *Making transformations*

Transformations modify how you draw to your graphical context. They do this by changing the grid upon which you're drawing by moving its origin, rotating, or resizing.

Why would you want to do these transformations?

- They can be useful for drawing photographs (or other images), because the transformations allow you to scale or rotate the picture.
- They can make it a lot easier to do certain types of mathematical drawing. For example, it's probably easier to draw a symmetric mathematical construct if you've got your origin in the center of the screen rather than up at the top left corner.
- They can allow you to flip your screen if you end up in a context (or using a function) with an inverse coordinate system.

CTM TRANSFORMATIONS

The simplest way to apply a transformation is to use one of the functions that modify the current transformation matrix (CTM), which is a matrix that's applied to all drawing done in your current graphical state. These functions are described in table 19.8.

Table 19.8 CTM transformation functions that allow you to change how you draw

Function	Arguments	Summary
CGContextRotateCTM	context, radian rotation	Rotates the grid
CGContextScaleCTM	context, x-scale, y-scale	Scales the grid
CGContextTranslateCTM	context, x-change, y-change	Moves the origin

There are two gotchas that you should watch for.

First, note that the ordering of translations is somewhat pickier than the order of color commands. You need to start your transformation *before* you add the relevant lines to your path, and you need to maintain it until *after* you paint that path.

Second, although these transformations can be applied in any sequence, order matters. Following are two transformation commands that could be applied together:

```
CGContextTranslateCTM(ctx, 100, 100);
CGContextRotateCTM(ctx, .25*M_PI);
```

These functions move a drawing 100 to the right and 100 down and rotate it by 45 degrees. Figure 19.2 shows the untransformed picture (which we've seen before), the results if these commands are applied with the translation before the rotation, and the results if they're applied in the opposite order.

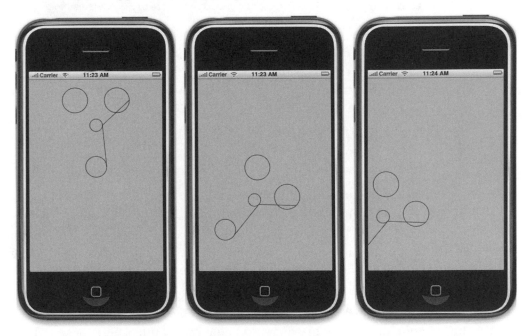

Figure 19.2 As these variant transformations show, order matters. The left picture is untransformed; the middle one is translated and then rotated; and the right one is rotated and then translated.

Clearly, you need to be careful and think about ordering when you're applying CTM transformations.

But CTM transformations aren't the only way to change your drawing space.

AFFINE TRANSFORMATIONS

Just as you can create a reusable path and then apply that to the context with the `CGContextAddPath` function, you can also create a reusable transformation matrix (using the affine transformation functions) and then apply that to the context with the `CGContextConcatCTM` function. This is managed by a set of six core functions, listed in table 19.9. Half of them create a new matrix, applying a transformation at the same time, and the other half apply a transformation to an existing matrix. The last function is the one that applies an affine transformation to your current graphical state.

Table 19.9   Affine transformations for creating reusable transforms

Function	Arguments	Summary
`CGAffineMakeRotation`	`radian rotation`	Makes an array with the rotation
`CGAffineMakeScale`	`x-scale, y-scale`	Makes an array with the scale
`CGAffineMakeTranslation`	`x-change, y-change`	Makes an array with the translation
`CGAffineTransformRotate`	`array, radian rotation`	Rotates the array
`CGAffineTransformScale`	`array, x-scale, y-scale`	Scales the array
`CGAffineTransformTranslate`	`array, x-change, y-change`	Translates the array
`CGContextConcatCTM`	`context, array`	Applies the transformation

The following code applies a rotation followed by a translation using a reusable affine matrix:

```
CGAffineTransform myAffine = CGAffineTransformMakeRotation(.25*M_PI);
CGAffineTransformTranslate(myAffine, 100, 100);
CGContextConcatCTM(ctx, myAffine);
```

Besides being able to create reusable affine transformations, you can also modify the transforms at a much lower level. Any affine transformation is constructed from a 3x3 matrix that is then multiplied across the individual vectors of your path using matrix multiplication. If you have specific needs, you can use the `CGAffineTransformMake` function to create a matrix by hand. Using it looks like this:

```
CGAffineTransform flip = CGAffineTransformMake(1,0,0,-1,0,0);
```

Information on how the matrix works and on some other functions can be found in the `CGAffine` reference.

The next sort of state you might want to change is one that makes fairly large-scale changes to your drawings: the clipping path.

19.4.3 *Setting clipping paths*

We already spoke about clipping paths in section 19.3. You create a path as usual, but then you clip it, rather than filling it or stroking it. Anything that you paint on the screen afterward (within that graphical state) will only appear if it's inside the clipping path.

For example, the following code causes later painting to only appear inside a large circle centered on the iPhone screen:

```
CGContextBeginPath(ctx);
CGContextAddArc(ctx,160,240,160,0,2*M_PI,1);
CGContextClip(ctx);
```

Figure 19.3 shows what a full-screen image might look like before the clipping occurred, and then after.

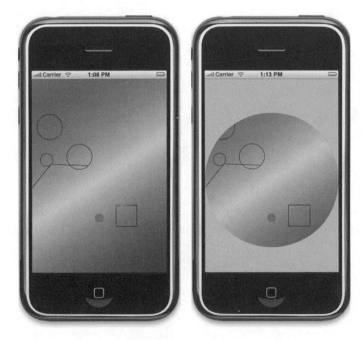

Figure 19.3   An example of a clipping path in use. The unclipped image is on the left, and the clipped image is on the right.

As with most of these Quartz functions, there are some opportunities for subtleties when using clipping paths. The CGContext reference offers a few additional functions for creating and modifying clipping paths.

So far we've discussed all the big-picture options for modifying your graphic state. There are many smaller things you can do too.

19.4.4 *Other settings*

There are a wide variety of additional settings that can be used as part of the graphic state. Table 19.10 lists many of the most interesting ones.

Table 19.10   A selection of other ways to change state

Function	Arguments	Summary
CGContextSetAlpha	context, alpha	Sets alpha transparency
CGContextSetBlendMode	context, CGBlendMode	Sets blending to one of almost 30 values, which specify how objects laid on top of each other interact with each other
CGContextSetFlatness	context, flatness	Defines the accuracy of curves
CGContextSetLineCap	context, CGLineCap	Defines how to draw the end of a line
CGContextSetLineDash	context, phase, lengths array, count	Describes how to draw dashes along a stroke
CGContextSetLineJoin	context, CGLineJoin	Defines how lines come together
CGContextSetLineWidth	context, width	Describes the width of a stroke
CGContextSetShadow	context, CGSize, blur	Sets a shadow behind all drawings
CGContextSetShadowWithColor	context, CGSize, blur, color	Sets a colored shadow behind all drawings

A number of more complex state changes can also be found in the CGContext class reference, but we've described the ones you're most likely to use in the course of an average program.

We're drawing to a close on the topic of graphical state, so let's step back for a moment and look at how graphical state works.

19.4.5   *Managing the state*

When you use any of the various functions that modify the graphical state, you're changing how you paint inside your current graphical context. The functions change the colors you're using, they transform your underlying grid, they clip the area you're allowed to paint within, or they make various smaller changes.

You can constantly reset these variables as your needs change, but this can get annoying. That's why you'll want to use the stack of states. It allows you to make a whole bunch of changes to state and then revert to a previous setup that you were happy with. We've already shown the two functions that do this in table 19.6.

Remember to save the state before you make a big change, such as adding a clipping path or running a whole bunch of graphical state functions. Then restore the state when you're done with that. If you want, you can even be clever and slowly build up a whole set of states in your stack, and move back through them appropriately.

You should now understand the three most important elements of drawing with Quartz: contexts, which specify *where* to draw; paths, which specify *what* to draw; and graphical states, which specify *how* to draw. There are numerous more advanced things you can do in Quartz, and although we're not going to get to all of them, the next section covers the most interesting ones.

19.5   Advanced drawing in Quartz

Quartz has a number of advanced capabilities that go beyond simple line drawings. In this section we're going to look at using gradients, images, and words.

19.5.1   Drawing gradients

Gradients are a core part of SDK design, because they're a clearly evident aspect of the standard user interface. Unfortunately there's no UIKit level class for creating gradients; instead, you have to fall back on Quartz.

There are two ways to create gradients inside Quartz: using a CGShadingRef object or a CGGradientRef object. As is often the case in Core Foundation functions, the difference is in complexity. CGGradientRef allows you to draw pretty simple gradients, and CGShadingRef requires you to define a CGFunctionRef object to precisely calculate how the colors in the gradient are displayed. As you've probably guessed, we're going to talk about CGGradientRef here and point you to the Apple class references for CGShadingRef.

Table 19.11 shows the important functions required to draw gradients with CGGradientRef.

Table 19.11   `CGColorSpace`, `CGGradient`, and `CGContext` **functions for drawing gradients**

Function	Arguments	Summary
CGColorSpaceCreateWithName	color space constant	Creates a color space by name
CGGradientCreateWithColors	color space, color array, location array	Creates a gradient using pregenerated colors
CGGradientCreateWithColorComponents	color space, color components array, location array, color count	Creates a gradient with an array of color parts
CGContextDrawLinearGradient	context, gradient, start CGPoint, end CGPoint, options	Draws a linear gradient

Table 19.11   `CGColorSpace`, `CGGradient`, and `CGContext` functions for drawing gradients *(continued)*

Function	Arguments	Summary
`CGContextDrawRadialGradient`	context, gradient, start center, start radius, end center, end radius, options	Draws a radial gradient
`CGColorSpaceRelease`	color space	Frees up a color space object
`CGGradientRelease`	gradient	Frees up a gradient object

Drawing a gradient is a four-step process:

1 Define your color space, which will usually be `kCGColorSpaceGenericRGB` for the iPhone.

2 Define your gradient by listing colors and where they appear in the gradient, from 0 to 1. There are two ways to do this. You can hand off an array of `CGColors` (which might be useful if you want to generate them using `UIColors`) or you can hand off a longer array that defines the colors using another method, such as RGBA.

3 Draw your gradient as a linear gradient (going from point to point) or a radial gradient (going from the center to the edge of a circle).

4 Free up your memory.

Listing 19.4 shows all the steps required to draw a three-color linear gradient that spans the entire iPhone screen.

Listing 19.4   Drawing a three-color linear gradient

```
CGColorSpaceRef myColorSpace =
   CGColorSpaceCreateWithName(kCGColorSpaceGenericRGB);
CGFloat components[12] = {1,0,0,1,
   0,1,0,1,
   0,0,1,1};
CGFloat locations[3] = {0,.5,1};

CGGradientRef myGradient =
   CGGradientCreateWithColorComponents(myColorSpace,
      components, locations, (size_t)3);
CGContextDrawLinearGradient(ctx, myGradient, CGPointMake(0,0),
   CGPointMake(320,480), (CGGradientDrawingOptions)NULL);

CGColorSpaceRelease(myColorSpace);
CGGradientRelease(myGradient);
```

This code steps through the steps we just listed, defining the color space, creating the parts of the gradient, drawing it, and cleaning up after it. As usual, you can find some more info on gradients in the `CGGradient` reference. For now, though, we're ready to move on to the next advanced category of Quartz work: images.

19.5.2 Drawing images

In the last chapter, we saw one way to work with images, using methods that largely hid the specifics of graphical contexts from you as a programmer. Now that you're fully immersed in Quartz, you can choose to use the Core Graphic functions instead.

THE IMAGE FUNCTIONS

There are two major Core Graphic functions for drawing, listed in table 19.12.

Table 19.12 A few image functions in Quartz

Function	Arguments	Summary
CGContextDrawImage	context, CGRect, image	Draws an image scaled to fit the rectangle
CGContextDrawTiledImage	context, CGRect, image	Draws an image scaled to fit the rectangle but filling the current clip region

These functions both require a `CGImageRef`, but remember that you can use the `CGImage` property of a `UIImage` to produce one. Alternatively, you can use the commands described in the `CGImage` reference, which offer more precise functionality, to create a new `CGImage`. Our suggestion is to go with what you know, which means using the `UIKit` methods, unless they can't do what you need.

There's one big gotcha to using the Quartz-related image-drawing functions: at the time of this writing, they produce a flipped image because they use Quartz's native coordinate system internally. We'll show you how to fix that momentarily.

DRAWING ON A BITMAP

Often you'll want to turn an image into a bitmap and modify it before displaying it on the screen, most frequently so that you can make multiple uses of the image. We're going to offer a quick example of crossing out a picture here.

Part of what's unique about this example is that you can do all your drawing work without ever showing the image to the user (unlike if you were drawing on a `UIView`), thus opening up the possibility of many image-editing functions. When you do decide to display your newly saved image, you'll see results like the image in figure 19.4.

Figure 19.4 You can change a `UIImage` without showing it to the user.

The code needed to accomplish this simple crossing-out is shown in listing 19.5.

Listing 19.5 Using bitmaps to edit images

```
UIImage *origPic = [UIImage imageNamed:@"pier.jpg"];      ❶ Creates
UIGraphicsBeginImageContext(origPic.size);                     context
CGContextRef thisctx = UIGraphicsGetCurrentContext();

// Image transformations Go Here       ❷ Transforms image
```

```
CGContextRotateCTM(thisctx, M_PI);
CGContextTranslateCTM(thisctx, -origPic.size.width, -origPic.size.height);
CGContextDrawImage(thisctx,CGRectMake(0,0,origPic.size.width,
    origPic.size.height),[origPic CGImage]);              ◄─┐

                                                            ❸  Draws
                                                               image
// Image modification goes here    ◄─┐
CGContextSetLineWidth(thisctx, 20);  │
CGContextBeginPath(thisctx);         └──  ❹  Draws on
CGContextMoveToPoint(thisctx, 0, 0);         image
CGContextAddLineToPoint(thisctx, origPic.size.width,origPic.size.height);
CGContextMoveToPoint(thisctx, 0, origPic.size.height);
CGContextAddLineToPoint(thisctx, origPic.size.width, 0);
CGContextSetStrokeColorWithColor(thisctx, [[UIColor redColor] CGColor]);
CGContextStrokePath(thisctx);

UIImage *newPic =                                          ❺  Saves
    UIGraphicsGetImageFromCurrentImageContext();    ◄─┘       image
UIGraphicsEndImageContext();
```

The process of modifying an image involves relatively few steps. You start off by creating your bitmap context ❶. Next, you apply any transformations that you want to use for the picture ❷. If you want to rotate or scale the original picture, here's where you do it. Likewise, you could use a combination of translations and the context size to easily crop an image. In this example, you flip the picture over by applying a rotation and a translation, to account for the fact that CGContextDrawImage produces an inverted picture. (We'll see an alternative way to do this in our next example.)

When your transformations are done, you can draw your image ❸, and then draw whatever you want on top of it ❹ (or modify it in some other way). Finally, you save the new picture ❺.

We'll return to the idea of drawing on pictures in section 19.6 (though we'll do it in a much more interactive way), but in the meantime we're ready to draw words.

19.5.3 *Drawing words*

Unlike Canvas, Quartz does support drawing words on top of your pictures. The functions required are pretty intricate, though, and we generally suggest using UILabel or other UIKit objects, and placing them on top of your Quartz objects. But if you need words inside Quartz (either because you're interweaving your words with other Quartz content or because you're adding words to a picture), you'll need to make use of the CGContext text options.

The majority of the text-related functions modify the graphical state, as described in table 19.13. The last two functions in the table draw your text.

Table 19.13   A variety of functions for drawing text in Quartz

Function	Arguments	Summary
CGContextSelectFont	context, font name, size, text encoding	Sets a font for the graphical state
CGContextSetTextDrawingMode	context, CGTextDrawingMode	Defines how to draw text in the graphical state

Table 19.13   A variety of functions for drawing text in Quartz *(continued)*

Function	Arguments	Summary
CGContextSetTextMatrix	context, affine transform	Places a transformation matrix in the graphical state for drawing *only* text
CGContextSetSetPosition	context, x, y	Sets where to draw in the graphical state
CGContextShowText	context, string, length	Draws the text at the current position
CGContextShowTextAtPoint	context, x, y, string, length	Draws the text at the specified position

You can find several other text-related functions in the CGContext reference. Most notably, if you need more control over your fonts (and particularly if you want to link up to UIFonts), you should use CGContextSetFont and CGContextSetFontSize instead of the CGContextSelectFont function that's noted here—but keep in mind that you can't use CGContextShowTextAtPoint when you set your font in this alternative way.

Listing 19.6 shows a simple example of printing text in Quartz.

Listing 19.6   Outputting text in Quartz

```
CGContextSelectFont (ctx, "Helvetica",20,kCGEncodingMacRoman);
CGContextSetTextDrawingMode(ctx, kCGTextFill);
CGAffineTransform flip = CGAffineTransformMake(1,0,0,-1,0,0);
CGContextSetTextMatrix(ctx, flip);
CGContextShowTextAtPoint(ctx, 20, 85, "A Quartz Example", 16);
```

The only thing of note here is the creation of the affine transformation matrix, flip. We've already pointed out that the text-drawing functions don't use the iPhone coordinate system at present. Instead, they're still stored in an inverted manner, so you need to flip them over to use them correctly. (We hope that this changes in some future release of the iPhone OS.)

The affine transformation shown here describes the matrix using the CGAffineTransformMake function. It effectively does the same thing as our two-part transformation in listing 19.5. In our view, it's a bit simpler, but less clear.

That's only the basics of using text, but it should be enough to get you started when you need to draw within Quartz.

19.5.4   *What we didn't cover*

Quartz 2D is a fully featured drawing and painting language that we can only briefly touch on in this chapter. Among the other topics that you might want to research if you're going to do more advanced work with Quartz are patterns, transparency layers, layer drawing, and PDF creation. As we've mentioned previously, Apple's "Quartz 2D Programming Guide" is an excellent introduction to these topics.

We're not quite done with Quartz yet. Before we finish up this chapter, we're going to put together a more fully featured example combining some of the Quartz lessons from this chapter with some of the photographic work we covered in chapter 18.

19.6   *Drawing on a picture: an example*

To put together the lessons we've covered, you're going to create a program that allows a user to load up a picture, draw on it, and then save the results. Figure 19.5 shows our intended result.

As usual, you'll start in Interface Builder, but you only have two simple things to do here:

1   Create a `UIButtonBar` with a single action-type button (which is one of the standard styles you can select for a button).
2   Link the existing `UIView` to a new `drawView` class (which should be a `UIView` subclass).

Figure 19.5   Photodraw can place drawings on pictures.

Once you get into Xcode, the programming will look a lot like the collage program in the last chapter, but with some nuances related to your greater understanding of Quartz.

You'll be doing your coding in two parts. The overall structure of the program will go in photodraw-ViewController.m and the drawing specifics will go in drawView.m.

19.6.1   *The photodraw view controller*

The view controller manages an image selector as well as several toolbar buttons, including the "action" button that you created in Interface Builder and Save and Cancel buttons that will appear later on. The code is shown in listing 19.7. We've omitted some of the view controller's overall structure and focused on the code that's involved when the user pushes the action button and activates `choosePic:`.

Listing 19.7   The important bits of a view controller for a photodraw program

```
-(IBAction)choosePic:(id)sender {                          Starts image
                                                        ① picker
    UIImagePickerController *myImagePicker =
        [[UIImagePickerController alloc] init];
    myImagePicker.delegate = self;
    myImagePicker.allowsImageEditing = NO;

    [self presentModalViewController:myImagePicker animated:YES];
}

- (void)imagePickerController:(UIImagePickerController *)picker
    didFinishPickingImage:(UIImage *)image
```

```
    editingInfo:(NSDictionary *)editingInfo {            ←        Finishes
                                                          ❷      image picker
    [self dismissModalViewControllerAnimated:YES];
    [picker release];

    [myTools setItems:[NSArray arrayWithObjects:
      [[UIBarButtonItem alloc]
        initWithBarButtonSystemItem:UIBarButtonSystemItemSave
        target:self action:@selector(savePic:)],
      [[UIBarButtonItem alloc]
        initWithBarButtonSystemItem:UIBarButtonSystemItemCancel
        target:self action:@selector(clearDrawing:)],
      nil] animated:YES];

    [self.view drawPic:image];
}
- (void)imagePickerControllerDidCancel:            ❸    Resolves image
    (UIImagePickerController *)picker {        ←           cancellation

    [self dismissModalViewControllerAnimated:YES];
    [picker release];
}                                          ❹    Saves
- (void)savePic:(id)sender {        ←            picture

    UIGraphicsBeginImageContext(self.view.bounds.size);
    [myTools removeFromSuperview];
    [self.view.layer renderInContext:UIGraphicsGetCurrentContext()];
    UIImage *finishedPic = UIGraphicsGetImageFromCurrentImageContext();
    UIGraphicsEndImageContext();

    UIImageWriteToSavedPhotosAlbum(finishedPic,self,
        @selector(exitProg:didFinishSavingWithError:contextInfo:),nil);
}
- (void)exitProg:(UIImage *)image didFinishSavingWithError:(NSError *)error
    contextInfo:(void *)contextInfo {          ←
                                               ❺    Ends program
    exit(0);
}
- (void)clearDrawing:(id)sender {          ←
                                           ❻    Clears drawing
    [self.view cancelDrawing];
}
```

This is a pretty simple snippet of code because it shows the view controller acting as a traffic cop, accepting input from controls and sending off messages to other objects, which is pretty much the definition of what a view controller should do.

For once, you don't have any setup in viewDidLoad:. Instead, the toolbar created in Interface Builder will initiate your program's actions. At startup, the user has only one choice, to click the action button and start the image picker ❶. When the picker returns, you modify the UIButtonBar to now give options for Save and Cancel, and then you send the picture off to drawView to be dealt with ❷. Alternatively, you clear the image picker away if the user canceled it ❸.

The save picture routine works the same way as the one you wrote in the collage program ❹. The only difference is that this one includes a callback, which ends the program after the saving is done ❺. The clear drawing method ❻, meanwhile, makes a call to the drawView object again.

To learn what's done with the initial picture, how drawing occurs, and what happens when the drawing is cleared, we need to look at this program's other major class.

19.6.2 *The photodraw view*

As we saw in the previous section, the view controller hands off three responsibilities to the view: displaying a picture, responding to touch events, and clearing the drawing. We're going to step through these functions one at a time.

Listing 19.8 shows what's done when a user picks an image.

Listing 19.8   Preparing a picture for drawing

```
- (void)drawPic:(UIImage *)thisPic {

    myPic = thisPic;
    [myPic retain];
    [self setNeedsDisplay];
}
```

This routine is simple: it saves the picture to an instance variable and then alerts the UIView that its CALayer must be drawn.

We are going to save the CALayer's drawRect: method for last, so we'll look now at how the drawView class interprets touch events. This is shown in listing 19.9.

Listing 19.9   Recording touch events

```
- (void) touchesBegan:(NSSet *)              ❶ Sets up touches
    touches withEvent:(UIEvent *)event {

    [myDrawing addObject:[[NSMutableArray alloc] initWithCapacity:4]];

    CGPoint curPoint = [[touches anyObject] locationInView:self];
    [[myDrawing lastObject] addObject:[NSNumber
        numberWithFloat:curPoint.x]];
    [[myDrawing lastObject] addObject:[NSNumber
        numberWithFloat:curPoint.y]];
}
- (void) touchesMoved:(NSSet *)touches       ❷ Displays touches
    withEvent:(UIEvent *)event {                temporarily

    CGPoint curPoint = [[touches anyObject] locationInView:self];
    [[myDrawing lastObject] addObject:[NSNumber
        numberWithFloat:curPoint.x]];
    [[myDrawing lastObject] addObject:[NSNumber
        numberWithFloat:curPoint.y]];
    [self setNeedsDisplay];
}
- (void) touchesEnded:(NSSet *)touches       ❸ Resolves
    withEvent:(UIEvent *)event {                touches

    CGPoint curPoint = [[touches anyObject] locationInView:self];
```

```
    [[myDrawing lastObject] addObject:[NSNumber
        numberWithFloat:curPoint.x]];
    [[myDrawing lastObject] addObject:[NSNumber
        numberWithFloat:curPoint.y]];
    [self setNeedsDisplay];
}
```

The overall concept here is pretty simple. You maintain an NSMutableArray called myDrawing as an instance variable. Within that, you create a number of NSMutable-Array subarrays, each of which contains an individual path. You set up a new subarray when a touch starts ❶ and then add the current point when the touch moves ❷ or ends ❸. The result is an array that contains a complete listing of all touches. But once again, we're going to have to wait a bit to see how that's drawn.

It's notable that we tell drawView to draw (via the setNeedsDisplay method) both when a touch moves *and* when it ends. That's because whenever the touch moves, you want to provide instant gratification by drawing what the user has sketched out so far. When the touch ends, you do the same thing.

Listing 19.10 shows the method that clears all current drawings. Its functionality is pretty obvious now that we know that the list of drawings is held as an array.

Listing 19.10   Clearing drawings

```
-(void)cancelDrawing {

    [myDrawing removeAllObjects];
    [self setNeedsDisplay];
}
```

At this point, your drawView object is maintaining two different instance variables: myPic contains the current picture and myDrawing contains an array of paths. Putting them together into a coherent whole just requires using some of the Quartz functions that we've discussed in the last two chapters. The results are shown in listing 19.11.

Listing 19.11   Drawing from user-created variables

```
- (void)drawRect:(CGRect)rect {

    float newHeight;
    float newWidth;

    if (!myDrawing) {
        myDrawing = [[NSMutableArray alloc] initWithCapacity:0];
    }
    CGContextRef ctx = UIGraphicsGetCurrentContext();

    if (myPic != NULL) {
        float ratio = myPic.size.height/460;
        if (myPic.size.width/320 > ratio) {
            ratio = myPic.size.width/320;
        }
        newHeight = myPic.size.height/ratio;
        newWidth = myPic.size.width/ratio;

        [myPic drawInRect:CGRectMake(0,0,newWidth,newHeight)];     ❶
```

```
        }
     if ([myDrawing count] > 0) {
        CGContextSetLineWidth(ctx, 5);
        for (int i = 0 ; i < [myDrawing count] ; i++) {
           NSArray *thisArray = [myDrawing objectAtIndex:i];

           if ([thisArray count] > 2) {
              float thisX = [[thisArray objectAtIndex:0] floatValue];
              float thisY = [[thisArray objectAtIndex:1] floatValue];

              CGContextBeginPath(ctx);          ❷
              CGContextMoveToPoint(ctx, thisX, thisY);     ❸

                for (int j = 2 ; j < [thisArray count] ; j+=2) {
                thisX = [[thisArray objectAtIndex:j] floatValue];
                thisY = [[thisArray objectAtIndex:j+1] floatValue];

                CGContextAddLineToPoint(ctx, thisX,thisY);     ❹
              }
              CGContextStrokePath(ctx);      ❺
           }
        }
     }
  }
```

The main bulk of this method is spent iterating through the information that you saved in other methods. There are five Quartz functions that do the drawing work. First, you draw the selected image. We've gone back to using the UIKit methods from the last chapter ❶, so that your image doesn't end up upside-down. Then, you start working through the myDrawing array. Each subarray results in your program beginning a new path ❷ and moving to the start ❸. As you move through the array, you add lines ❹. Finally, when a subarray is complete, you stroke the path ❺.

The result allows for drawing simple lines on a picture, which can then be saved, as we saw back in the view controller.

But is it possible to do more with this example? As usual, the answer is Yes.

19.6.3 *Expanding on the example*

If you wanted to expand this example into a more complete application, there are several routes you could take.

The first and most obvious expansion would be to select a color before drawing a line. The hard part here would be to create a color picker, though you could make a stand-alone class that you could then reuse elsewhere. With that in hand, it would be simple to add a color variable to your line arrays, probably by always saving it as the 0 element of a subarray.

The program might also benefit from a more sophisticated line-drawing algorithm that tosses out nearby points and smoothes the lines into curves, removing some of the sharp edges that show up in the current program.

In any case, that ends our look at Quartz 2D. There's lots more you can learn, but you should have the foundation that you need to move forward.

There are two other ways that you can draw using the SDK: Core Animation and OpenGL. We don't have the space in this introductory book to give full-length attention to either, but we'll at least introduce them and show you where to go for more information, beginning with Core Animation.

19.7 An introduction to Core Animation

Core Animation is a fundamental technology on the iPhone. It's what manages all the nifty scrolls, pivots, zoom-ins, zoom-outs, and other bits of animation that make up the iPhone user interface. As you've already seen, many UIKit classes give you an option to use animation or not, usually by having an `animated:` argument as part of a method.

Core Animation is also tightly integrated with Quartz. As you've seen, each UIView is linked up to a graphical layer called the CALayer, which is the *Core Animation* layer. Though you've just used it to depict simple graphics and images so far, you can also use it to manage more complex changes.

But you don't *have* to use Quartz at all to create animations. There's a CALayer behind every UIView, and because almost everything on the iPhone is built on a UIView, you can animate your existing UIViews, possibly including pictures that you've loaded into UIImageViews. For example, figure 19.6 shows how you could use Core Animation to show an approaching plane, by moving its UIImageView and turning it opaque as it approaches.

Figure 19.6 A jet approaches, thanks to Core Animation.

This is the example that we're going to show later in this section, using two different means to create the animation.

19.7.1 The fundamentals of Core Animation

When we speak of animation using Core Animation, what we're talking about is changing the properties of the CALayer and then smoothly animating those property changes. The CALayer class reference lists which properties can be animated; they include `anchorPoint`, `backgroundColor`, `opacity`, `position`, `transform`, and several others. This means that you can use Core Animation to animate the position of an object, its color, its transparency, and also its `CGAffine` transformations.

Before we get further into Core Animation, we want to talk about its fundamentals—those terms and ideas that you'll meet throughout this section:

- *Layer*—This is where animation occurs. You always have one `CALayer` hooked up to every `UIView`, accessible via the `layer` property. You can call up additional layers with a `[CALayer layer]` class message and then add them to your existing `CALayer` with the `addSublayer:` method. Adding layers in this way will result in inverted coordinate systems. Each layer can be individually animated, allowing for complex interactions between numerous animated properties. On the iPhone, you might find it as easy to create a more complex animation by creating multiple `UIKit` objects (most likely multiple `UIImageViews`) and animating each one.
- *Implicit animation*—This is the simplest type of animation. You tell the `UIView` that it should animate, and then you change properties.
- *Explicit animation*—This is an animation created with `CABasicAnimation` that allows you to more explicitly define how the property change animates.
- *Key frame animation*—This is an even more explicit type of animation, where you define not only the start and end of the animation, but also some of the frames in between.

You can also create much more complex animations, such as redefining how implicit animations work, collecting animations into transactions, and building complex animation layer hierarchies. For more information, look at the "Core Animation Programming Guide" and the "Core Animation Cookbook," both available from Apple.

19.7.2   *Getting started with Core Animation*

To use Core Animation, make sure that you add Quartz Core, the framework required for animation, to your project. You'll also want to include QuartzCore/QuartzCore.h, the main header file for Core Animation.

With that done, you're now ready to try out the two simplest types of animation: a simple implicit animation and an explicit animation.

19.7.3   *Drawing a simple implicit animation*

Implicit animations are the simplest type of animation, because they just require starting an animation block and then changing `CALayer`-level properties. Listing 19.12 shows a simple example of this, involving a `UIImageView` called `plane` that contains a clipart picture of a plane. The image starts out at the top-left corner of the screen with 25 percent opacity and moves downward while growing more opaque.

Listing 19.12   A simple implicit animation

```
[UIView beginAnimations:nil context:NULL];
CGAffineTransform moveTransform
```

```
  = CGAffineTransformMakeTranslation(200, 200);
[plane.layer setAffineTransform:moveTransform];
plane.layer.opacity = 1;
[UIView commitAnimations];
```

Between them, beginAnimations:context: and commitAnimations define an animation block.

Within the block, you set two properties to animate. setAffineTransform: is a special CALayer method that allows the setting of its transform property using an affine transform matrix, which you're already familiar with; opacity is a more obvious property.

As soon as you close out the block, the animation begins. Your plane will move and grow more distinct. That's all there is to it!

But sometimes an implicit animation won't give you as much control as you want. That's where explicit animations come in.

19.7.4 *Drawing a simple explicit animation*

When you're working with explicit animations, instead of defining a bunch of changes to a CALayer and executing them all, you define animations one by one using the CABasicAnimation class. Each of these animations can have its own value for duration, repeatCount, and numerous other properties. You then apply each animation to a layer separately, using the addAnimation:forKey: method.

Listing 19.13 executes an animation similar to the one shown in listing 19.12, but with more control.

Listing 19.13   A simple explicit animation

```
CABasicAnimation *opAnim = [CABasicAnimation
  animationWithKeyPath:@"opacity"];
opAnim.duration = 3.0;
opAnim.fromValue = [NSNumber numberWithFloat:.25];
opAnim.toValue= [NSNumber numberWithFloat:1.0];
opAnim.cumulative = YES;
opAnim.repeatCount = 2;
[plane.layer addAnimation:opAnim forKey:@"animateOpacity"];

CGAffineTransform moveTransform
  = CGAffineTransformMakeTranslation(200, 200);
CABasicAnimation *moveAnim = [CABasicAnimation
  animationWithKeyPath:@"transform"];
moveAnim.duration = 6.0;
moveAnim.toValue= [NSValue valueWithCATransform3D:
  CATransform3DMakeAffineTransform(moveTransform)];
[plane.layer addAnimation:moveAnim forKey:@"animateTransform"];
```

This example is definitely longer than our implicit animation example, but you get to define the two animations with separate durations, which is the first step to creating a more beautiful and better-controlled animation. Note that you also make use of yet another way to change an affine transform matrix into a Transform3D matrix of the type used by Core Animation: the CATransform3DMakeAffineTransform function.

The code does include a bit of a kludge: to keep the plane opaque through the last 3 seconds, it keeps counting opacity up cumulatively, making it climb from 1.0 to 1.75 the second time through. A better solution would be to have created three key frames for opacity: .25 at 0 seconds, 1.00 at 3 seconds, and 1.00 at 6 seconds. *That's* why you might want to use a key-frame animation of the sort we alluded to at the start of this section, rather than a basic animation.

These simple methods for using Core Animation can take you far. Look through the CALayer class reference for everything that you're allowed to animate. For more details, read though the two Apple guides we pointed out.

Before we leave graphics entirely behind, there's one other toolkit that we want to briefly touch on: OpenGL.

19.8   *An introduction to OpenGL*

OpenGL is SGI's standardized 2D and 3D graphical drawing language. The iPhone more specifically uses OpenGL ES, or OpenGL for Embedded Systems, which features a reduced API for use on devices like mobile phones. For full information on using OpenGL, you should pick up a book on the topic or read Apple's "OpenGL ES Framework Reference," which links to the most important documents available from Apple. We're going to cover some of the general information you'll need to access OpenGL through the iPhone OS.

The iPhone manages OpenGL through EAGL, a class that interfaces between the iPhone's views and OpenGL's drawing functions. It allows for the writing of OpenGL functions onto an EAGLView, which is the CAEAGL layer of a UIView, showing the same layer-based paradigm we met when using Core Animation.

To simplify your programming of OpenGL projects, Xcode supplies a standard template to use, which sets up all the OpenGL defaults for you. It's the OpenGL ES Application template, the only Xcode template that we have yet to examine. This template includes all the basic setup of OpenGL, which is pretty extensive. That includes the setup of a timer, the creation of frame buffers, and the code needed to draw something. To do basic OpenGL programming, all you have to do is write your code into the drawView method of the EAGLView class.

Rather than giving a completely insufficient overview of this enormous library, we'll instead point you toward a few bits of sample code. The OpenGL template comes complete with a rotating square as an example. There are also three OpenGL samples currently available from Apple: GLGravity shows simple OpenGL rendering related to accelerometer output, GLSprite demonstrates texturing, and GLPaint explores another way to allow finger painting.

These examples should be sufficient to get you started if you already have a strong basis in OpenGL and need to see how it's integrated into the iPhone.

19.9   *Summary*

Graphics are one of the most important elements for making your iPhone projects look great. Not only does the iPhone OS support high-quality graphics, but it also gives you a wide variety of options, depending on the needs of your program.

Quartz 2D will be your main workhorse for most graphical programs. If you're already familiar with the Canvas library for the web, you'll see that Quartz is quite similar. You'll be able to draw paths and use many graphical state variables to modify exactly how that path is painted. This chapter includes a pretty extensive look at Quartz.

Core Animation is an expansion to Quartz that was created for the iPhone. You've already seen it integrated into numerous programs native to the iPhone, and now you can use it yourself. Core Animation is built around the idea of automated animations: you tell it the endpoints, and Core Animation fills in the rest for you. Again this is much as you may have seen on the web, with the WebKit's various styles of implicit and explicit animation. This chapter covers the basics of how to use the simpler forms of Core Animation.

OpenGL is a whole new graphics library that has been imported into the iPhone, much as SQLite is a third-party library that Apple made available to iPhone users. The difference here is that Apple has made OpenGL easier to use, thanks to the creation of the EAGL framework. Though this chapter suggests how to get started with OpenGL, the topic is large enough that you'll need to pick up a book on OpenGL to fully explore the topic.

With graphics covered, there's one last topic that we need to complete our SDK toolkit: the internet. How do you access the net and how do you make use of various protocols that will allow you access to the ever-growing social network of the web? That's the topic of our last chapter, bringing this book full circle.

The web: web views
and internet protocols

This chapter covers

- Using web views
- Parsing XML
- Accessing other protocols

We started this book with a look at the web. Chapters 3 through 8 offered an extensive discussion of building iPhone web apps using HTML, CSS, JavaScript, and the dynamic programming language of your choice. As we said at the time, web development is one of two major ways that you can program for the iPhone, the other being the SDK that we've spent the last ten chapters on.

We've generally suggested web apps as the proper platform for creating internet-related programs. This chapter will present some solutions for when that's not the case. Even if you're depending heavily on the web, there are numerous reasons that you might want to program using the SDK. You might want to make use of its more extensive graphic capabilities. You could be designing something of sufficient complexity that you want to use a well-organized object-oriented environment. You might want to monetize your app without having to depend on ads. For whatever reason, you've decided to design an SDK web app, not an HTML-based web app, and now you need to know how to do so.

In this chapter, we're going to cover the major ways to access the internet from the SDK. You can do so in a variety of ways, and we'll outline their hierarchy in our first section.

20.1 The hierarchy of the internet

internet programming involves a hierarchy of protocols. At the lowest level, you have the sockets that you use to connect from one computer to another. Above them are a variety of more sophisticated technologies, such as FTP, Bonjour, and HTML. HTML is a critical protocol, represented on the iPhone by both low-level access and the high-level `UIWebView`. Recently an increasing number of protocols have been built on top of HTML, forming what we call the *social network*.

This hierarchy of internet protocols is shown in figure 20.1, along with iPhone OS classes of note.

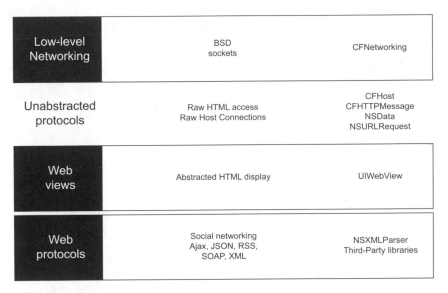

Figure 20.1 internet protocols are arranged in a hierarchy.

In this chapter, we're going to cover all of these protocols, starting with the lowest level, but our real focus will be on the higher-level internet and social network protocols, because they're the protocols that are best supported by the iPhone, and they're the ones you're most likely to want to interact with.

20.2 Low-level networking

We've opted not to pay much attention to BSD sockets and the lower-level networking classes, because we expect they'll be of little interest to most iPhone programmers. If you need to work with BSD sockets, you should look at Apple's "Introduction to CFNetwork Programming Guide."

If you need to work with the lower-level protocols, CFNetwork provides a variety of classes that you'll find useful. You can find more information on them in the "Networking & internet" topic in the Apple docs. In particular, the "CFNetwork Framework Reference" will give you an overview of the various classes. Among the classes are CFFTPStream, which lets you communicate with FTP servers, and CFNetServices, which gives you access to Bonjour—Apple's service discovery protocol. There are also two low-level HTTP-related classes, CFHTTPMessage and CFHTTPStream. We're going to leave these classes alone, as our HTML work will be related to the higher-level NSURL, NSURLRequest, UIWebView, NSMutableURLRequest, and NSURLConnection classes.

Rather than skipping over these low-level and unabstracted protocols entirely, we'll take a look at one of them, CFHost. It's the easiest to work with and perhaps the most immediately useful.

20.2.1 *The CFHost class*

CFHost allows your program to request information about an internet host, such as its name, its address, and whether it's reachable. Listing 20.1 shows a sample of how to determine whether a host name exists or not.

Listing 20.1   A simple host name lookup

```
-(IBAction)reportStatus:(id)sender {

    CFStreamError errorTest;

    if (myInput.text) {

        CFHostRef myHost = CFHostCreateWithName(kCFAllocatorDefault,
            (CFStringRef)myInput.text);        ❶

        if (myHost) {
            if (CFHostStartInfoResolution(myHost, kCFHostAddresses,
                &errorTest)) {       ❷

                myOutput.text = [myInput.text stringByAppendingString:
                    @" COULD be resolved."];
            } else {
                myOutput.text = [myInput.text stringByAppendingFormat:
                    @" could NOT be resolved (Error: %i).",
                        errorTest.error];
            }
        }
        CFRelease(myHost);
    }
}
```

Our sample method, reportStatus:, is activated by a button push. It reads a host name from a UITextField called myInput and reports out to a UITextView called myOutput.

All uses of the CFHost commands follow the same pattern. First you create a CFHostRef object with CFHostCreateCopy, CFHostCreateWithAddress, or CFHost-CreateWithName ❶. Then you use CFHostStartInfoResolution to request a certain type of information, which can be kCFHostAddresses, kCFHostNames, or kCFHost-Reachability ❷. This example omits a final step where you retrieve your information with CFHostGetAddressing, CFHostGetNames, or CFHostReachability—something that wasn't necessary here because the point was to see if the request for an address resolved correctly at all.

You can find more information on these functions, and on how to use a callback function to make the host resolution asynchronous, in the CFHost reference.

We consider this look at low-level networking—and CFHost—an aside, meant only to hint at what is possible if you must do lower-level networking work. Now we'll move on to higher-level HTML-related network work that's more likely to be the focus of your iPhone network programming. The first thing you'll need to know is how to use the iPhone's URL objects.

20.3 Working with URLs

With HTTP being the basis of most iPhone internet programming, it shouldn't be a surprise that URLs are a foundational technique for internet-based programming. You'll use them whether you're calling up UIImageViews, accessing content by hand, or parsing XML. As a result, we're going to spend a bit of time on the two fundamental URL classes: NSURL and NSURLRequest.

20.3.1 Creating an NSURL

An NSURL is an object that contains a URL. It can reference a web site or a local file, like any URL can. You've used it in the past to access Apple's stock page and to load up local media files for play.

> **NSURL and CFURLRef**
> NSURL is a toll-free bridge to CFURL, making an NSURL * and a CFURLRef equivalent. We took advantage of this in chapter 18 when dealing with the MPMoviePlayerController and with sounds. Whenever you need to create a CFURLRef, you can do so using the standard methods for NSURL creation that are described in this chapter.

As noted in the NSURL class reference, there are numerous methods that you can use to create an NSURL. The most important ones are listed in table 20.1.

Table 20.1 A variety of NSURL creation methods

Method	Summary
fileURLWithPath:	Creates a URL from a local file path
URLWithString:	Creates a URL from a string; equivalent to initWithString:
URLWithString:relativeToURL:	Adds a string to a base URL; equivalent to initWithString:relativeToURL:

Once you've got an NSURL in hand, you can do any number of things with it:

- You can pass it on to functions that require a bare NSURL, as was the case with those media functions in chapter 18.
- You can query its properties to easily break down the URL to its parts. As usual, you can find a complete list of properties in the Apple reference, but properties like baseURL, fragment, host, path, port, and query might be particularly useful.
- You can use the NSURL to load up a UIWebView.

The first two possibilities require only the use of an NSURL, but when you're working with a UIWebView, you must first create an NSURL and then turn it into an NSURLRequest.

20.3.2 Building an NSURLRequest

The NSURLRequest class contains two parts: a URL and a specific policy for dealing with cached responses. As noted in table 20.2, there are four ways to create an NSURL-Request, though we expect that you'll usually fall back on the simple factory method, requestWithURL:.

Table 20.2   The related NSURLRequest init methods

Method	Summary
requestWithURL:	Creates a default request from the URL; equivalent to initWithURL:
requestWithURL:cachePolicy:timeoutInterval:	Creates a request with specific caching choices; equivalent to initWithURL: cachePolicy:timeoutInterval:

By default, an NSURLRequest is built with a caching policy that's dependent upon the protocol, and a timeout value of 60 seconds, which should be sufficient for most of your programming needs. If you need to get more specific about how things are loaded, you can call requestWithURL:cachePolicy:timeoutInterval:, giving it an NSURLRequestCachePolicy for the policy and an NSTimeInterval for the timeout.

You can also create a more interactive NSURLRequest by using the NSMutableURL-Request class, which allows you to more carefully form and modify the request that you're sending. We'll talk about this in section 20.6, when we examine how to send POST requests from an iPhone.

The NSURLRequest will get you through most web page work. As with the NSURL, there are a few different things that you can do with an NSURLRequest. You can hand it off to a UIImageView, or you can use it to read in the contents of a web page, to later manipulate it by hand.

20.3.3 Manipulating HTML data by hand

To read the contents of a web page manually, you need to access an NSURLRequest's properties. Table 20.3 lists some of the most important ones, though, as usual, more information can be found in the class reference.

Table 20.3 `NSURLRequest` **can give access to a page's content**

Property	Summary
`allHTTPHeaderFields`	Returns an `NSDictionary` of the header
`HTTPBody`	Returns an `NSData` with the body
`valueforHTTPHeaderField:`	Returns an `NSString` with the header

The catch with these properties is that you can only work with well-defined HTML pages. Most notably, the `NSURLRequest` properties can't read fragments, such as would be generated by Ajax or JSON, nor can they parse other sorts of content, such as XML or RSS.

Other ways to read HTTP content

If you're not reading data that meets the HTML protocol, you can't use `NSURLRequest`'s properties to access the data. Instead, you must fall back on other functions that let you read in data from an `NSURL`.

We're already met functions that read data that follows other protocol specifications, such as the `MPMoviePlayerController` and the sound players from chapter 18. Similarly, in this chapter we'll talk about an XML parser. All of these classes can read directly from a URL.

If you need to capture raw data that isn't set in HTML, the best way to do so is with an init or factory method that reads from a URL, such as `NSData`'s `dataWithContentsOfURL:`. We'll look at an example of that in the last section of this chapter.

You may also discover that you need a more interactive way to deal with HTML data. In this case, you'll probably use an `NSURLConnection` object; but as with the `NSMutableURLRequest`, we're going to save that for later, because you'll typically only need to use it when you're POSTing information to a web page rather than just retrieving it.

For the moment, we're going to put all of these complexities aside and instead look at how to display straight HTML data using the SDK's `UIWebView`.

20.4 *Using UIWebView*

One of the easiest ways to connect up to the internet is to use the `UIWebView` class, which gives you full access to web pages of any sort. In some ways, this class is of limited utility, because it largely duplicates Safari, and Apple isn't interested in approving applications that just duplicate their existing technology. But there are clearly situations where you'll want a program to be able to refer to some specific web pages, and that's what `UIWebView` is for.

The class is easy to use—we included it in simple examples way back in chapters 11 and 12. The only real complexity is in building an `NSURL` or `NSURLRequest` object to get your web view started, but that process follows the methods we've already seen.

20.4.1 *Calling up the web view*

There are two main ways to fill a web view once you've created it, as listed in table 20.4. Most frequently, you'll start with an NSURLRequest, which you must have created using the two-step process that we described in the previous section, but you can also load a web view with an NSURL and an NSString. A few other init methods can be found in the class reference.

Table 20.4   Methods for loading `UIWebView`

Method	Summary
loadHTMLString:baseURL:	Loads a page from a URL and a string
loadRequest:	Loads a page from an NSURLRequest

Assuming you use the more common NSURLRequest method, you can put all the lessons you've learned so far together, which is just what you did back in chapter 11 when you created your first UIWebView:

```
[myWebView loadRequest:
    [NSURLRequest requestWithURL:
        [NSURL URLWithString:url]]];
```

Once you've got a UIWebView, you can start working with it. There are five UIWebView methods and properties of particular note, which are summarized in table 20.5.

Table 20.5   Some sterling `UIWebView` options

Method/Property	Type	Summary
detectsPhoneNumbers	Property	Boolean that determines whether phone numbers become links
goBack	Method	Moves back a page; check canGoBack property first
goForward	Method	Moves forward a page; check canGoForward property first
reload	Method	Reloads the current page
scalesPageToFit	Property	Boolean that determines whether the page is zoomed into a viewport and whether user zooming is allowed

The most exciting options in our mind are the goBack, goForward, and reload methods, which can give you some control over how the UIWebView moves among pages. Similarly, the loadRequest: method can be continually rerun if you want to move a user through multiple pages, treating the UIWebView more like a web slideshow than a browser.

WARNING   In our opinion, the scalesPageToFit property does not work correctly at the current time. It always scales the page as if the UIWebView were full screen, and it leaves a less than optimal view if you create a small UIWebView, as we will do in our next example. We expect this to be resolved in a future version of the SDK.

As we wrote in chapter 12, the biggest gotcha in using a `UIWebView` is that you can't load a URL straight from Interface Builder. Expect to always use the `NSURL-NSURLRequest-loadRequest:` process that we've laid out here to load up pages into your web views.

20.4.2 Managing the web view delegate

There's one other element that we didn't discuss when we talked about web views in chapters 11 and 12: you can set a delegate to manage a few common responses. You must follow the `UIWebViewDelegate` protocol, which lists four methods, described in table 20.6.

Table 20.6 Managing `UIWebViews` with delegate methods

Method	Summary
`webView:shouldStartLoadWithRequest:navigationType:`	Called prior to content loading
`webViewDidStartLoad:`	Called after content begins loading
`webViewDidFinishLoad:`	Called after content finishes loading
`webView:didFailLoadWithError:`	Called after content fails to load

Together with the `UIWebView` methods, these delegate methods give you considerable power. You could use them to load alternative web pages if the preferred ones don't load. Or, continuing our slideshow analogy, you could use them to continuously load new pages when old ones finish. All those possibilities highlight the ways that you might be able to use the `UIWebView` as more than just a Safari clone.

20.4.3 Thumbnails: a web view example

As we've previously stated, `UIWebViews` are pretty easy to set up, and we're not going to spend a lot of time on a coding sample. Listing 20.2 presents a simple example that creates a set of web page thumbnails, similar to the startup page of the Google Chrome browser. It uses delegates first to get rid of `UIWebViews` that don't load, and later to zoom in on the one the user selects.

It should be initially created in Interface Builder by laying out four `UIWebViews`. Make sure that they're set to scale, and set their delegates to be the view controller.

Listing 20.2 A thumbnail web viewer

```
- (void)viewDidLoad {          ⟵┐  Sets up web
    [super viewDidLoad];         ❶  views

    webArray = [[NSArray alloc]
        initWithObjects:webView1,webView2,webView3,webView4,nil];

    NSString *paths = [[NSBundle mainBundle] resourcePath];
    NSString *filePath = [paths
        stringByAppendingPathComponent:@"weblist.txt"];
    NSString *webList = [NSString stringWithContentsOfFile:filePath];
    NSArray *webListArray = [webList componentsSeparatedByString:@"\n"];

    for (int i = 0 ; i < [webArray count] ; i++) {
```

```
        [[webArray objectAtIndex:i] loadRequest:
          [NSURLRequest requestWithURL:
            [NSURL URLWithString:
               [webListArray objectAtIndex:i]]]];
    }
}

- (void)webView:(UIWebView *)webView
    didFailLoadWithError:(NSError *)thiserror {

    NSLog(@"Web Thumbs Error: %@",thiserror);
    if (thiserror.code == -1003) {
      [webView removeFromSuperview];
    }
}

- (void)webViewDidFinishLoad:(UIWebView *)webView {

    if (webView.canGoBack == YES) {
       for (int i = 0 ; i < [webArray count] ; i ++) {
          if ([webArray objectAtIndex:i] != webView) {
             [[webArray objectAtIndex:i] removeFromSuperview];
          } else {
             webView.frame = [[UIScreen mainScreen] bounds];
          }
       }
    }
}
```

② Resolves errors

③ Zooms active view

To start with, you read a set of (exactly) four URLs from a file and use the `NSString` method `componentsSeparatedByString:` to turn them into an `NSArray` that you use to seed your web views **①**. After that, it's a question of responding to delegation messages.

The `webView:didFailLoadWithError:` method **②** shows off some valuable techniques for both debugging and error management. `NSLog` is what you want to use when you want to do a `printf`-style reporting of runtime variables. It'll output to `/var/log/system.log` when you run it inside the iPhone Simulator.

Within a `UIWebView`, there are two error codes that will come up with some frequency: -1003 is "Can't find host" and -999 is "Operation could not be completed." This example ignores -999 (which usually means that the user clicked a link before the page finished loading), but in the case of a -1003 failure, you dismiss the web view.

Finally, the `webViewDidFinishLoad:` method **③** zooms in on an individual web view (dismissing the rest) once a user clicks on a link. Realistically, this should occur whenever the user touches the web view, but we wanted to show the `UIWebView` delegate methods, so we chose this slightly more circuitous route.

And that's it—a simple web thumbnail program, as shown in figure 20.2. It could be improved by giving the

Figure 20.2 The thumbnail program removes web views that fail to load.

Debugging

We haven't talked about debugging your SDK program much in this book, primarily for reasons of space. Here's a short overview of our favorite techniques:

Xcode itself provides the best debugging. Pay careful attention to autocompletion of words and note when an expected autocompletion doesn't occur, because that usually means you didn't set a variable correctly.

The warnings and errors that appear on compilation should always be carefully considered.

During Simulator runtime, we suggest keeping an eye on /var/log/system.log on your Mac. You can do this by opening a terminal and running `tail -f /var/log/system.log`. You can log to the system log by hand with `NSLog`. If a program crashes at runtime, you'll usually see an error here, and then you can go to Xcode to step back through a trace to see exactly where the crash occurred.

Finally, once you're done with your program you should run it through Instruments to check for memory leaks.

For more information, take a look at the "Xcode Debugging Guide," "Debugging with GDB," and the "Instruments User Guide," Apple articles which contain comprehensive explanations of those subjects.

user the ability to manage the selected URLs and by polishing up the way the user selects an individual page (including an option to return to the thumbnail page afterward). For our purposes, though, it does a great job of demonstrating some of the intricacies of the `UIWebView`.

Before we finish with web views entirely, we're going to look at one more example. Back in chapter 17, we talked about how Core Location would be better served once we got into the world of the internet. We're going to look at the first of two Core Location internet examples.

20.4.4 *Google Maps: a Core Location example*

Google Maps should be an ideal way to show off Core Location, because it's already built into the iPhone and because it can take longitude and latitude as URL arguments. Unfortunately, the reality falls a little short of that ideal at the time of this writing.

There are two problems. Most notably, as a programmer, you have no access to the proprietary Google Maps interface built into the iPhone. Instead, you have to use a `UIWebView` to show the map results (which is why we're covering this example in this section). That leads to the second problem, which is that Google Maps often sizes itself in weird ways when displayed in a `UIWebView`. This is probably due to some combination of `UIWebView` always assuming that it will appear full screen—which we've already discussed—and Google Maps trying to do the right thing when it detects that you're on the iPhone. We expect Google Maps' presentation will slowly improve through future releases of the SDK.

DISPLAYING GOOGLE MAPS WITH A WEB VIEW

Listing 20.3 shows a simple example of how to call up Google Maps in a web view called myWeb using a Location Manager–derived latitude and longitude. It additionally does a search for a type of business listed in a UITextField called myEntry.

Listing 20.3   Calling up Google Maps from the SDK

```
- (void)locationManager:(CLLocationManager *)manager
  didUpdateToLocation:(CLLocation *)newLocation
  fromLocation:(CLLocation *)oldLocation {

  if (myEntry.text) {
    [myLM stopUpdatingLocation];
    [myActivity stopAnimating];

    NSMutableString *googleSearch = [NSMutableString
      stringWithFormat:@"http://maps.google.com?q=%@&sll=%f,%f",
      myEntry.text,newLocation.coordinate.latitude,
      newLocation.coordinate.longitude];
    [googleSearch replaceOccurrencesOfString:@" " withString:@"+"
      options:NSCaseInsensitiveSearch range:NSMakeRange(0,
      [googleSearch length])];

  [myWeb loadRequest:[NSURLRequest requestWithURL:[NSURL
    URLWithString:googleSearch]]];
  }
}
```

The functionality here is quite simple. All the hard work is done in the middle two lines, which set up the googleSearch string. When you pass Google Maps a q argument, you're sending it a word to search on, and when you give it an sll argument, you're defining a latitude and longitude. As a result, when handed this search string, Google will put together a map that shows businesses of type q near the designated coordinates. The following line of code then does some minimal work to urlencode the query.

This sample code is sufficient if you're playing around with a program for your own use, and more importantly it shows how easy it is to mix Core Location and a UIWebView, but what would you do if you wanted to make a production-ready app that supported Google Maps? The answer is, you'd do a fair amount more work—more than we can cover here in anything but an outline.

HYBRIDIZING GOOGLE MAPS WITH A WEB APP

In order to create a top-grade Google Maps application, you'd need to fall back on one of the lessons we covered way back in chapter 2: you must create an integrated client-server program that uses both web development and the SDK. Here's how it would work:

1   Sign up for a Google Maps API key, as described at http://code.google.com/apis/maps/.

2   Use that key to create a static map on your web server that creates a Google Map based on the arguments it's sent.

3 Set values like `viewport` to make sure that your map displays well on an iPhone, conquering the main problem with accessing Google Maps at this time.

4 Have your iPhone app call up your specialized Google Maps page, rather than Google's.

We don't think this sort of work will be necessary forever, but, for now, it's a powerful example of exactly why you might want to create a hybridized program using both iPhone web development and the SDK.

There's a Google Code project that presumes this sort of setup and includes a sophisticated iPhone-side interface for viewing the map. It's called iPhone Google Maps and can currently be found here:

http://code.google.com/p/iphone-google-maps-component

Having now covered web views in some depth, we're ready to begin looking at higher-level protocols, starting with the one best supported by the iPhone: XML.

20.5 *Parsing XML*

XML (Extensible Markup Language) is a generalized markup language whose main purpose is to deliver data in a well-formed and organized way. It has some similarities to HTML, and an XML version of HTML has been released, called XHTML.

Because of XML's popularity on the internet, the iPhone SDK includes its own XML parser, the `NSXMLParser` class. This is an event-driven API, which means that it will report start and end tags as it moves through the document, and you must take appropriate action as it does.

XML and files

When using `NSXMLParser`, you'll probably immediately think about reading data taken from the internet, but it's equally easy to read XML from your local files. You create a path to the file, and then use `NSURL`'s `fileURLWithPath:` method, as we've seen elsewhere in this book.

An XML file can be a nice intermediary step between saving data in plain text files and saving it in databases, which were two of the options we saw in chapter 16. Although you're still using files, you can do so in a well-organized manner. We'll see an example of this in section 20.5.3.

Running the `NSXMLParser` involves setting it up, starting it running, and then reacting to the results. We'll cover that process in the rest of this section. For more information on any of these topics, we suggest reading Apple's "Event-Driven XML Programming Guide for Cocoa," but we'll provide a tutorial on the basics, starting with the parser class.

20.5.1 *Starting up NSXMLParser*

In order to get started with the NSXMLParser, you need to create it, set various properties, and then start it running. The most important methods for doing so are listed in table 20.7.

Method	Summary
initWithContentsOfURL:	Creates a parser from an NSURL
initWithData:	Creates a parser from an NSData
setDelegate:	Defines a delegate for the parser
parse	Starts the parser going

Table 20.7 Methods to get your NSXMLParser going

Not listed are a few additional setters that allow the parser to process namespaces, report namespace prefixes, and resolve external entities. By default, these properties are all set to NO; you shouldn't need them for simple XML parsing.

20.5.2 *Acting as a delegate*

There are approximately 20 delegate methods for NSXMLParser. They are all optional: you only need to write delegates for things that you're watching for. We're not going to cover the delegate methods for mapping prefixes, comments, external entities, and many other things of somewhat less importance.

Instead, we're going to look at the five most critical methods that you'll need to use whenever you're parsing XML. These are the methods that report the start and end of elements, the contents inside, when the XML parsing has ended (unrecoverably!) with an error, and when the XML parsing has ended because it's all done. These are listed in table 20.8.

Table 20.8 The five most important NSXMLParser delegate methods

Method	Summary
parser:didStartElement:namespaceURI: qualifiedName:attributes:	Reports the start of an element and its attributes
parser:foundCharacters:	Reports some or all of the content of an element
parser:didEndElement:namespaceURI: qualifiedName:	Reports the end tag of an element
parserDidEndDocument:	Reports the end of parsing
parser:parseErrorOccurred:	Reports an unrecoverable parsing error

Generally, when you're parsing XML, you should take the following steps as you move through elements:

1 When you receive the didStartElement: method, look at the NSString to see what element is being reported, and then prepare a permanent variable to save its content, to prepare your program to receive the information, or both. Optionally, look at the NSDictionary passed by the attributes: handle and modify things accordingly.

2 When you receive the foundCharacters: method, save the contents of the element into a temporary variable. You may have to do this several times, appending the results to your temporary variable each time, because there's no guarantee that all of the characters will appear in one lot.

3 When you receive the didEndElement: method, copy your temporary variable into your permanent variable, take an action based upon having received the complete element, or both.

4 Optionally, when you receive parserDidEndDocument:, do any final cleanup.

Beyond that, the parser:parseErrorOccurred: method should call up an NSAlert or otherwise alert the user to the problem. As we noted, this is only for an unrecoverable problem: the user won't be able to do anything about it without modifying the original XML itself.

In order to show how the NSXMLParser can be used, our next example involves writing a simple RSS reader. Building an RSS reader on your own will allow you to walk through the basic functionality of NSXMLParser using an XML type that's widely available on the internet for testing.

20.5.3 *Building a sample RSS reader: an XML example*

Now that you understand the basics of XML, you're ready to put together a sample program that uses NSX-MLParser in two ways: first to read a text file and then to read an RSS feed. The results will be output to a hierarchy of tables. The first level of the hierarchy will show all the possible feeds, and the second level will show the contents of individual feeds. An example of the latter sort of page is shown in figure 20.3.

THE TOP-LEVEL TABLE

To start this project, you'll need to create a Navigation-Based Application, which will provide you with the navigator and initial table setup needed to get this project started. In a more advanced program, you'd give users the opportunity to create a settings file for what RSS feeds they want to read, but for the purposes of this example, create an XML settings file called rss-feeds.xml by hand, using the following format:

Figure 20.3 RSS feeds can easily be placed in table views.

```
<rdf:RDF xmlns:rdf="http://www.w3.org/1999/02/22-rdf-syntax-ns#"
   xmlns="http://purl.org/rss/1.0/"
   xmlns:dc="http://purl.org/dc/elements/1.1/">
  <feed title="RPGnet News" url="http://www.rpg.net/index.xml" />
  <feed title="RPGnet Columns" url="http://www.rpg.net/columns/index.xml"
      />
</rdf:RDF>
```

For each entry, create a singular `<feed>` element and include `title` and `url` attributes.

Once you've added `rssfeeds.xml` to your project, you're ready to write the code for the top-level table, which will parse your local XML file and give your user the option to select one of the RSS feeds. Listing 20.4 displays this code, which appears in the main view controller.

Listing 20.4   Reading an XML text file

```
- (void)viewDidLoad {          ◁──┐ ❶ Parses
   [super viewDidLoad];              XML file

   self.title = @"RSS Feeds";
   rssList = [[NSMutableArray alloc] initWithCapacity:1];

   NSString *paths = [[NSBundle mainBundle] resourcePath];
   NSString *xmlFile = [paths
      stringByAppendingPathComponent:@"rssfeeds.xml"];
   NSURL *xmlURL = [NSURL fileURLWithPath:xmlFile isDirectory:NO];

   NSXMLParser *firstParser = [[NSXMLParser alloc]
      initWithContentsOfURL:xmlURL];
   [firstParser setDelegate:self];
   [firstParser parse];

}

- (void)parser:(NSXMLParser *)parser
   didStartElement:(NSString *)elementName
   namespaceURI:(NSString *)namespaceURI
   qualifiedName:(NSString *)qualifiedName          ❷ Reads attribute
   attributes:(NSDictionary *)attributeDict {    ◁──┘   elements

   if ([elementName compare:@"feed"] == NSOrderedSame) {
      [rssList addObject:[[NSDictionary alloc] initWithObjectsAndKeys:
         [attributeDict objectForKey:@"title"],@"title",
         [attributeDict objectForKey:@"url"],@"url",
         nil]];
   }
}
- (void)parserDidEndDocument:(NSXMLParser *)parser {   ◁─── ❸ Cleans up
   [parser release];                                         parser
}

- (NSInteger)numberOfSectionsInTableView:
   (UITableView *)tableView {   ◁──┐
                                     ❹ Sets up
   return 1;                           sections
}
```

```
-  (NSInteger)tableView:(UITableView *)tableView
   numberOfRowsInSection:(NSInteger)section {

   return [rssList count];
}
-  (UITableViewCell *)tableView:(UITableView *)tableView
   cellForRowAtIndexPath:(NSIndexPath *)indexPath {

   static NSString *CellIdentifier = @"Cell";

   UITableViewCell *cell = [tableView
      dequeueReusableCellWithIdentifier:CellIdentifier];
   if (cell == nil) {
      cell = [[[UITableViewCell alloc] initWithFrame:CGRectZero
   reuseIdentifier:CellIdentifier] autorelease];
   }

   cell.text = [[rssList objectAtIndex:indexPath.row]
      objectForKey:@"title"];
   cell.accessoryType = UITableViewCellAccessoryDisclosureIndicator;

   return cell;
}
-  (void)tableView:(UITableView *)tableView
   didSelectRowAtIndexPath:(NSIndexPath *)indexPath {

   rssViewController *nextController =
      [[rssViewController alloc] initWithURL:
         [[rssList objectAtIndex:indexPath.row] objectForKey:@"url"]];
   nextController.title = [[rssList objectAtIndex:indexPath.row]
      objectForKey:@"title"];

   [self.navigationController pushViewController:nextController
      animated:YES];
   [nextController release];
}
```

5 Sets up rows

6 Sets up cells

7 Calls up RSS view

This example begins by reading in XML from a file **1**. The result is a lot more pleasing than trying to read raw text, as in the thumbnail example earlier in this chapter, so we'll suggest encoding simple preference files as XML in the future.

Because we designed a simple XML format, where the information is encoded as attributes, you only have to watch one delegate method, didStartElement: **2**. Here you add the information to rssList, an NSMutableArray, for use later. The only other thing you have to do with your XML parser is clean it up when you're done **3**.

The next few functions are standard table view work, as you define the sections **4**, rows **5**, and cells **6** using the rssList array you created. Finally, you define what happens when the user selects a row **7**, and that is to call up a brand new type of object, the rssViewController.

THE RSS TABLE

The rssViewController is a subclass of the UITableViewController that will display an RSS feed if initialized with a URL. Listing 20.5 shows the complete contents, much of which is similar to listing 20.4. The biggest differences are in the XML parsing, because an RSS feed is a much more complicated XML format, even when you're only using minimal information from it, as is the case here.

Listing 20.5   Creating a table from an RSS feed

```
- (id)initWithURL:(NSString *)url {        ◄──❶ Parses RSS feed
    if (self = [super init]) {
        feedList = [[NSMutableArray alloc] initWithCapacity:0];
        NSXMLParser *nextParser = [[NSXMLParser alloc]
            initWithContentsOfURL:[NSURL URLWithString:url]];
        [nextParser setDelegate:self];
        [nextParser parse];
    }
    return self;
}

- (void)parser:(NSXMLParser *)parser
    didStartElement:(NSString *)elementName
    namespaceURI:(NSString *)namespaceURI
    qualifiedName:(NSString *)qualifiedName        ❷ Preps for
    attributes:(NSDictionary *)attributeDict {   ◄──┘  reading

    if ([elementName compare:@"item"] == NSOrderedSame) {
        currentItem = [[NSMutableDictionary alloc] initWithCapacity:0];
    } else if (currentItem != NULL) {
        currentContents = [[NSMutableString alloc] initWithCapacity:0];
    }
}

- (void)parser:(NSXMLParser *)parser            ❸ Reads
    foundCharacters:(NSString *)string {    ◄──┘  content
    if (currentContents && string) {
        [currentContents appendString:string];
    }
}

- (void)parser:(NSXMLParser *)parser
    didEndElement:(NSString *)elementName
    namespaceURI:(NSString *)namespaceURI        ❹ Finishes
    qualifiedName:(NSString *)qName {    ◄──┘  reading

    if ([elementName compare:@"item"] == NSOrderedSame) {
        [feedList addObject:currentItem];
        [currentItem release];
    } else if (currentItem && currentContents) {
        [currentItem setObject:currentContents forKey:elementName];
        currentContents = nil;
        [currentContents release];
    }
}

- (void)parserDidEndDocument:(NSXMLParser *)parser {
    [parser release];
}

- (NSInteger)numberOfSectionsInTableView:(UITableView *)tableView {
    return 1;
```

```
}

- (NSInteger)tableView:(UITableView *)tableView
  numberOfRowsInSection:(NSInteger)section {

  return [feedList count];

}

- (UITableViewCell *)tableView:(UITableView *)tableView
  cellForRowAtIndexPath:(NSIndexPath *)indexPath {

  static NSString *CellIdentifier = @"Cell";
  UITableViewCell *cell = [tableView
  dequeueReusableCellWithIdentifier:CellIdentifier];
  if (cell == nil) {
     cell = [[[UITableViewCell alloc]
        initWithFrame:CGRectZero
        reuseIdentifier:CellIdentifier]
           autorelease];
  }

  if ([[feedList objectAtIndex:indexPath.row] objectForKey:@"title"]) {
     cell.text = [[feedList objectAtIndex:indexPath.row]
        objectForKey:@"title"];
  }
  if ([[feedList objectAtIndex:indexPath.row] objectForKey:@"link"]) {
     cell.accessoryType = UITableViewCellAccessoryDisclosureIndicator;
  }
  return cell;

}

- (void)tableView:(UITableView *)tableView
  didSelectRowAtIndexPath:(NSIndexPath *)indexPath {
```

❺ Calls up web view

```
  UIWebView *thisInfo = [[UIWebView alloc] init];
  [thisInfo loadRequest:[NSURLRequest requestWithURL:
     [NSURL URLWithString:[[feedList objectAtIndex:indexPath.row]
        objectForKey:@"link"]]]];
  thisInfo.scalesPageToFit = YES;

  UIViewController *thisVC = [[UIViewController alloc] init];
  thisVC.view = thisInfo;
  thisVC.title = [[feedList objectAtIndex:indexPath.row]
     objectForKey:@"title"];

  [self.navigationController pushViewController:thisVC animated:YES];

  [thisInfo release];
  [thisVC release];

}
```

The difference in this new table view starts with the fact that you've got a custom init function that allows you to start an XML parser running on an RSS feed ❶. In a more polished application, you'd check for the feed's existence, but for this example you can dive right in.

Because this XML file is more complex than the previous one, you can't do all your work in didStartElement: ❷. Instead, you use this method as part of a systemic examination of the XML content, by preparing variables, creating a dictionary to hold the contents of a complete RSS item, and initializing a string to hold each individual element.

In foundCharacters: ❸, you have to keep appending data to the current element's string, as we promised. The XML parser *will* break the data from an individual element into multiple strings, so you have to be careful about this.

When you're done ❹, you can add your string to the element's dictionary, and when the element is done, you can add the dictionary to the array of RSS contents that you're maintaining.

From here on, most of the table work is pretty similar to the previous example. You read back through your master array to fill in the contents of the table. The only thing of note comes in the last method ❺, when a user clicks on a table row. At this point, you call up a UIWebView so that the user can hop straight to the RSS feed item he is interested in.

Before we finish with XML entirely, we want to look at one more Core Location example, using GeoNames to read in altitude, as we promised we would in chapter 17.

20.5.4 *Altitude redux: a Core Location example*

GeoNames, which can be found at geonames.org, offers a variety of web services related to location. It can give you information on postal codes, countries, addresses, and more. A complete listing of their web services can be found at http://www.geonames.org/export/ws-overview.html.

Most of GeoNames' information is returned either in XML or JSON format, as you prefer. We're going to look at their XML interface here. Table 20.9 shows off some of the XML-based GeoNames information that you may find particularly useful.

Table 20.9 GeoNames searches allowable with coordinate information

Information	Summary
findNearestIntersection	Returns nearest street intersection in the US
gtopo30	Returns altitude of location or -9999 for sea
srtm3	Returns altitude of location or -32768 for sea
timezone	Returns not only time zone info, but also the current time

We're going to use gtopo30 to follow through on our promise from chapter 17 to look up the altitude from GeoNames based upon the Location Manager's results. This project requires a somewhat complex chaining together of multiple delegate-driving classes, as shown in figure 20.4.

The bare skeleton of the code needed to make this work is shown in listing 20.6.

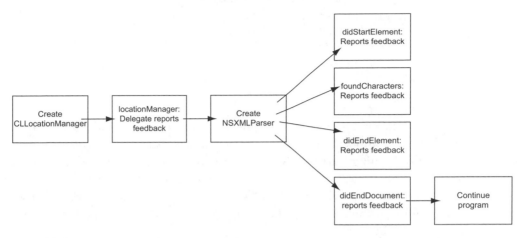

Figure 20.4 Complex SDK programs can chain multiple classes.

Listing 20.6 Deriving altitude from GeoNames

```
- (void)locationManager:(CLLocationManager *)manager
  didUpdateToLocation:(CLLocation *)newLocation
  fromLocation:(CLLocation *)oldLocation {          ◁─┐  ❶ Prepares
                                                        XML
  [myLM stopUpdatingLocation];
  [myActivity stopAnimating];

  NSString *gnLookup = [NSString stringWithFormat:
    @"http://ws.geonames.org/gtopo30?lat=%f&lng=%f&style=full&type=XML",
    newLocation.coordinate.latitude,newLocation.coordinate.longitude];
  NSXMLParser *gnParser = [[NSXMLParser alloc]
    initWithContentsOfURL:[NSURL URLWithString:gnLookup]];
  [gnParser setDelegate:self];
  [gnParser parse];
}

- (void)parser:(NSXMLParser *)parser
  didStartElement:(NSString *)elementName
  namespaceURI:(NSString *)namespaceURI
  qualifiedName:(NSString *)qualifiedName          ❷ Watches for
  attributes:(NSDictionary *)attributeDict {    ◁─┘  gtopo30

  if ([elementName compare:@"gtopo30"] == NSOrderedSame) {
    gnAlt = [[NSMutableString alloc] initWithCapacity:4];
  }
}

- (void)parser:(NSXMLParser *)parser            ❸ Saves
  foundCharacters:(NSString *)string {   ◁─┘      altitude

  if (gnAlt && string) {
    [gnAlt appendString:string];
  }
}

- (void)parser:(NSXMLParser *)parser
```

```
didEndElement:(NSString *)elementName
namespaceURI:(NSString *)namespaceURI        ❹  Writes
qualifiedName:(NSString *)qName {            ◄─┘   altitude

  if ([elementName compare:@"gtopo30"] == NSOrderedSame) {
    altLabel.text = [NSString stringWithFormat:@"%@ m.",gnAlt];
  }
}
```

In general, this is a pretty simple application of lessons you've already learned. It's also an interesting application of the internet to Core Location.

The only thing particularly innovative comes in the Core Location delegate ❶, where you create a GeoNames URL using the format documented at the GeoNames site. Then you watch the start tags ❷, content ❸, and end tags ❹, and use those to derive altitude the same way that you pulled out XML information when you were reading RSS feeds.

As we mentioned in chapter 17, the result should be an altitude that's much more reliable than what the iPhone can currently provide, unless you're in a tall building, in an airplane, or hang gliding.

To date, all of our examples of web parsing have involved simple GET connections, where you can encode arguments as part of a URL. That won't always be the case on the web, and before we leave web pages behind, we're going to return to some basics of URL requests and look at how to POST information to a web page when it becomes necessary.

20.6   POSTing to the web

Many web pages will allow you to GET or POST information interchangeably. But there will also be situations when that is not the case, and you're instead forced to POST (and then to read back the response manually). In this section, we're going to look at both how to program a simple POST, and how to do something more complex, like a form.

20.6.1   POSTing by hand

When you need to POST to the web, you'll need to fall back on some HTML-related low-level commands that we haven't yet discussed in depth, including NSMutableURL-Request (which allows you to build a piecemeal request) and NSURLConnection (which allows you to extract information from the web).

In general, you'll follow this process:

1 Create an NSURL pointing to the site that you'll POST to.
2 Create and encode the data that you plan to POST, as appropriate.
3 Create an NSMutableURLRequest using your NSURL.
4 Use the NSMutableURLRequest's addValue:forHTTPHeaderField: method to set a content type.
5 Set the NSMutableURLRequest's HTTPMethod to POST.
6 Add your data to the NSMutableURLRequest as the HTTPBody.
7 Create an NSURLConnection using your NSMutableURLRequest.

8 Either immediately capture the return using a synchronous response, or set up a delegate to receive the data as it comes, as defined in the `NSURLConnection` class reference.

9 Parse the `NSData` you receive as you see fit.

For a simple synchronous response, listing 20.7 shows how to put these elements together.

Listing 20.7 A simple `POSTing` example

```
NSURL *myURL = [NSURL URLWithString:@"http://www.example.com"];

NSMutableURLRequest *myRequest = [NSMutableURLRequest
    requestWithURL:myURL];
[myRequest setValue:@"text/xml" forHTTPHeaderField:@"Content-type"];
[myRequest setHTTPMethod:@"POST"];
[myRequest setHTTPBody:myData];

NSURLResponse *response;
NSError *error;
NSData *myReturn = [NSURLConnection sendSynchronousRequest:myRequest
    returningResponse:&response error:&error];
```

A large number of steps are required to move from the URL through to the data acquisition, just as there were when creating a URL for a simple `UIWebView`, but once you have them down, the process is pretty easy. The hardest part, as it turns out, is often getting the data ready to `POST`.

This code will work fine for posting plain data to a web page. For example, you could use it with the Google Spell API found at http://www.google.com/tbproxy/spell to send XML data and then read the results with `NSXMLParser`.

Things can get pretty tricky if you're doing more intricate work than that, such as `POSTing` form data.

20.6.2 *Submitting forms*

Sending form data to a web page follows the same process as any other `POSTed` data, and reading the results works the same way. The only tricky element is packaging up the form data so that it's ready to use.

The easiest way to work with form data is to create it using an `NSDictionary` or `NSMutableDictionary` of keys and values, because that matches the underlying structure of HTML forms. When you're ready to process the data, you pass the dictionary to a method that turns it into `NSData`, which can be sent as an `NSMutableURLRequest` body. Once you've written this method the first time, you can use it again and again.

Listing 20.8 shows how to turn a dictionary of `NSStrings` into `NSData`.

Listing 20.8 Creating form data

```
- (NSData*)createFormData:(NSDictionary*)myDictionary
    withBoundary:(NSString *)myBounds {

  NSMutableData *myReturn = [[NSMutableData alloc] initWithCapacity:10];
```

```
    NSArray *formKeys = [dict allKeys];
    for (int i = 0; i < [formKeys count]; i++) {

        [myReturn appendData:
          [[NSString stringWithFormat:@"--%@\n",myBounds]
            dataUsingEncoding:NSASCIIStringEncoding]];

        [myReturn appendData:
          [[NSString stringWithFormat:
            @"Content-Disposition: form-data; name=\"%@\"\n\n%@\n",
                [formKeys objectAtIndex:i],
                [myDictionary valueForKey:[formKeys objectAtIndex: i]]]
            dataUsingEncoding:NSASCIIStringEncoding]];

    }

    [myReturn appendData:
      [[NSString stringWithFormat:@"--%@--\n", myBounds]
        dataUsingEncoding:NSASCIIStringEncoding]];

    return myReturn;
}
```

There's nothing particularly notable here. If you have a sufficiently good understanding of the HTML protocol, you can easily dump the dictionary elements into an NSData object. The middle appendData: method is the most important one, because it adds both the key (saved in an NSArray) and the value (available in the original NSDictionary) to the HTML body.

Back outside the method, you can add the data to your NSMutableURLRequest just as in listing 20.7, except the content type will look a little different:

```
NSMutableURLRequest *myRequest = [NSMutableURLRequest
    requestWithURL:myURL];
NSString *myContent = [NSString stringWithFormat:@"multipart/form-data;
    boundary=%@",myBounds]
[myRequest setValue:myContent forHTTPHeaderField:@"Content-type"];
[myRequest setHTTPMethod:@"POST"];
[myRequest setHTTPBody:myReturn];
```

Some other types of data processing, such as file uploads, will require somewhat different setups, and you'd do well to look at HTML documentation for the specifics, but the general methods used to POST data will remain the same.

With POSTing out of the way, we've now covered all of the SDK's most important functions related to the internet. But there's one other topic that we want to touch upon before we close this chapter—a variety of internet protocols that you can access through third-party libraries.

20.7 *Accessing the social web*

Since the advent of web 2.0, a new sort of internet presence has appeared. We call it the *social web*. This is an interconnected network of web servers that exchange information based on various well-known protocols. If you're building internet-driven programs, you may wish to connect up to this web so that your iPhone users can become a part of it.

20.7.1 *Using web protocols*

In order to participate in the social web, clients need to speak a number of protocols, most of them built on top of HTML. These include Ajax, JSON, RSS, SOAP, and XML. Here's how to use each of them from your iPhone:

- *Ajax*—Ajax is something that—as it turns out—you can largely ignore on the iPhone. It's usually used as part of a client-server setup, with HTML on the front side, but the iPhone uses an entirely different paradigm. You can dynamically load material into labels or text views, and you can dynamically call up websites using the XML or HTML classes we've discussed. There's no need for Ajax-type content as long as you have good control over what a server will output. You just need to remember some of the lessons that Ajax teaches, such as downloading small bits of information, rather than a whole page.

- *JSON*—JSON is perhaps the most troublesome protocol to integrate. It's quite important as a part of the social web, because it's one of the standardized ways to download information from a web site. It also depends on your iPhone being able to understand JavaScript—which it doesn't (unless you do some fancy work with DOM and the WebKit, which are beyond the scope of this section). Fortunately, there are already two JSON toolkits available: JSON Framework and TouchJSON. We'll look at an example of the latter shortly.

- *RSS*—At the time of this writing, we're not aware of any RSS libraries for the iPhone. But as we've already demonstrated in this chapter, it's quite easy to parse RSS using an XML parser.

- *SOAP*—SOAP isn't as popular as most of the other protocols listed here, but if you must use it, you'll want a library. There are a number of SOAP libraries written for Objective-C (though not necessarily for the iPhone), including SOAP Client and ToxicSOAP.

- *XML*—XML is, as we've already seen, fully supported by the iPhone OS, but if you don't like how the default parser works and want an alternative, you should look at TouchXML.

These libraries should all be easy to find with simple searches on the internet, but table 20.10 lists their current locations as of this writing.

Library	Location
JSON Framework	http://code.google.com/p/json-framework/
TouchJSON	http://code.google.com/p/touchcode/
Soap Client	http://code.google.com/p/mac-soapclient/
ToxicSOAP	http://code.google.com/p/toxic-public/
TouchXML	http://code.google.com/p/touchcode/

Table 20.10 Download sites for social protocol libraries

Because of its importance to the social web, we're going to pay some additional attention to JSON, using the TouchJSON library.

20.7.2 Using TouchJSON

For our final example, we're going to return to Core Location one more time, because GeoNames offers a lot of JSON information. You're going to use GeoNames to display the postal codes near a user's current location. Figure 20.5 shows our intended result by highlighting the postal codes near Apple headquarters, the location reported by the iPhone Simulator.

In order to get to this point, you must first install this third-party library and make use of it.

INSTALLING TOUCHJSON

To integrate TouchJSON into your project, you must download the package from Google and move the source code into your project. The easiest way to do this is to open the TouchJSON download inside Xcode and copy the Source folder to your own project. Tell Xcode to copy all the files into your project as well. Afterward, you'll probably want to rename the copied folder from Source to TouchJSON.

Then you need to include the header CJSONDeserializer.h wherever you want to use TouchJSON.

Figure 10.5   **It's easy to extract data using TouchJSON.**

USING TOUCHJSON

In order to use TouchJSON, you pass the CJSONDeserializer class an NSData object containing the JSON code. Listing 20.9 shows how to do so. In this example, this work occurs inside a location manager delegate. It's part of a program similar to our earlier GeoNames example, but this time we're looking up postal codes with a JSON return rather than altitudes with an XML return.

Listing 20.9   Using TouchJSON

```
- (void)locationManager:(CLLocationManager *)manager
    didUpdateToLocation:(CLLocation *)newLocation
    fromLocation:(CLLocation *)oldLocation {

  [myLM stopUpdatingLocation];
  [myActivity stopAnimating];

  NSString *gnLookup = [NSString
      stringWithString:@"http://ws.geonames.org/findNearbyPostalCodesJSON
        ?lat=37.331689&lng=-122.030731"];

  NSData *gnData = [NSData dataWithContentsOfURL:
      [NSURL URLWithString:gnLookup]];        ❶
  NSError *error = nil;
```

```
NSDictionary *dictionary = [[CJSONDeserializer deserializer]
    deserializeAsDictionary:gnData error:&error];      ❷

if(error) {
    postalLabel.text = [NSString stringWithFormat:@"Error: %@",
        [[error userInfo] objectForKey:@"NSLocalizedDescription"]];
} else {
    NSMutableString *postCodes = [NSMutableString
        stringWithString:@"Nearby post codes are:\n\n"];

    for (int i = 0 ;
        i < [[dictionary objectForKey:@"postalCodes"] count] ;
        i++) {

        [postCodes appendFormat:@"%@ (%@)\n",
            [[[dictionary objectForKey:@"postalCodes"] objectAtIndex:i]
                objectForKey:@"postalCode"],
            [[[dictionary objectForKey:@"postalCodes"] objectAtIndex:i]
                objectForKey:@"placeName"]];
    }
    postalLabel.text = postCodes;
}
}
```

To access the JSON results, you first retrieve the data from a URL using the datawith-ContentsOfURL: method ❶, which was one of the ways we suggested for retrieving raw data earlier in the chapter. Then you plug that NSData object into the CJSONDeserializer ❷ to generate an NSDictionary containing the JSON output.

The TouchJSON classes are much easier to use than the XML parser we met earlier in this chapter. All you need to do is read through the arrays and dictionaries that are output. The downside is that the resulting dictionary may take up a lot of memory (which is why the XML parser didn't do things this way), so be aware of that if you're retrieving particularly large JSON results.

Absent that concern, you should be on your way to using JSON and creating yet another link between your users' iPhones and the whole world wide web.

20.8 Summary

"There's more than one way to do it."

That was the slogan of Perl, one of the first languages used to create dynamic web pages, and today you could equally use that slogan to describe the iPhone, one of the newest cutting-edge internet devices.

We opened this book by talking about the two different ways that you could write iPhone programs: using web technologies and using the SDK. That also highlighted two different ways that you could interact with the internet: either as an equal participant—a web-based member of the internet's various peer-to-peer and client-server protocols—or as a pure iPhone client that runs its own programs and connects to the internet via its own means.

We've said before that each programming method has its own advantages, and we continue to think that web development is often a better choice when you're interacting with the internet already, but when you need to use other SDK features, the SDK offers some great ways to connect to the web.

As we've seen in this chapter, you have easy and intuitive access to the social web—that conglomeration of machines that's connected via various public protocols. You should have no trouble creating projects that use the HTML and XML protocols, and even further flung protocols like JSON and SOAP are usable thanks to third-party libraries. That'll cover most programmers' needs, but for those of you who need to dig deeper, the SDK has you covered there too, thanks to Core Foundation classes.

In coming full circle, returning to the web technologies that opened this book, we're also bringing this book to its end. But read on (if you haven't already) for some helpful appendixes that list numerous SDK objects, talk more about getting your projects ready for market, and point toward other resources.

appendix A:
iPhone OS class reference

After this book, your main resource for learning more about the iPhone should be the references at developer.apple.com. To help you find documents that might interest you, this appendix lists the major classes in the UIKit and Foundation hierarchies that you might want to know more about, excluding classes that only appear as a part of another class.

A.1   *UIKit framework classes*

The UIKit framework contains those classes most tightly connected to the iPhone, including all of the graphical classes you use to make up pages. A partial listing appears as table A.1. It's current as of iPhone OS 2.1, and will probably be mostly correct when you read this, but the UIKit does sometimes change between releases.

Table A.1   A listing of the most important User Interface classes

Class	Parent	Summary
UIActionSheet	UIView	A pop-up window that includes options; similar to a UIAlertView
UIActivityIndicatorView	UIView	An indeterminate progress display
UIAlertView	UIView	A pop-up window that includes options; similar to a UIActionSheet
UIApplication	UIResponder	The main source for application information and control
UIButton	UIControl	A push button
UIColor	NSObject	A color output class

Table A.1   A listing of the most important User Interface classes *(continued)*

Class	Parent	Summary
UIControl	UIView	An abstract class that is parent to many user controls
UIDatePicker	UIControl	A wheeled date-selection device
UIDevice	NSObject	A class that holds info about the iPhone itself
UIEvent	NSObject	A container for touches; part of the event model
UIFont	NSObject	A font output class
UIImage	NSObject	A non-displaying image holder
UIImagePickerController	UINavigationController	A modal controller for image selection
UIImageView	UIView	An image display that holds one or more UIImage objects
UILabel	UIView	A small, non-editable text display
UINavigationController	UIViewController	A hierarchical controller; often linked with a UITableView-Controller to produce hierarchical menus
UIPageControl	UIControl	A toolbar for navigating among pages using dots
UIPickerView	UIView	A wheel-based selection mechanism
UIProgressView	UIView	A determinate progress display
UIResponder	NSObject	An abstract class that defines all classes that can receive and respond to events
UIScreen	NSObject	A class containing an iPhone's entire screen
UIScrollView	UIView	A parent class for views with multiple pages of content
UISearchBar	UIView	A text-input mechanism specialized for searches
UISegmentedControl	UIControl	A control for making one of several choices
UISlider	UIControl	A control for setting discrete values

Table A.1   A listing of the most important User Interface classes

Class	Parent	Summary
UISwitch	UIControl	A control for selecting binary values
UITabBarController	UIViewController	A controller for moving among multiple screens
UITableViewController	UIViewController	A controller for displaying tables of content; often linked with a UINavigationController
UITextField	UIControl	A control for inputting short text
UITextView	UIScrollView	A display for text of any size
UITouch	NSObject	An individual touch on the iPhone's screen
UIView	UIResponder	The abstract class that lies at the core of most UIKit objects
UIViewController	UIResponder	A simple view controller
UIWebView	UIView	A Safari-like web browser
UIWindow	UIView	The root for the view hierarchy

A.2   *Foundation framework classes*

Foundation framework classes, whose names begin with *NS*, are almost as important as the UI classes because they represent foundational variable types, like strings and numbers. Table A.2 only lists the major classes that have some relevance to the sort of work you've done in this book; for more, look at Apple's developer site under "Core Services" frameworks.

Table A.2   A listing of the most important Foundation classes

Class	Parent	Summary
NSArray	NSObject	An array
NSAutoreleasePool	NSObject	A memory-management class
NSBundle	NSObject	A pointer toward a project's file system home
NSCharacterSet	NSObject	Methods for managing characters
NSCountedSet	NSMutableSet	An unordered collection of elements
NSData	NSObject	A wrapper for a byte buffer
NSDictionary	NSObject	An associative array
NSError	NSObject	Encapsulated error information

Table A.2   A listing of the most important Foundation classes *(continued)*

Class	Parent	Summary
NSFileHandler	NSObject	
NSFileManager	NSObject	
NSIndexPath	NSObject	
NSLog	NSObject	
NSMutableArray	NSArray	
NSMutableCharacterSet	NSCharacterSet	
NSMutableData	NSData	Data that can be changed
NSMutableDictionary	NSDictionary	A dictionary that can be changed
NSMutableSet	NSSet	A set that can be changed
NSMutableString	NSString	A string that can be changed
NSMutableURLRequest	NSURLRequest	A URL request that can be changed
NSNotificationCenter	NSObject	A notification manager
NSNumber	NSValue	A way to encapsulate many types of numbers
NSObject	N/A	The root class for Cocoa Touch
NSString	NSObject	A class for various sorts of string storage and manipulation
NSURL	NSObject	A simple URL object
NSURLRequest	NSObject	A URL plus a cache policy
NSValue	NSObject	A simple container for data
NSXMLParser	NSObject	An XML parser

A.3   *Other classes*

The UI and NS classes should contain most of the objects you use when programming.

We've also covered several other frameworks throughout this book, including the Address Book framework (chapter 16), the Address Book UI framework (chapter 16), the Core Location framework (chapter 17), the Core Audio framework (chapter 18), the Media Player framework (chapter 18), the Core Graphics framework (chapters 18 and 19), the Quartz Core framework (chapter 19), the OpenGL ES framework (chapter 19), and the CFNetwork framework (chapter 20). Finally, you may wish to pay some attention to the Core Foundation framework, which we've used (as infrequently as possible) throughout part 4 of this book.

appendix B:
External sources and references

What follows are web resources that we suggest for continuing your exploration of iPhone development.

B.1 General resources

Site	URL	Summary
The Apple Blog	http://theappleblog.com	General Apple blog, including some iPhone discussion
iPhone Atlas	http://www.iphoneatlas.com	IPhone news blog
iPhone Dev Forums	http://www.iphonedevforums.com	Forums for SDK or web discussion
iPhone in Action	http://iphoneinaction.manning.com/	The authors' blog for this book; we'll keep you up to date with new links of interest and occasionally cover some of the topics that we didn't cover in this book
iPhone in Action on Magnolia	http://ma.gnolia.com/people/iPhoneInAction/	The authors' listing of links of note

B.2 Web app resources

Site	URL	Summary
Apple Developer Connection	http://developer.apple.com/webapps/	The official Apple site for developer resources; requires ADC login
iPhoneWebDev	http://www.iphonewebdev.com/	The authors' own site, complete with examples and webdev discussion list
WebKit Open Source Project	http://webkit.org	WebKit home, including the Surfin' Safari blog

B.3   SDK resources

Site	URL	Summary
Apple Developer Site	http://developer.apple.com/iphone/	The official Apple site for developer resources; requires ADC login
Apple Developer Forums	https://devforums.apple.com/community/iphone	Official Apple forums for SDK discussion, including betas; requires ADC login
Apps Amuck	http://www.appsamuck.com/	Thirty-one programs with source code in 31 days
Cocoa Dev Central	http://cocoadevcentral.com/	A hub of Objective-C and Cocoa information
Cocoa Is My Girlfriend	http://www.cimgf.com/	News and tutorial blog
Cocoa Samurai	http://cocoasamurai.blogspot.com/	Cocoa and iPhone discussion
Furbo.org	http://furbo.org/	General blog that's mostly iPhone discussion
iDevKit	http://idevkit.com/	Forums and news
iPhone Dev SDK	http://www.iphonedevsdk.com/	Forums
iPhone Development	http://iphonedevelopment.blogspot.com/	Blog with extensive original content
iPhone Development Central	http://www.iphonedevcentral.org/	Online iPhone tutorials
Lap Cat Software Blog	http://lapcatsoftware.com/blog/	Coding blog that's mostly about iPhone and Cocoa
Mobile Orchard	http://www.mobileorchard.com/	A news blog
Safe from the Losing Fight	http://www.losingfight.com/blog/	A blog about Macs with some emphasis on iPhones

B.4   Other technologies

Site	URL	Summary
JavaScript.com	http://www.javascript.com/	A comprehensive JavaScript site
SQLite	http://www.sqlite.org/	The official SQLite site
W3C XML	http://www.w3.org/XML/	The official XML site

appendix C:
Publishing
your SDK program

All of your programming will be for naught if you don't sign up for the iPhone Developer Program with Apple. This is a multistep process that can take quite some time, so make sure to get it all in hand well before you want to upload your program to the iPhone App Store.

C.1 Signing up with Apple

To get started, you must register as a developer at developer.apple.com/iphone/ program. When you register, you'll be asked for some basic information about what you'll be developing and you'll need to sign Apple's Terms & Conditions for working with the iPhone. You've probably already done this step, as it was required to get access to the SDK and the online documentation.

Sometime afterward—maybe in a few hours, maybe in a few weeks—you'll get a call from Apple confirming your signup information and giving you the OK for the program. They'll then send you an email that'll allow you to finish your registration. At this point, expect to pay a fee, currently $100 (standard) or $300 (enterprise), to become a full-fledged developer. The standard program allows for distribution via the iPhone App Store, while the enterprise program allows distribution of in-house applications to over 500 employees.

C.2 Compiling to the iPhone

The first advantage of being a registered iPhone developer is that you'll be able to compile programs directly to your iPhone. This is fairly critical for certain types of testing. As we've seen in this book, features like altitude detection, volume control, and the accelerometer don't work correctly when tested in the iPhone Simulator.

To compile to an iPhone, you must create a provisioning profile, which is a multi-step process. You'll need to use some new tools that will appear under a Program Portal link at the top of developer.apple.com once you've finished your signup and paid your fee. Apple has a complete "iPhone Developer Program Portal User Guide" that explains how to use everything here, but we're going to outline the main steps:

1 *Add team members (admin)*—If you registered as a company, you can add additional team members under the Team tab. The initial creator of a team will be the Team Agent, who has the highest-level powers in the Developer Program; other users will be Team Administrators or Team Members. From here, individual members can set themselves up to compile to their iPhones, with some steps requiring interventions from Team Admins.

2 *Create a certificate signing request (member)*—This is the first step required to generate the certificate you'll need to sign (and thus run) applications on your iPhone. You create a certificate signing request (CSR) inside Keychain Access on your Mac and then upload it from the Certificates > Development tab; a Team Admin must then approve it.

3 *Download a certificate (member)*—Once your Admin (who might be you) has approved your CSR, you can download a certificate. From Certificates > Development, download the WWDR Intermediate Certificate and double-click to install it. Afterward, download your developer certificate and double-click to install it.

WARNING Your certificate will now be permanently installed in your keychain. However, if you rebuild your machine or move to a new machine, you'll lose it. To avoid this, be sure you export the private key associated with your developer certificate. You can then import it on a different machine, and redownload the two certificates from Apple. If you fail to do this, a Team Admin may need to revoke your certificate so that you can create a new one.

4 *Add devices (admin)*—Add any devices (iPhones or iPods) that you want to build on using the Devices tab.

5 *Create an app ID (admin)*—Each application needs an app ID, which controls its access to devices. For the purposes of testing, you'll probably just use one general wildcard ID that you create by appending a wildcard (.*) to your app ID bundle identifier.

6 *Create a provisioning profile (admin)*—A provisioning profile is a unique combination of multiple developer certificates, multiple iPhone device IDs, and a single app ID. It's what ties your iPhone to your overall development profile and what allows you to actually run programs. You create a provisioning profile from the Provisioning > Development tab, at which point you'll be asked to enter the three elements that make it up.

7 *Download a provisioning profile (member)*—Download the profile from Devices and drag it your Xcode dock icon or the organizer window of Xcode.

Though the setup can be a bit extensive for an admin, once the initial work is done, a member can just create a CSR, download a certificate, and download a provisioning profile. From that point, the member can choose to compile onto a device rather than to the iPhone Simulator by changing the pop-up window at the top-left of Xcode.

C.3　*Preparing for distribution via the iPhone App Store*

Preparing your program for distribution via the iPhone App Store follows much the same process as preparing your programs for testing on iPhones, except that the steps can only be undertaken by the Team Agent.

1 *Create a certificate (agent)*—As before, you must upload a CSR, but here you should create a certificate from the Certificates > Distribution tab, rather than Certificates > Development.

2 *Create a provisioning profile (agent)*—Create a provisioning profile in the Provisioning > Distribution tab. It will usually be an App Store profile. As before, drag your new profile to Xcode.

3 *Prepare to compile (anyone)*—Create a new Distribution configuration that uses the distribution provisioning profile. Update other info in the configuration, update your Info.plist as appropriate, and then build.

4 *Prepare media (anyone)*—Prepare a 57x57 PNG home screen icon, a 512x512 JPG/ TIF large application icon, and a full-screen screenshot, as well as other information required by the iPhone App Store.

5 *Upload (agent)*—Go to the Distribution tab and run iTunes Connect.

6 *Wait (everyone)*—It'll take a bit of time for your application to be approved and go on sale.

These procedures may well change over time, but for now, this should be what you need to get your program from your desktop to the iPhone App Store.

There are also two alternative ways to distribute your software: enterprise distribution lets you distribute an in-house application to employees within your company, and ad hoc distributions let you distribute to up to 100 other iPhone users by email or a website. Both are explained further in Apple's documentation.

index

Symbols

@end 200
@implementation 195, 200
@interface 200
@property 200
@selector syntax 254
@synthesize 200
*.framework 192
*.h 192
*.m 192
*.mm 192
#import 158, 200
#import directive 193
#include command 158

Numerics

3G network 6

A

ABAddressBook 314
ABAddressBookCopyArrayOfAll
    People 317
ABAddressBookCreate 317
ABCreateMutableCopy 315
ABGroup 314
ABMultiValue 314
ABMultiValueCopyLabelAt-
    Index 315
ABMultiValueCopyValueAt-
    Index 315
ABMultiValueGetCount 315

ABMultiValueReplaceLabelAt-
    Index 315
ABMultiValueReplaceValueAt-
    Index 315
ABMutableMultiValue 314
ABNewPersonViewController
    314, 318, 321
ABPeoplePickerNavigation-
    Controller 314, 318–321
ABPerson 314, 317
ABPersonViewController
    314, 318, 322
ABRecord 314
ABRecordCopyCompositeName
    317
ABRecordCopyValue 315, 317
ABRecordRef 321
ABRecordSetValue 315
absolute values, problems 34
absolutes 39–40
ABUnknownPersonView-
    Controller 314, 318, 322
accelerometer 8, 11, 18, 75
    filtering, and 328
    force, measuring 327
    gestures, and 333–335
    gravity, measuring 328–329
    iPhone Simulator, and 326
    movement, and 326–333
    movement, basic,
        detecting 331
    orientation, and 325–326
    rhythm 331
    shared action 328
Accelerometer Graph 330
    shake characteristics 333

accelerometer:didAccelerate:
    327–328, 335
Accessibility 21
Accessory view 274
Action target listed as nil 254
Actions in use 261
ad hoc distribution 431
addAnimation:forKey: 393
addColorStop 113
addEventListener 71
addObserver:selector:name:
    object: 262
Address Book 313–322
    and Core Foundation 321
    Apple tutorial 313
    classes 314
    contacts, extracting 317
    framework 313
    getter functions 315
    include files 313
    individuals, extracting 317
    memory management 321
    people-picker view
        controller 318
    properties, accessing 314–316
    querying 316–318
    setter functions 315
    UI framework 313, 318–322
    view controllers 318, 321
addSubview 198
addTarget:action:forControl-
    Events: 255
addTarget:action:forControl-
    Events: method 254
affine transformation, in
    Quartz 378

Ajax 419
    introduction 88
    reloading pages 152
allHTTPHeaderFields 401
allTouches 244
altitude
    Core Location, and 340
    iPhone Simulator, and 340
    monitor 342
    original iPhone, and 340
always-on Internet 10, 83
animation 64
    applying to layer 393
    block, defining 393
    explicit 392–393
    implicit 392
    in Canvas 121
    key frame 392
    key frame vs. basic 394
animationDuration 346
animationImages 346
animationRepeatCount 346
Apache, using locally 143
app delegate database 312
app ID, creating 430
App Store, preparing app for
        distribution 431
Apple
    Blog 427
    Developer Connection 427
    Developer Forums 428
    Developer Site 428
    docs 32
    registering as developer
        with 429
application bundle 297
application delegate 194
    linking a new class 202
    tasks 194
applicationDidFinishLaunching
        194–195, 202, 268, 313
Apps Amuck 428
arc 107
arcTo 107–108
arrayForKey 297
Assisted GPS (A-GPS) 8
attribute, iUI 93
Attributes tab 245, 267
audio file ID 363
Audio File Services 360
    Celestial 360
Audio File Stream Services 360
audio queue 363
    AudioQueueFlush 364
    AudioQueuePause 364

AudioQueuePrime 364
AudioQueueReset 364
AudioQueueStart 364
AudioQueueStop 364
buffer 363
callback 363
functions 364
playing from 363
recording 364
Audio Queue Services 360, 362
Audio Session Services 360
Audio Toolbox 361
    framework 361
AudioServicesAddSystemSound
    Completion 361
AudioServicesCreateSystem-
    SoundID 361
AudioServicesDisposeSystem-
    SoundID 361
AudioServicesPlaySystemSound
    361
AudioServicesRemoveSystem-
    SoundCompletion 361
autocompletion 204
Autoresize 133
autoresizesSubviews 228
autoresizingMask 228
    properties 228
    UIViewAutoresizingFlexible-
        BottomMargin 229
    UIViewAutoresizingFlexible-
        Height 229
    UIViewAutoresizingFlexible-
        LeftMargin 229
    UIViewAutoresizingFlexible-
        RightMargin 229
    UIViewAutoresizingFlexible-
        TopMargin 229
    UIViewAutoresizingFlexible-
        Width 229
    UIViewAutoresizingNone 229

B

back button 82
backButton 86
Background 57
bandwidth 151
bar button items 276
becomeFirstResponder 242, 250
beginAnimations:context: 393
beginIgnoringInteractionEvents
    UIApplication method 250
beginPath 106
Bezier curve 108

bezierCurveTo 107, 109
Bibeault, Bear 99
bitmap
    drawing on 383
    drawing to 370
bitmask 253
blueButton 92
Bluetooth 8
<body> 105
Bonjour, accessing 398
bookmarklet 150–151
bottom bar 46
    eliminating 69
Box 57
Brisbin, Jon 150
browsing paradigm 83
BSD socket 397
bubble 14, 47
bundle 293
    accessing 298
    application 293, 297
    framework 293
    settings 293
button 86, 92, 286
    iUI 92

C

C 18, 155
.c (source) files 158
C Concept 155
C# 18
C++ 18
CABasicAnimation 393
CAEAGL 394
CALayer 348, 391
    anchorPoint 392
    backgroundColor 392
    opacity 392
    position 392
    properties 391
    transform 392
camera 8
    resource usage 350
canBecomeFirstResponder 251
canResignFirstResponder 251
Canvas 18, 77, 197
    animation 121
    basic setup 104
    Bezier curves 108
    <body> 105
    browsers 104
    clipping 114–116
    clipping paths 116
    color styles 112

Canvas *(continued)*
compatibility 103
composition, modifying 114–116
curve commands 107
drawing order 121
enabling 103
first steps 103–105
global variables 115
gradient styles 112
graphics, accessing 103
image commands 118
images 117–120
integrating with Dashcode 141
Internet Explorer, and 104
line styles 114
onload attribute 105
path commands 106
paths 105–109
pattern commands 119
patterns 117–120
rectangles 110
restoring shapes 116–117
shape functions 110
shapes 110–111
controlling 114–115
drawing order 110
state stacking 117
styles 112–114
text 117, 119–120
transformations 116–117
transparency 114–116
viewport metatag 105
<canvas> 57, 103
Canvas graphic library 24
carbon-based frameworks 157
CATransform3DMakeAffine-Transform 393
cc 158
Celestial 360
certificate
creating 431
downloading 430
signing request, creating 430
CFBundle 293
CFFTPStream 398
CFHost 398–399
CFHTTPMessage 398
CFHTTPStream 398
CFNetServices 398
CFNetwork 398
CFStringRef 157
CFURLRef 399
CGAffine 378, 385

CGColorSpaceCreateWithName 381
CGColorSpaceRelease 382
CGContextAdd 371–372
CGContextBeginPath 371
CGContextClearRect 374
CGContextClip 373
CGContextClosePath 373
CGContextConcatCTM 378
CGContextDraw 381–383
CGContextFillPath 373
CGContextFillRect 374
CGContextMoveToPoint 372
CGContextRestoreGState 375
CGContextRotateCTM 377
CGContextSaveGState 375
CGContextScaleCTM 377
CGContextSelectFont 384
CGContextSetAlpha 380
CGContextSetBlendMode 380
CGContextSetFillColorWith-Color 376
CGContextSetFlatness 380
CGContextSetFont 385
CGContextSetLine 380
CGContextSetRGBFillColor 376
CGContextSetRGBStrokeColor 376
CGContextSetSetPosition 385
CGContextSetShadow 380
CGContextSetShadowWithColor 380
CGContextSetStrokeColorWith Color 376
CGContextSetTextDrawing-Mode 384
CGContextSetTextMatrix 385
CGContextShowText 385
CGContextShowTextAtPoint 385
CGContextStrokePath 373
CGContextStrokeRect 375
CGContextTranslateCTM 377
CGGradientCreateWithColors 381
CGGradientRef 381
CGGradientRelease 382
CGPath 373
CGPathCloseSubpath 373
CGPathCreateMutable 373
CGPathMoveToPoint 373
CGPDFContextCreate 369
CGRectMake 197
CGShadingRef 381
changedTouches 71

chrome 81
action buttons 82
adjusting the chrome 69
CJSONDeserializer 420–421
class controls 204
clearRect 110
click 48
client-server development 27
client-side database 66, 68
clientX 72
clientY 72
clip 114, 116
clipping path
closing 373
setting, in Quartz 379
clipping, in Canvas 114–116
CLLocation 336–337
altitude 337
coordinate 337
timestamp 337
CLLocationManager 336
location 337
startUpdatingLocation 337
stopUpdatingLocation 337
CLLocationManagerDelegate 336–337
didFailWithError: 337
didUpdateToLocation: 337
fromLocation: 337
locationManager: 337
closeFile 300
Cocoa Dev Central 428
Cocoa Is My Girlfriend 428
Cocoa Samurai 428
Cocoa Touch
Plugin 208
Quartz, and 367
Cocoa's Foundation framework 157
code folding 204
collage 348
example 351–356
temporary image view 354
view 355
view controller 351–354
collageView 351
collageViewController 351
color
setting, in Quartz 375
styles, in Canvas 112
color space 375
color stop 113
colspan 43
Column Layout 140

columns 41
commitAnimations 393
compile at runtime 158
compile in Xcode 192
compiler directive 158
compiling 158
   to iPhone 429
complex structures 159
composition, in Canvas 114–116
contentForMenuWithParent 310
contentsAtPath 300
context
   graphical 348
   in Quartz 367–371
controlChange 339
coordinate system 368
Core Animation 391–394
   explicit animation 392–393
   fundamentals 391
   implicit animation 392
   key frame animation 392
   layer 391
Core Audio frameworks 360
Core Foundation 157
   and Quartz 367
   and the Address Book 321
Core Graphics
   drawing simple images 347
   functions for 383
Core Location 335–342
   GeoNames example 414–416
   Google Maps example 405–407
   Internet, and 342
   iPhone Simulator, and 335
   TouchJSON example 420
   using altitude 340–342
   using location and
      distance 337–340
Core Services
   framework classes 425
CoreGraphics 202
countForMenuWithParent 310
create new class, steps to 199
CREATE TABLE 67
createLinearGradient 113
createPattern 118
createRadialGradient 113
creating
   new objects as subviews of
      your tab bar controller 266
   nextResponder function in a
      subclass 251

tab bar controller by
   hand 266
tab bar through a
   template 267
CSS 12, 16, 39
   gradients 76
   masks 77
   optimizing 44
CSS positioning 40
   absolute 40
   fixed 40
   relative 40
   static 40
CTM
   transformation 376
   transformation functions 377
current transformation matrix
   (CTM). See CTM
curve commands 107
custom audio structure 363
customizableViewControllers
   271
cut and paste 14

D

Dashcode 24, 32
   action buttons 134–135
   alternative to iUI 101
   anatomy 126
   Attributes inspector 129
   Behaviors inspector 129
   Browser 140
   Browser template 126, 135
   Canvas 140
   canvas 127
   Code Library 130
   Custom template 126
   Debugger 130
   Edge-to-Edge List 140
   Fill & Stroke inspector 129
   graphical orientation
      gauge 132
   handlers, writing 135
   history 125
   input from objects 134
   inspector 126, 128
      tabs 129
   integrating
      with Canvas 141
      with existing libraries
         140–142
      with iUI 141
      with WebKit 140
   integration, deep 142

introduction 125–130
iPhoto interface 130
JavaScript integration 128
Library 126
library 129
library parts 132–134
listController 136
lists 136
main screen 126
Metrics inspector 129
navigator 126, 128
   Application Attributes 128
   Home Screen Icon 128
   object hierarchy 128
   Share 128
object Behavior tab 134
orientation code 133
outputting to a part 132–134
Parts Library 129
photos 130
Podcast template 126
program, writing 131–140
project
   deploying 130
   running 130
   saving 131
   starting 126
Quartz Composer 140
QuickTime 140
resizing objects 133
Rounded-Rectangle List 140
RSS template 126
Run button 130
Share button 130
source code window 126, 128
Stack Layout 140
stackLayout 136–139
   building outside 137
   creating 136
   getAllViews 137
   getCurrentView 137
   manipulating 137
   methods 137
   populating 137
   setCurrentView 137
   setCurrentViewWith-
      Transition 137
   variable views 139
tab bar 138
templates 126
   choosing 126
   SDK equivalents 126
Text inspector 129
top bar 126–127
transitions 138

Dashcode *(continued)*
  User Guide 125, 138
  Utility template 126
  viewports 133
  widgets 125
data paradigm 83
data, non-HTML, capturing 401
database 65
  building navigation menu
    from 306–313
  loading a database 65
  transaction 65
  transaction handlers 66
database view controller 311
DatabaseViewController 307
dataHandler 66
datawithContentsOfURL 421
dealloc 201, 310
deallocate 157
debugging
  with desktop browser
    144–149
  with Firefox 146–148
  with iPhone Debug 150
  with iPhone Simulator 148
  with iPhone Web
    Developer 150
  with Safari 144
  with your iPhone 149–151
declarations 156
default element 87
delegate connection 217
delegateClassName 194
delegation 217, 241
dereference 157
design element 83
  always-on Internet 83
  input 84
  location awareness 84
  orientation awareness 84
  output 84
  power consciousness 84
desktop Internet, browsing
  paradigm 83
detectsPhoneNumbers 402
developer
  registering with Apple 429
development
  web resources 427
Development Methodologies 23
deviceDidRotate 326
device-width 35
dialog, iUI 89
didEndElement 409

didReceivedMemoryWarning
  157
didRotateToInterface-
  Orientation 230
didStartElement 409
directory, accessing 299
dismissModalViewController-
  Animated 281
display text using the SDK 257
distance, and Core
  Location 337
distribution
  ad hoc 431
  enterprise 431
  via iPhone App Store 431
doc lookup 205
doc sets 205
documents directory 297
  files retrieving 299
DOM Inspector 146
dot shorthand 198
do-while 159
drawAsPatternInRect 348
drawAtPoint 348
drawAtPoint:blendMode:alpha:
  348
drawImage 118
drawInRect 348
drawInRect:blendMode:alpha:
  348
drawRect 348, 356, 368
Drosera 145
dynamism 20

E

EAGL 394
EAGLView 394
EDGE network 6
editing window 204
endGeneratingDeviceOrientation-
  Notifications 326
endIgnoringInteractionEvents
  250
energy consciousness 152
enterprise distribution 431
enumerated 253
Error Console 145
error.code 66
error.message 66
errorHandler 66
event 21, 70, 240, 242
  regulating 242
  unhandled 241

event encoding 242
event handlers 71
eventreporter 245
exclusiveTouch 251
  UIView property 250
executeSql 65
  dataHandler 66
  errorHandler 66

F

Facebook 50
Fettig, Abe 150
<fieldset> 90
file
  copying 299
  manipulating 300–301
  NSData, writing to 300
  NSString, writing to 301
  opening 297, 300–303
  retrieving from Documents
    directory 299
file manager 300
fileHandleForReadingAtPath
  299–300
fileHandleForWritingAtPath
  300
filesaver example 302–303
fileURLWithPath 399, 407
fill 106
fillRect 110
fillStyle 112
fillText 119
filtering
  and accelerometer 328
  movement 329
  high pass 330
Firebug 146–147
  console object 148
  console.assert 148
  console.dir 148
  console.dirxml 148
  console.log 148
  console.profile 148
  console.trace 148
  debugging with 148
Firefox 56
  add-ons 146
  CSS-related functions 146
  debugging with 146–148
  DOM Inspector 146
  download sites 146
  Error Console 146
  Firebug 147–148
  Firebugsites 146

Firefox *(continued)*
  forms functions 146
  Web Developer 146
first responder 241–242
FirstViewController class 267
fixed 40
Flash 7, 37
flick 14, 47
  two-finger 14
flipside controller 223, 279
  creating preferences 289
  RootViewController 279
  Using 280
FlipsideViewController 279
font types 40
force
  measuring with
    accelerometer 327
  start 331
  stop 331
form, submitting 417
formfield.onclick 38
formfield.onmousedown 38
formfield.onmousemove 38
formfield.onmouseout 38
formfield.onmouseover 38
formfield.onmouseup 38
Foundation 202
foundation framework
    classes 425
foundCharacters 409
frame 197
framesets 49
framework bundle 293
framework classes 423
  foundation 425
frameworks 21
Frameworks folder 202
free() 157
freemium 26
FTP server, communicating
    with 398
function 156, 197
function prototypes 158
Furbo.org 428

G

gcc 158
GeoNames 414–416
  findNearestIntersection 414
  gtopo30 414
  srtm3 414
  timezone 414
  TouchJSON example 420

gesture 14, 70
  and accelerometer 333–335
  gesturechange 70, 74
  gestureend 70
  gesturestart 70, 74
  recognizing 333
getAllViews 137
getCurrentView 137
getDistanceFrom 340
getElementById 118
g-force 327
global variable states 117
globalAlpha 112, 114–115
globalCompositeOperation 114
  values 115
globalization 20
Gmail 50
goBack 402
goForward 402
Google 11
Google Maps
  Core Location example
    405–407
  hybridizing 406
googleSearch 406
goto 159
GPS 8, 11, 18
  accessing 335
  built-in 336
  location aware 11
gradient
  color stop 113
  drawing, in Quartz 381–382
  styles, in Canvas 112
graphic state
  methods 379
  setting in Quartz 375–381
graphical context 348
  in Quartz 367
  methods 369
  stack 368
graphical interface 81–83
  back button 82
  choices, separating 82
  chrome 81
    action buttons 82
  guidelines 81
  look and feel 81
  sliding menus 82
  viewports 82
  zooming 82
graphical state, managing 380
gravity, checking 328
grayButton 93

H

.h (or header) files 158
Hahlo twitter client 20
Harte, Erwin 150
Hello, World! 196
helloworldxcAppDelegate.h 193
helloworldxcAppDelegate.m
  193
Hewitt, Joe 84, 147
hidesWhenStopped 342
hitTest:withEvent:
  UIView method 250
HTML 16
  data, manipulating by
    hand 400
  displaying with
    UIWebView 401–407
HTTP_USER_AGENT 44
HTTPBody 401

I

IBAction 209, 212, 217, 255–256
IBOutlet 195–196, 209, 212,
  246, 267–268
  coding with 216
  connecting to Interface
    Builder object 215
  creating 246
  declaring 215
icon 269
identifier 72
Identity tab 246
iDevKit 428
image 345–347
  blending 348
  collage example 351–356
  data types 345
  drawing, in Quartz 383–384
  duplicating 118
  in Canvas 117–120
  layering 347
  modifying 383
  modifying in UIKit 347
  transparency 348
image picker 349
image property 346
Image View object 210
imageNamed 345
imagePickerController:did-
  FinishPickingImage:
  editingInfo: 349
imagePickerControllerDid-
  Cancel 349, 354

imageWithCGImage 345
imageWithContentsOfFile
    345-299
imageWithData 345
includes 191
info button 17
Info.plist 192
initWithCoder 218, 234
initWithContentsOfFile
    300
initWithContentsOfURL 408
initWithData 408
    encoding 301
initWithFile
    310
initWithFrame 197, 346
initWithImage 346
initWithNibName 279
initWithParentid
    Menu 311
initWithStyle 289, 291
initWithTitle:image:tag:
    method 269
input 84
    user, accepting 286
    user, controls 286
insertSubview 280
Inspector window 245
integerForKey 297
integerForMenuWithParent 310
integrated development
    methodologies 25
Interface Builder 207–210, 245,
    266, 268
    .xib files, accessing 219
    advanced topics 217–219
    anatomy 207–209
    Connections panels 216
    delegate connections 217
    external object, creating 218
    first project 210–214
    IBAction 209, 217
    IBOutlet 209, 214–217
        coding with 216
        declaring 215
    Image View object 210
    image, adding 213
    inspector 209
    inspector window 211–213
        Attributes tab 211
        Class Actions 213
        Class Identity 213
        Class Outlets 213
        Connections tab 212
        Identity 213

Identity tab 212
    Size tab 212
Label object 210
Library 208
linking objects 260
main display window 208
nib document window 207
object
    connecting to IBOutlet 215
    creating 210
    inializing 218
    initWithCoder 218
    linking with Xcode 214
    manipulating 211
proxies 207
proxy, creating 218
resizing views 229
simulating in 210
UI elements 208
Web View object 210
windows 207
interfaceOrientation 229, 325
Internet 9
    always-on 9
    host, requesting information
        about 398
    programming, hierarchy 397
Internet Explorer 56
    and Canvas 104
IP address, locating 144
iPhone
    browser specifications 7
    built-in programs,
        preferences 288
    chrome 45
        Mobile Safari web
            browser 45
    compatible 23
    compiling to 429
    events 47
    incompatible 23
    multi-touch-capable
        capacitive touch screen 13
    optimized 24
        CSS 44
        iPhone events 47
    profiling for 151
    UI, creating 81–84
    vibrating 362
    viewport 32
    web apps 24
    web resources 427
iPhone Atlas 427
iPhone Dev Forums 427

iPhone Dev SDK 428
iPhone Development 428
iPhone Development
    Central 428
iPhone friendly 24, 36
    columns are too big 43
    common problems 41
    graphics are too small 41
    words are too small 42
iPhone in Action, blog 427
iPhone SDK 16
iPhone Simulator 130, 144,
    148, 192
    and altitude 340
    and Core Location 335
    and volume controls 359
iPhone Specifications 4
    802.11g 6
    input 5
    Streaming bit rate 6
    Visual output 5
    Wireless network
        connectivity 6
iPhone Web Developer
    debugging with 150
    getting bookmarklets
        from 151
iPhoneWebDev 427
iPhoto interface via
    Dashcode 130
iPod Touch 5
isFirstResponder 251
isIgnoringInteractionEvents
    UIApplication method 250
isSourceTypeAvailable 350
iUI 18, 24, 84
    Ajax 90, 97
        improving data listings 97
    alternatives 101
    attributes 93
        hideBackButton= 93
        orient= 93
        selected= 93
        target= 96
        toggled= 93
        type=cancel 93
        type=submit 93
    back end 94–95
    blueButton 92
    blueButton class 86
    button 92
    button class 86
    button classes 90
    classes 86

iUI *(continued)*
  code, organizing 95
  compressing 98
  data listings, improving 97
  default element 87
  developing with 85–94
  dialog class 86
  dialogs 89
  download sites 84
  example, integrating PHP 94
  external links 88
  grayButton 93
  grayButton class 86
  group class 86
  home page 96
  integrating with
    Dashcode 141
  integrating with other
    libraries 99–100
  jQuery, and 99
  leftButton 92
  leftButton class 86
  license 85
  lists 87
  look and feel 98
  organizational approaches 96
  page navigation 96
  panel class 86
  panels 91
  row class 86
  rows 91
  search 89
    with Ajax 90
  tabbed interface 96
  table view 87
  tips and tricks 95–99
  toggle class 86
  toggles 92
  toolbar 86
  toolbar class 86
  WebKit, and 100
  whiteButton 93
  whiteButton class 86
iUi example, color-selector 85

J

J2ME 18
Java 7, 37
JavaScript 16, 18
  and energy use 152
Javascript Events 37
  Drag 37
  Hover 37

JavaScript.com 428
jQuery
  and iUI 99
  iPhone package 99
    disableTextSizeAdjust 99
    enableTextSizeAdjust 99
    hideURLbar 99
    orientchange 99
    version 99
jQuery in Action 99
JSON 419

K

Katz, Yehuda 99
Kernighan 155
key window 241
keyboard 18, 242, 258–259
  Done key 258
  getting rid of 242
  Return key 259
Konquerer 56

L

labeledwebview 199
Lap Cat Software Blog 428
leftBarButtonItem 276
leftButton 92
Library directory 297
life cycle, monitoring 230
line styles, in Canvas 114
lineCap 114
lineJoin 114
lineTo 106
lineWidth 114
Link 58
links 39
list, iUI 87
listController 136
  methods 136
loadHTMLString:baseURL: 402
loadRequest 402
loadURL 201
loadView 230, 268
location
  and Core Location 337
  classes 336
  CLLocation 336
  CLLocationManager 336
location awareness 84
locationInView 243, 249
locationManager:didUpdateTo-
    Location:fromLocation: 339

look and feel 01
lookupSingularSQL 308

M

.m and .mm files 158
Mac OS X, setting up Apache
    web server 143
macro 158
main.m 193
mainBundle 299
MainWindow.xib 192, 207, 226
makefile 158
malloc() 157
manageTouches 248
Marquee 58
<marquee> 57
media
  detection, CSS 44
  loading from files 359
media player 357
  difficulties with 359
  framework 357–360
  integrating 359
  invoking 357
  notifications 357
  volume 359
memory management 157
  in Address Book 321
  with tables 234
menu, sliding 82
metatag 36
method 197
Metrics inspector 133
microphone, and audio
    queue 364
mirrored development 26
miterLimit 114
mixed development 26
mobi 10
Mobile Orchard 428
Mobile Safari 16, 63, 145
mobile web standards 10
mobileOK 10
modal view 281
  controllers 281
modalViewController 227
monetization 21
mousedown 48, 69
mousemove 47–48
mouseout 47–48
mouseover 47–48
mouseup 48, 69
mousewheel 48–49

movement
  and accelerometer 326–333
  casual, force 331
  checking 330
  filtering out 329
  forceful, force of 331
  slightly forceful, force of 331
moveTo 106
movieDidLoad 358
MPMoviePlayerContentPreload
    DidFinishNotification 357
MPMoviePlayerController 357
MPMoviePlayerPlaybackDid-
    FinishNotification 357
MPMoviePlayerScalingMode-
    DidChangeNotification 357
MPVolumeSettingsAlertHide
    359
MPVolumeSettingsAlertIsVisible
    359
MPVolumeSettingsAlertShow
    359
MPVolumeView 357, 359
multiple pages of screens 267
multipleTouchEnabled 251
  UIView property 250
multi-touch screen 11
music 5
MVC 240, 242, 245–246, 249,
    268, 270

N

navigate using tables 277
navigation controller 271, 273
  activating 277
  adding a title 274
  adding actions 275
  adding the links 274
  anatomy of 272
  difference from Tab Bar 272
  minimal configuration 272
  navigating backward 278
  navigating forward 276
  other methods and
    properties 278
  other types of navigation 278
  using 276
navigation menu 304
  building from database
    306–313
navigation paradigm 278
Navigation-Based
    Application 203–204
  template 273

navigationController 227
navigators and databases 278
Neal, Jonathan 99
nested message 196
Network Timeline 145, 151
networking, low-level 397–399
nextResponder 247
  UI Responder method 250
Nib Document window 246
Nib file 192
  vs. .xib 207
Nib window 274
nil 254
nonatomic 195
notifications 262
NSArray 425
  creating table content 233
  list of the view controllers 268
NSAutoreleasePool 193, 425
NSBundle 293, 425
NSCharacterSet 425
NSCountedSet 425
NSData 300, 425
  initWithContentsOfFile 300
  writeToFile atomically 300
NSDictionary 277, 417, 425
  creating table content 233
NSDocumentDirectory 299
NSError 425
NSFileHandler 426
NSFileManager 300, 426
  contentsAtPath 300
NSHandle 300
  closeFile 300
  fileHandleForReadingAtPath
    300
  fileHandleForWritingAtPath
    300
  readsDataofLength 300
  readsDataToEndOfFile 300
NSIndexPath 235, 426
NSLibraryDirectory 299
NSLog 426
NSMutableArray 233, 426
NSMutableCharacterSet 426
NSMutableData 426
NSMutableDictionary 417, 426
NSMutableSet 426
NSMutableString 426
NSMutableURLRequest
    398, 400, 416–417, 426
NSNotification 262
NSNotificationCenter 262, 426
NSNotificationQueue 262
NSNumber 426

NSObject 199, 224, 426
NSSearchPathForDirectoriesIn-
    Domains 299
NSSet 244, 248
NSString 157, 300, 402, 426
  initWithData encoding 301
  stringWithContentsOfFile
    encoding error 300
  writeToFile atomically
    encoding error 301
NSURL 201, 398, 402, 426
  creating 399
  fileURLWithPath: 399
  URLWithString: 399
  URLWithString:relative-
    ToURL: 399
NSURLConnection
    398, 401, 416
NSURLRequest 201, 398,
    402, 426
  allHTTPHeaderFields 401
  building 400
  HTTPBody 401
  init methods 400
  properties 401
  requestWithURL: 400
  requestWithURL:cachePolicy:
    timeoutInterval: 400
  valueforHTTPHeaderField:
    401
NSUserDefaults 292
  objectForKey 292
  resetStandardUserDefaults
    292
  setObjectForKey 292
  standardUserDefaults 292
NSValue 426
NSXMLParser
    407, 409–416, 426
  delegate methods 408
  initWithContentsOfURL: 408
  initWithData: 408
  parse 408
  parser:didEndElement:names
    paceURI:qualifiedName:
    408
  parser:didStartElement:
    namespaceURI:qualified-
    Name:attributes: 408
  parser:foundCharacters: 408
  parser:parseErrorOccurred:
    408
  parserDidEndDocument: 408
  setDelegate: 408
  starting 408

numberOfRows 136
numberOfSections 237
numberOfSectionsInTableView
    235

O

object
    creation, abstracting 225
    registers to receive notice 262
objectForKey
    292, 297
Objective-C 16, 18, 156–157
onblur 38
onchange 38
onclick 38
oncontextmenu 38
ondblclick 38
onerror 38
onfocus 38
onkeydown 38
onkeypress 38
onkeyup 38
onload 38
onload attribute 105
onmousedown 38
onmouseenter 38
onmouseleave 38
onmousemove 38
onmouseout 38
onmouseover 38
onmouseup 38
onreset 38
onresize 38
onscroll 38, 47–48
onselect 38
onsubmit 38
OOP (object-oriented
    programming) 159
OOP concept 159
    class 159
    framework 159
    inheritance 159
    message 159
    method 159
    object 159
    subclass 159
OpenAL 360, 364
openDatabase 65
OpenGL 394
    EAGL 394
    standard template 394
OpenGL ES Application
    203–204
Organizer 205

orientation 12
    and accelerometer 325–326
    interfaceOrientation 325
    notification 325
    of a view, checking 229
    precise, determining 328
    property 325
    UIDevice 325
orientation awareness 11, 84
output 84
Overriding
    hitTest:withEvent: 251

P

package 293
page control 286
pageTitle 86
pageX 72
pageY 72
panel, iUI 91
parentViewController 227
parse 408
parser:didEndElement:namespace-
    URI:qualifiedName: 408
parser:didStartElement:name-
    spaceURI:qualifiedName:
    attributes: 408
parser:foundCharacters: 408
parser:parseErrorOccurred:
    408–409
parserDidEndDocument
    408–409
path 371
    Canvas 105–109
    commands 106
    drawing, in Quartz 371–375
    finishing 372
    reusable, creating 373
pattern, in Canvas 117–120
Pederick, Chris 146
peoplePickerNavigation-
    Controller
    shouldContinueAfter-
    SelectingPerson 320
peoplePickerNavigation
    ControllerDidCancel 320
Perl 18
phase 243
    property 249
photo
    accessing 349–350
    taking 349
photo album, saving to 350

PHP 18
    and iUI 94
picker view 286
pinch 14, 48
plist editor 295
pointer 157
popToRootViewController-
    Animated 278
popToViewController:Animated:
    278
popViewControllerAnimated
    278
portrait mode 33
POST 416–418
power consciousness 11, 84, 152
<pre> 57
preference
    of built-in iPhone
        programs 288
    user, creating 288–293
        creating hierarchical
            settings 296
        creating settings 295
        dictionary of values 295
        editing settings 294
        maintaining 288–297
        saving 291–293
        saving in a database 291
        saving in a file 291
        saving NSUserDefaults 291
        settings, accessing 296
        using system settings
            293–297
preferences page
    drawing 289–291
    select list 291
    switches 291
PreferenceSpecifiers 295
prepareRow 136
presentModalViewController
    animated 320
presentModalViewController:
    animated: method 281
preventDefault 71
previousLocationInView 243
principalClassName 193
procedural language 159
profiling 151
program
    distribution, via iPhone App
        Store 431
project_Prefix.pch 192
project.app 192
projectAppDelegate 194

provisioning profile
  creating 430–431
proxy
  creating in Interface
    Builder 218
  in Interface Builder 207
PSChildPaneSpecifier 295–296
PSGroupSpecifier 295
PSMultiValueSpecifier 295
PSSliderSpecifier 295
PSTextFieldSpecifier 295
PSTitleValueSpecifier 295
PSToggleSwitchSpecifier 295

Q

quadraticCurveTo 107–108
Quartz 367
  advanced drawing 381–386
  affine transformation 378
  bitmap, drawing to 370
  clipping path, setting 379
  Cocoa Touch, and 367
  color space 375
  color, setting 375
  context 367–371
  coordinate system 368
  Core Foundation, and 367
  gradients, drawing 381–382
  graphic state, setting 375–381
  images, drawing 383–384
  paths 367
    drawing 371–375
    finishing 372
    reusable 373
  program example 386–391
  rectangle, drawing 374
  state 367
  text, drawing 384
  transformation 376–378
  UIView, drawing to 369
  words, drawing 384–385
Quartz 2D, UIImage 345

R

radian 107
readsDataToEndOfFile 300
recording 364
rectangle
  drawing 374
  drawing in Canvas 110
regulating events 250
relative 40

release 201
reload 402
report coordinates 249
requestWithURL 400
requestWithURL:cachePolicy:
    timeoutInterval: 400
resetGPS 342
resetStandardUserDefaults 292
Resig, John 131
resignFirstResponder 242, 251
resort array 136
resourcePath 299
Responder 244
responder 241
  chain 241, 247
restore 117
results.insertId 66
results.rows.item 66
results.rows.length 66
results.rowsAffected 66
retain 195
returnKeyType 258
rhombus function 111
rightBarButtonItem 276
Ritchie, Dennis 155
Root.plist 294
  editing 294
Root.strings 294
RootViewController 274
rotate 117
rotate animation 63
row, iUI 91
RSS (Really Simple
    Syndication) 20, 419
RSS reader 409–414
rssViewController 411
Ruby on Rails 18

S

Safari 7
  debugging with 144
  development tools 144
  Drosera 145
  Error Console 145
  Network Timeline 145, 151
  User Agent 145
  Web Content Guide for
    iPhone 32
  Web Inspector 145
Safe from the Losing Fight 428
save 117
scale 117
scalesPageToFit 402
screenX 72

screenY 72
scroll bars 14
SDK 10, 19, 82
  docs 32
  grid starting point 197
  native apps 25
  programmer 18
  programming 17
  resources 428
  web apps 25
search bar 287, 316
search, iUI 90
searchBarSearchButtonClicked
    259, 316
secondController.tabBarItem.
    badgeValue 270
Secrets of the JavaScript Ninja 131
segmented control 287
select list 291
select menus 45
sendAction:to:forEvent: 254
sendAction:to:fromSender:
    forEvent:
    UIApplication method 254
sendActionsForControlEvents
    254
setAffineTransform 393
setCurrentView 137
setCurrentViewWithTransition
    137
setDelegate 408
setObjectForKey 292, 297
Settings page tools, creating 294
shake 333–335
shape
  drawing in Canvas 110–111
  drawing order 110
  function, writing 110
sharedApplication 193, 251
shouldAutorotateToInterface-
    Orientation 228
simplicity 19
SKDatabase 307
SKMenu 307
slider 287
sliding menu 82
SOAP 419
social web, accessing 418–421
sophistication 21
sound
  complex, playing 362
  playing manually 360–364
  simple, playing 361
source availability, checking 350
special control events 253

SQL 65
  transaction handler 66
SQL responses
  error.code 66
  error.message 66
  results.insertId 66
  results.rows.item 66
  results.rows.length 66
  results.rowsAffected 66
SQLite 65, 157, 428
  API commands 306
  database, setting up 304–305
    accessing 305
    creating from command
      line 304
  documentation 303
  framework, adding 305
  include file 305
  limitations 303
  using 303–313
sqlite3_close 306
sqlite3_column_int 306
sqlite3_column_string 306
sqlite3_exec 306
sqlite3_finalize 306
sqlite3_get_table 306
sqlite3_open 306
sqlite3_prepare 306
sqlite3_step 306
sqlite3.h 305
stack, in graphical context 368
stackLayout 136–139
  building outside 137
  creating 136
  getAllViews 137
  getCurrentView 137
  manipulating 137
  methods 137
  populating 137
  setCurrentView 137
  setCurrentViewWith-
    Transition 137
  variable views 139
standardUserDefaults 292
start force 331
startAnimating 346
state 76
  in Quartz 367
  stack 117
statelessness 37
static 40
status bar 46
stop force 331
stopAnimating 346

stringByAppendingPath-
  Component 299
stringForKey 297
stringWithContentsOfFile
  encoding error 300
stroke 106
stroke width 114
strokeRect 110
strokeStyle 112
strokeText 119
style, creating in Canvas
  112–114
stylesheets, minimizing 151
subpage, back button 82
subscribe 205
subvert 242
subviews 201
superclass lookup 205
superview 241
SVG 37
switch 287
symbolic constants 159
synthesized 195
System Sound Services 361
  AudioServicesAddSystem-
    SoundCompletion 361
  AudioServicesCreateSystem-
    SoundID 361
  AudioServicesDisposeSystem-
    SoundID 361
  AudioServicesPlaySystem-
    Sound 361
  AudioServicesRemoveSystem-
    SoundCompletion 361
  functions 361
  vibrating iPhone 362

T

tab bar
  adding more tabs 267
  allow users to customize tab
    bar 271
  badge 270
  building a tab bar
    interface 267
  connecting views 268
  customization 271
  customizing items 270
  delegate 265
  icon image size 269
  initWithTabBarSystemItem:
    tag: method 270
  item 265
  modifying the buttons 269

tabBarController:didEnd-
  CustomizingViewControllers:
  changed: 271
tabBarController:didSelect-
  ViewController: 271
  title 270
  view controller 265
  where title is found in
    Xcode 270
Tab Bar Application 203–204
tab bar controller 267
  delegate 265
  setup example 268
tabBarController 227
tabBarController:didEnd-
  CustomizingViewControllers:
  changed: 271
tabBarController:didSelectView
  Controller: 271
tabBarItem 227
table
  cell properties 235
  cell, selecting 238
  cells 234
  cells, accessories 236
  content, creating 233
  interface, building 233–238
  memory management 234
  sections 236
table view controller 231, 238
  anatomy 231
  cell properties 235
  cell, selecting 238
  creating 231
  subclasses 233
  table cells 234
    accessories 236
  table content 233
  table interface, building
    233–238
  table sections 236
table view controller
  subview 247
tableView
  cellForRowAtIndexPath 312
  didSelectRowAtIndexPath 312
tableView:cellForRowAtIndex-
  Path: 233, 235, 238
tableView:didSelectRowAtIndex-
  Path: 238, 277
tableView:numberOfRows: 237
tableView:numberOfRowsIn-
  Section: 235
tableView:titleForHeaderIn-
  Section: 237

tap 14
/change 48
/click 47
/nothing 47
double 14, 48
tapCount 243, 249
target 71–72
target= 88
targetTouches 71
tempImageView 351, 354
templates for iPhone users 49
Text 58
text
drawing, in Quartz 384
in Canvas 117–120
text field 287
text view 18
TextField/Slider mashup 260
TextFieldDelegate 303
textViewDidEndEditing 259
thumbnails 403–405
timestamp 243–244
T-Mobile 10
tmp directory 297
toggle 92
toggleView method 279
toolbar 287
iUI 86
touch 70–71, 241–242
touch data 249
Touch events 244, 246
Touch Phase 249
Touch properties
clientX 72
clientY 72
identifier 72
pageX 72
pageY 72
screenX 72
screenY 72
target 72
touchcancel 70
touchend 70
touches and gestures 70
touchesBegan:withEvent:
244, 247
touchesCancelled:withEvent:
244
touchesEnded:withEvent: 244
touchesForView 244
touchesForWindow 244
touchesMoved:withEvent: 244
TouchJSON 419–420
installing 420

touchmove 70
touchstart 70
transaction 65
handler 66
transactionError 66
transactionSuccess 66
transformation
CTM 376
in Canvas 116
in Quartz 376–378
reusable 378
transition 61
in Dashcode 138
transitioning transforms 62
transitions 61
translate 117
transparency 348
in Canvas 114–116
tricks 204
turning off default behavior 71
two-finger flick/not
scrollable 48
two-finger flick/scrollable 48
two-fingered gestures 48
type casting 156
type preference, describing 295
typing 156

U

UIAcceleration 327
and gravity 327
and movement through
space 327
parsing 327
UIAccelerometer 326
UIActionSheet 423
UIActivityIndicatorView 423
UIAlertView 423
UIApplication 193, 241, 423
regulation 251
sendAction:to:fromSender:for
Event: 254
sharedApplication 193
UIApplicationDelegate
protocol 194–195
UIApplicationMain 193
UIBarButtonItem 276, 278
modifying the look 278
UIBarButtonItems 272
UIButton 252, 286, 423
UIColor 196, 198, 423
UIControl 252, 424
accepting user input 286

UIControl events
editing events 253
objects events 253
slider event 253
touch events 253
UIControlEventEditingDidEnd
259
UIDatePicker 286, 424
UIDevice 325, 424
orientation values 229
UIDeviceOrientationDidChange
Notification 326
UIDeviceOrientationFaceDown
325
UIDeviceOrientationFaceUp
325
UIDeviceOrientationLandscape
Left 325
UIDeviceOrientationLandscape
Right 325
UIDeviceOrientationPortrait
325
UIDeviceOrientationPortrait-
UpsideDown 325
UIEvent 242, 244, 424
allTouches 244
touchesForView 244
UIEvent reference 243
UIEventtimestamp 244
UIFont 424
UIGraphicsBeginImageContext
368–369
UIGraphicsEndImageContext
368–369
UIGraphicsGetCurrentContext
369
in Quartz 368
UIGraphicsGetImageFrom-
CurrentImageContext 369
UIImage 300, 424
Core Graphics, and 347
drawAsPatternInRect: 348
drawAtPoint: 348
drawAtPoint:blendMode:
alpha: 348
drawInRect: 348
drawInRect:blendMode:alpha:
348
factory methods 345
imageNamed: 345
imageWithCGImage: 345
imageWithContentsOfFile:
345
imageWithData: 345
instance methods 347

UIImage *(continued)*
  loading 345
  modifying 383
  size restriction 345
UIImagePickerController
    349, 424
  sources 349
UIImagePickerControllerSource
    TypeCamera 349
UIImagePickerControllerSource
    TypePhotoLibrary 349
UIImagePickerControllerSource
    TypeSavedPhotosAlbum
    349
UIImageView 346, 424
  animationDuration 346
  animationImages 346
  animationRepeatCount 346
  drawing 346
  frame, setting 347
  image property 346
  initializing 346
  layering 347
  properties and methods 346
  resizing 347
  startAnimating 346
  stopAnimating 346
UIImageWriteToSavedPhotos-
    Album 350
UIInterfaceOrientation-
    LandscapeLeft 229
UIInterfaceOrientation-
    LandscapeRight 229
UIInterfaceOrientationPortrait
    229
UIInterfaceOrientationPortrait
    UpsideDown 229
UIKit 196, 202, 350
  framework 193
  framework classes 423
  image, modifying in 347
UILabel 198, 245, 257, 424
UINavigationBar 223, 272, 274
UINavigationController
    222, 272, 274, 424
UINavigationItem 272, 274, 276
  View controller,
    navigationItem 276
UIPageControl 252, 286, 424
UIPickerView 286, 424
UIProgressView 424
UIResponder 241, 424
  regulation 250
  touchesBegan:withEvent: 244

touchesCancelled:withEvent:
    244
touchesEnded:withEvent: 244
touchesMoved:withEvent: 244
UIReturnKeyDone 258
UIScreen 424
UIScrollView 424
UISearchBar 257, 259, 287, 424
  not a child of UIControl 252
UISegmentedControl
    252, 287, 424
UISlider 252, 257, 259, 287, 424
  action method query the
    slider's properties 259
  maximumValue 259
  minimumValue 259
  value 259
UISwitch 252, 287, 425
UITabBar 222
  vs UIToolBar 267
UITabBarController 222, 425
UITabBarControllerDelegate
  protocol 271
UITabBarDelegate protocol 265
UITableView 222, 231
UITableViewCell 231, 233
UITableViewCellAccessory-
    Checkmark 236
UITableViewCellAccessoryDetail
    DisclosureButton 236
UITableViewCellAccessory-
    DisclosureIndicator 236, 274
UITableViewCellAccessoryNone
    236
UITableViewController
    222, 231, 273, 425
  RootViewController 274
UITableViewDataSource
    231, 233
UITableViewDelegate 231
UITableViewGrouped 289, 291
UITextField 242, 245–246, 252,
    257, 287, 425
  prime control for entering
    text 257
UITextFieldDelegate 258
UITextFields 245
UITextView 242, 257, 287, 425
UIToolBar 287
  vs UITabBar 267
UITouch 242, 425
  locationInView 243
  phase 243
  previousLocationInView 243
  reference 243

tapCount 243
timestamp 243
view 243
window 243
UITouchPhaseBegan
    243, 247, 249
UITouchPhaseCancelled 243
UITouchPhaseEnded 249
UITouchPhaseMoved 243
UITouchPhaseStationary 243
UIView 199, 222, 225, 425
  autoresizesSubviews 228
  autoresizingMask 228
    properties 228
    UIViewAutoresizingFlexible
      BottomMargin 229
    UIViewAutoresizingFlexible
      Height 229
    UIViewAutoresizingFlexible
      LeftMargin 229
    UIViewAutoresizingFlexible
      RightMargin 229
    UIViewAutoresizingFlexible
      TopMargin 229
    UIViewAutoresizingFlexible
      Width 229
    UIViewAutoresizingNone
      229
  drawing to 369
  properties 211
  regulation 251
UIViewController 222, 272,
    279, 425
  descendent of NSObject 224
  didRotateToInterface-
    Orientation: 230
  life-cycle events 230
  loadView 230
  properties 228
  viewDidLoad 230
  viewWillAppear: 230
  viewWillDisappear: 230
  willRotateToInterface-
    Orientation:duration: 230
UIViewControllers 272
UIWebView 12, 398,
    401–407, 425
  detectsPhoneNumbers 402
  goBack 402
  goForward 402
  loading, methods 402
  reload 402
  scalesPageToFit 402
UIWebViewDelegate 403
  protocol 202

UIWindow 241, 425
unions 159
URL bar 46
  eliminating 69
URL, working with 399–401
URLWithString 399
URLWithString:relativeToURL:
  399
User Agent 145
user input
  accepting 286
  controls 286
user preference
  creating 288–293
  creating settings 295–296
  dictionary of values 295
  editing settings 294
  maintaining 288–297
  saving 291–293
    in a database 291
    in a file 291
    using NSUserDefaults 291
  settings, accessing 296
  using system settings 293–297
USER_AGENT 43
utility 17
Utility Application 203–204, 279
  flipside controller 289
  preferences page,
    creating 289

V

ValueChanged event 253
valueforHTTPHeaderField 401
variable controls 205
variables 156
vibrating 362
view 82, 241, 243
  objects, as subviews of view
    controller 225
  orientation, checking 229
  resizing 228
    in Interface Builder 229
  rotating 228
  subviews 224
view controller 245, 265
  .xib files 219, 226
  accessing related objects 227
  anatomy 223
  as MVC controller 227
  collage 351–354
  controlling views and
    subviews 226
  creating 224

database 311
family 222
flipside controller 223
  in Address Book 318
interface, building 225–226
interfaceOrientation 229
introduction 223–231
life cycle, monitoring 230
life-cycle events 226
MainWindow.xib 226
modalViewController 227
multipage 223
navigationController 227
parentViewController 227
people picker 318–321
properties 227
proxy 246
single-page 223
standard 226
subviews 225
tabBarController 227
tabBarItem 227
table cells 234
  accessories 236
table content 233
table interface, building
  233–238
table sections 236
table view 231–238
  anatomy 231
  cell properties 235
  cell, selecting 238
  creating 231
  subclasses 233
  using 238
using 226–231
using the navigation-
  Controller property 277
view property 227
  orientation, checking 229
  resizing 228
  rotating 228
  setting up with Interface
    Builder 226
  setting up with Xcode 225
View-Based Application
  template 224
viewexViewController.xib 226
view hierarchy 247
view property 227
View-Based Application
  203, 245
  template 224

viewControllers 268
viewDidLoad 225, 230, 268, 289,
  348, 353
viewexViewController.xib 226
viewport 12, 33
  default pixel size 33
  device-width 35
  height 35
  initial-scale 35, 43
  local changes 34
  maximum-scale 35
  meta tag 33
  metatag multiple values 36
  minimum-scale 35
  sitewide changes 34
  user-scalable 35
  width 35
viewport metatag 105
viewport tags 23
viewWillAppear 230
viewWillDisappear 230
volume
  alerts 359
  controls, and iPhone
    Simulator 359
  view 359

W

W3C Mobile Web Initiative 10
W3C XML 428
WAP (Wireless Application
  Protocol) 10
web
  data-centric view 83
  POSTing to 416–418
  protocols 419
  protocols, download sites 419
  social, accessing 418–421
  vs. SDK 7
web app resources 427
web clip 46
  PNG size 47
Web Developer 146
web developer 18
web development 17
Web Development
  Advantages 19
web docs 32
Web Inspector 145
web page
  default resolution, 980
    pixels 12
  developing locally 144

web page *(continued)*
  redesign example
    Facebook 51
    Gmail 50
  testing locally 144
  testing on iPhone 144
web resources 427
  for development 427
web view
  calling up 402
  delegate 403
  example 403–405
  Google Maps example
    405–407
Webimage App Delegate 208
WebKit 18, 24, 32, 34, 56
  accessing an event 70
  accessing gestures 74
  accessing touches 72
  event properties 71
  gradients 76
  integrating with
    Dashcode 125, 140
  iPhone-specific CSS
    properties 59
  iUI, and 100
  masks 76
  new CSS elements 57
  new HTML elements 57
  recognize orientation
    changes 75
WebKit database 65
  event properties 71
  gesture properties 74
WebKit events
  gesturechange 70
  gestureend 70
  gesturestart 70
  orientationchange 75
  touchcancel 70
  touchend 70
  touchstart 70
WebKit Open Source
  Project 427
WebKit transforms
  Rotate 60
  Scale 60
  ScaleX 60
  ScaleY 60
  Skew 60

SkewX 60
SkewY 60
Translate 60
TranslateX 60
TranslateY 60
webkit-animation 64
webkit-animation-duration 64
webkit-animation-iteration-
  count 64
webkit-animation-name 64
webkit-background-size 57
webkit-border-bottom-left-
  radius 57
webkit-border-bottom-right-
  radius 57
webkit-border-image 57
webkit-border-radius 57
webkit-border-top-left-radius 57
webkit-box-shadow 58
webkit-gradient 76
webkit-keyframes 64
webkit-marquee-direction 58
webkit-marquee-increment 58
webkit-marquee-repetition 58
webkit-marquee-speed 58
webkit-tap-highlight-color
  58–59
webkit-text-fill-color 58
webkit-text-size-adjust 45, 58–59
webkit-text-stroke-color 58
webkit-text-stroke-width 58
webkit-touch-callout 58–59
webkitTransform 74
webkit-transform 59
webkit-transform-origin 60
webkit-transition properties 61
webkit-transition shorthand
  property 61
webkit-transition-delay 61
webkit-transition-duration 61
webkit-transition-property 61
webkit-transition-timing-
  function 61
webView:didFailLoadWithError:
  403
webView:shouldStartLoadWith-
  Request:navigationType: 403
webViewDidFinishLoad 403
webViewDidStartLoad 403
while 159

whiteButton 93
widget 125
willRotateToInterface-
  Orientation:duration: 230
window 243
  as a property 195
window.onscroll 38
window.orientation 75, 134
  values 75
window.scrollTo(0, 1) 46
Window-Based Application 203
  template 266
WML (Wireless Markup
  Language) 10, 37
word drawing, in Quartz
  384–385
Workspace 205
writeToFile
  atomically 300
    encoding error 301

X

Xcode 158, 190
  adding frameworks 202
  alternate templates 203
  templates 207
XHTML mobile profile
  documents 37
.xib file 207, 219, 245
  creating 219
  for view controller 219
  loading through view
    controller 219
  main 219
  multiple files 219
  templates 219
  vs. .nib 207
XML 409–414
  Core Location example
    414–416
  error 409
  parsing 407–416
  reading from files 407
XSLT 37

Z

zooming 35
  turning off 82